Career Mathematics

Merwin J. Lyng

L. J. Meconi

Earl J. Zwick

Houghton Mifflin Company/Boston

Atlanta   Dallas   Geneva, Ill.   Palo Alto   Princeton   Toronto

Authors

Merwin J. Lyng, Professor of Mathematics, Mayville State University, Mayville, North Dakota. Dr. Lyng teaches mathematics and mathematics education courses to students majoring in mathematics, business, and education.

L. J. Meconi, Professor of Mathematics Education, University of Akron, Akron, Ohio. Dr. Meconi teaches mathematics and mathematics methods courses in the college of education.

Earl J. Zwick, Professor of Mathematics, Indiana State University, Terre Haute, Indiana. Dr. Zwick teaches mathematics and mathematics education courses and supervises student teachers.

Editors	**Assembly and Technical Art**	**Book Design**
Thomas W. Dart	Dave Hunter	Michael van Elsen
Fran Seidenberg		

ISBN 0-395-48342-5

ABCDEFGHIJ-JD-976543210/898

Consultants

Arthur Contois, Chairman, Mathematics Department, Waukegan West High School, Waukegan, Illinois. Mr. Contois is a teacher and supervisor of mathematics staff and curriculum at Waukegan West High School and an instructor of mathematics at the College of Lake County.

June Danaher, Chief of Sciences and Mathematics Section, Maryland State Department of Education, Baltimore, Maryland. Ms. Danaher is responsible administratively for science, mathematics, health education, and environmental education programs. Her content responsibilities are in the area of mathematics.

Dennis Scott, Mathematics Teacher, Waynedale High School, Apple Creek, Ohio. Mr. Scott teaches vocational mathematics, consumer mathematics, and geometry courses for the Southeast Local Schools of Apple Creek, Ohio.

Graham Blundon, Assistant Superintendent, Deer Lake Integrated School Board, Deer Lake, Newfoundland. Mr. Blundon is responsible for the administration of district curriculum and inservice programs.

Paula Chabanais, President, Paula Chabanais & Associates Ltd., Toronto, Ontario. Her company provides advice and assistance in all areas of print communication to a wide variety of clients in Canada and the United States.

Allan McDermott, Consultant for Technological Studies, The York Region Board of Education, Aurora, Ontario. Mr. McDermott assists teachers and principals with the interpretation and implementation of curriculum policies and Ministry of Education guidelines for technological subjects taught from grade 7 through high school.

Douglas T. Owens, Associate Professor of Mathematics Education, University of British Columbia, Vancouver, British Columbia. Dr. Owens gives teacher education and graduate courses in mathematics education. He also does research into students' in-depth understanding of mathematical concepts.

Walter Szetela, Associate Professor of Mathematics Education, University of British Columbia, Vancouver, British Columbia. Dr. Szetela teaches elementary and secondary mathematics methods courses. He specializes in the area of problem solving with research and teaching activities.

Contents

Part 1   Reading Industrial Measurements

Part 2   Algebra in the Workplace

Part 3   Applied Geometry

Part 4   Occupational Simulations

Part 5   Trigonometry and Statistics at Work

Chapter 1
Reading Industrial Measurements

After completing this chapter, you should be able to:

1. Read several kinds of industrial meters.
2. Make precise measurements of length.
3. Read technical drawings.

Reading Meters

1-1 Electric Meters

Many industries depend on electricity as a source of power. The amount of electricity they use is measured by an electric meter in kilowatt hours (kW·h). The diagram below shows the four dials of a common electric meter. Industries using very large amounts of electricity might have five dials on their meters.

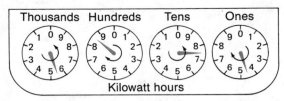

Thousands Hundreds Tens Ones

Kilowatt hours

This meter is read by noting the position of the pointer on each dial, starting at the far left. When the pointer is between two figures, use the smaller number. If it is between 9 and 0, use the 9. This meter is read as 5874 kW·h and means 5874 kW·h of electricity have been used.

Example

Skill(s) _____ 2 _____

According to the meter below, how many kilowatt hours of electricity were used for the month of May?

May 1

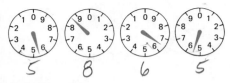

5   8   6   5

June 1

7   5   0   2

Solution

Find the difference between the two meter readings.

Present meter reading:	7502
Previous meter reading:	−5865
	1637

For the month of May, 1637 kW·h were used.

EXERCISES

Read the electric meters.

A 1.

9 8 2 4

2.

1 3 8 6

3.

6 6 4 5

4.

5 6 3 2

B 5.

1 7 0 0 6

6.

October 15	November 15

3 1 1 4

5 7 8 6

a. What is the latest meter reading? NOVEMBER 5786
b. What is the previous meter reading? OCTOBER 3114
c. How many kilowatt hours of electricity were used from October 15 to November 15? 2672

7.

January 1	February 1

2 6 1 8

6 1 4 9

a. What is the latest meter reading? FEBRUARY 6149
b. What is the previous meter reading? JANUARY 2618
c. How many kilowatt hours of electricity were used from January 1 to February 1? 3531

8. If the November 7 reading was 50,132 for the meter below, how many kilowatt hours were used for the month? 12,425

December 7

6 2 5 5 2

9. How many kilowatt hours were used from August 5 to September 5? 5233

August 5	September 5

4 3 8 2   9 6 5

10. How many kilowatt hours were used from April 10 to May 10? 1870

April 10	May 10

0 3 8 5   2 2 6 5

11. If 43,675 was the meter reading on November 3 for the meter below, how many kilowatt hours of electricity were used for the month? 24838

December 3

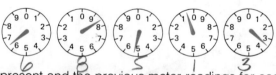

6 8 5 1 3

12. Sketch the present and the previous meter readings for each.

	a.	b.	c.	d.
Previous Reading	2175	3784	22,549	17,638
Present Reading	7325	9578	31,577	48,926

C 13. What are the usage charges for each electric-bill code if the cost per kilowatt hour is 5.6¢?

	Code	Previous Reading	Present Reading	Usage	Charges	
a.	057	6628	7435	807	?	45.19
b.	062	4067	5930	1863	?	104.33
c.	085	5293	8865	3572	?	200.03

1-2 Tachometers

Automobiles, motorcycles, boats, trains, and planes usually have a precision instrument called a tachometer in their dashboards or control panels to measure the speed in revolutions per minute (r/min). To read a tachometer, take the reading on the dial shown by the pointer and multiply by 100.

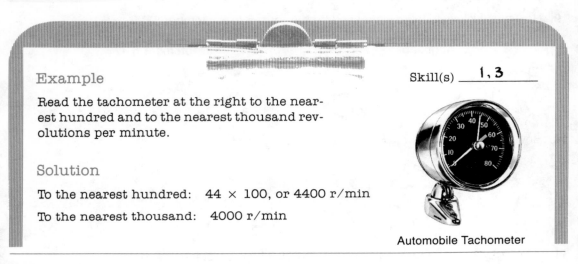

Example

Read the tachometer at the right to the nearest hundred and to the nearest thousand revolutions per minute.

Skill(s) ___1, 3___

Solution

To the nearest hundred: 44 × 100, or 4400 r/min

To the nearest thousand: 4000 r/min

Automobile Tachometer

EXERCISES

Read the tachometer to the nearest hundred revolutions per minute.

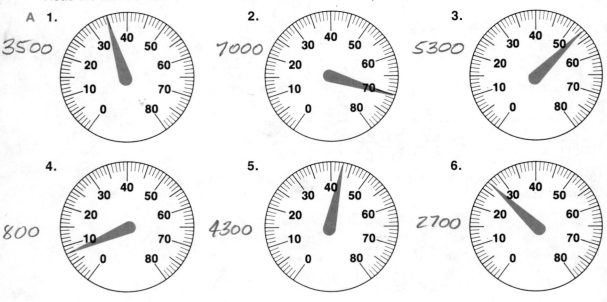

A 1. 3500 **2.** 7000 **3.** 5300

4. 800 **5.** 4300 **6.** 2700

B 7. Sketch a tachometer showing a reading of 2700 r/min.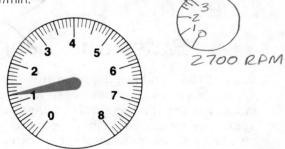

2700 RPM

C 8. To read a tachometer like the one at the right, you must multiply the number pointed on the dial by 1000. The smallest division on the tachometer represents 0.1.   1200

How many revolutions per minute are shown on the tachometer at the right? (Give your answer to the nearest hundred.)

Self-Analysis Test

1. How many kilowatt hours of electricity were used from May 1 to June 1?   595 KW·H

May 1   June 1

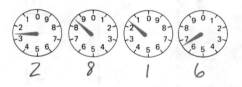   

2 8 1 6   3 4 1 1

2. How many kilowatt hours of electricity were used from December 5 to January 5?   4180 KW·H

December 5   January 5

0 3 8 6   4 5 6 6

Read each tachometer to the nearest hundred revolutions per minute.

3.

1200 R/MIN

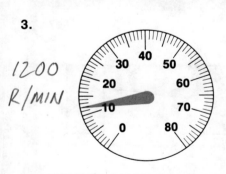

4.

5600 R/MIN

Reading Rules

1-3 Rules and Lengths

Most of the numbers involved in technical and scientific work are *approximate*, as they have been arrived at through some process of measurement. For example, a measure of a rope's length can be *approximated* as 5 m, or more precisely as 5.3 m (before being rounded off). Yet, neither measurement is the rope's *exact* length.

When approximating measurements in inches, you can use a rule showing eighths, tenths, sixteenths, or even thirty-seconds. When the tenths scale is used, the measurement can be written as a fraction or decimal.

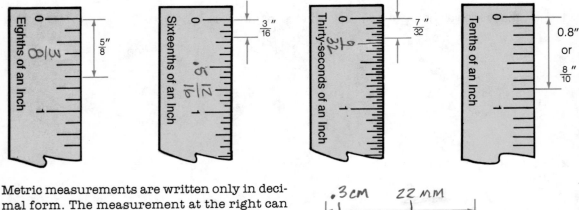

Metric measurements are written only in decimal form. The measurement at the right can be approximated as 4 cm, or more precisely as 3.9 cm.

Example 1

Measure the length of dimension T in the diagram at the right to the nearest sixteenth of an inch.

Solution

Dimension T measures $1\frac{1}{16}$ in.

Example 2

Draw a line $\frac{7}{16}$ in. long

Solution

Use an inch rule showing
sixteenths of an inch.

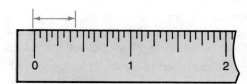

EXERCISES

A **1.** What length is shown by 25 divisions on an inch rule showing tenths? $2\frac{1}{2}$

2. Using an inch rule showing thirty-seconds, determine what length is shown
by: $\frac{7}{32}$ $\frac{35}{32} = 1\frac{3}{32}$ $\frac{67}{32} = 2\frac{3}{32}$
a. 7 divisions.   **b.** 35 divisions.   **c.** 67 divisions.

3. Using a metric rule showing millimeters, determine what is the length in
centimeters between: .3 cm   1.0 cm   2.9
a. 3 divisions.   **b.** 10 divisions.   **c.** 29 divisions.

B Draw a line to the given length. SEE PAGE 6 FOR ANSWERS

4. $\frac{3}{8}$ in.   **5.** $\frac{12}{16}$ in.   **6.** $\frac{9}{32}$ in.   **7.** 0.5 in.   **8.** 3.7 in.   **9.** 22 mm   **10.** 0.3 cm

11. Measure the lengths shown in each diagram below to the nearest millimeter.

12. Measure each length to the nearest sixteenth of an inch.

11.   **a.**

AB = 5mm
BC = 22 mm
CD = 52mm

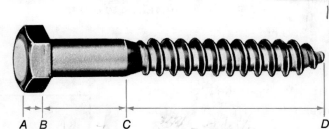

12.
$AB = \frac{3}{16}$
$BC = \frac{14}{16} = \frac{7}{8}$
$CD = \frac{33}{16} = 2\frac{1}{16}$

11.   **b.**

AB = 76 mm
CD = 6 mm
EF = 10 mm
GH = 9 mm
JK = 39 mm

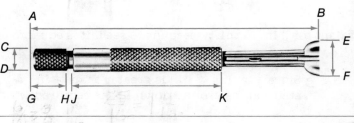

12.
$AB = \frac{48}{16} = 3''$
$CD = \frac{4}{16} = \frac{1}{4}$
$EF = \frac{6}{16} = \frac{3}{8}$
$GH = \frac{6}{16} = \frac{3}{8}$
$JK = \frac{25}{16} = 1\frac{9}{16}$

7

1-4 Equivalent Measures

Many jobs require that we work with **equivalent measures** as we approximate measurements. The rule at the right allows approximations to be made in both the U.S. and metric systems.

Two examples of equivalent measures are illustrated. The upper U.S. scale of the rule shows that dimension F is approximated as $\frac{10}{16}$ in. or in **simplified form** as $\frac{5}{8}$ in. Since both fractions are equal, $\frac{5}{8}$ in. and $\frac{10}{16}$ in. are equivalent measures.

In the lower metric scale, dimension G is approximated as 12 mm. This measurement is equivalent to 1.2 cm.

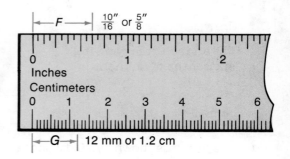

Example 1

Skill(s) _10, 31_

A piece of cord measures 3.5 m long. What is the length of the cord in centimeters?

Solution

Since 1 m = 100 cm, the conversion factor for changing meters to centimeters is 100. Multiply the measurement by 100.

$$3.5 \text{ (m)} = 100 \times 3.5 \text{ (cm)}$$
$$= 350 \text{ (cm)} \qquad \leftarrow 350 \text{ cm is equivalent to } 3.5 \text{ m.}$$

The piece of cord is 350 cm long.

Example 2

A metal strip $\frac{11}{32}$ in. wide is needed to fasten two parts of a blower. Can a strip $\frac{3}{8}$ in. wide be ground down to the required width?

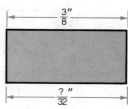

Solution

Find the equivalent for $\frac{3}{8}$ in. in thirty-seconds to see if it is equal to or greater than $\frac{11}{32}$ in. Multiply the numerator and denominator of $\frac{3}{8}$ by 4.

$$\frac{3 \times 4}{8 \times 4} = \frac{12}{32} \qquad \leftarrow \text{Since } \frac{3}{8} \text{ in. (or } \frac{12}{32} \text{ in.) is greater than } \frac{11}{32} \text{ in., a } \frac{3}{8} \text{ in.}$$
$$\text{strip can be ground down to the required width.}$$

EXERCISES

A Express each length shown
at the right in eighths and in
sixteenths of an inch.

1. A $2\frac{7}{8}$
2. B $2\frac{3}{8}$
3. C $1\frac{3}{8}$
4. D $\frac{7}{8}$

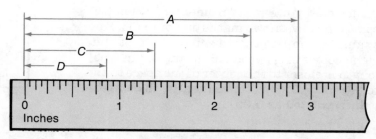

Measure the lengths of the following segments. Then write each length in
millimeters, centimeters, and decimeters.

5. 86 MM, 8.6 cm, & .86 DM
6. 133 MM, 13.3 cm & 1.33 DM
7. 118 MM, 11.8 cm & 1.18 DM

B 8. The square head of a machine bolt measures
$\frac{24}{32}$ in. across. Will a $\frac{3}{4}$ in. open-end wrench fit
this bolt head? YES

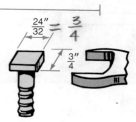

9. The owner's manual for a power lawn mower calls for an adjusting screw
not less than $\frac{7}{8}$ in. long. Will a screw $\frac{29}{32}$ in. long fit? YES

10. A rectangular piece of plywood is 120 cm wide and 240 cm long. What
are the dimensions in meters? ① 1.20 m ② 2.40 m

11. The hole in a flat washer measures $\frac{8}{32}$ in. Can
this washer be used on a $\frac{1}{4}$ in. bolt? YES

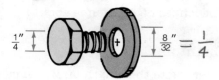

12. A plumber needs pieces of copper tubing in the following lengths: 96 cm,
134 cm, 108 cm, and 83 cm. How many meters of tubing does the plumber
need in all? 4.21 m

13. A computer programmer is designing a program that prints a word using
letters of the following widths: 5 mm, 2 mm, 3 mm, and 2 mm. What is the
length of the word in centimeters? 1.2 cm

C 14. A package contains one dozen flower bulbs. The bulbs are to be planted
2 dm apart. Are there enough bulbs to plant a row 2 m long? Are there
any bulbs left? YES   YES

1-5 Interpolating Lengths

Sometimes your measurement of a length comes *in between* two marks on your rule. For example, the length of segment *XY*, at the right, comes in between 0.6 in. and 0.7 in. The length is closer to the center of the space than to 0.6 in. or 0.7 in. Consequently, the length is **interpolated** as 0.65 in.

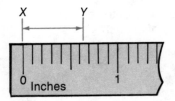

Example 1

Skill(s) ___10___

Interpolate the length of *AB*.

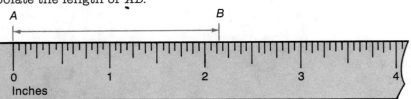

Solution

Since point *B* falls halfway between $\frac{2}{16}$ in. and $\frac{3}{16}$ in., interpolate the length of *AB* in thirty-seconds of an inch.

$\frac{2}{16}$ in. = $\frac{4}{32}$ in. $\frac{3}{16}$ in. = $\frac{6}{32}$ in.

Halfway between $\frac{4}{32}$ in. and $\frac{6}{32}$ in. is $\frac{5}{32}$ in.
The length of *AB* then is about $2\frac{5}{32}$ in.

Example 2

Interpolate the length of *MN* in centimeters.

Solution

The length *MN* is about halfway between 0.8 cm and 0.9 cm.
The length *MN* is about 0.85 cm.

EXERCISES

U.S. System

A Interpolate each length in thirty-seconds of an inch.

1. $C \, 3\frac{27}{32}$ **2.** $D \, 3\frac{1}{32}$

3. $E \, 2\frac{7}{32}$ **4.** $F \, 1\frac{21}{32}$

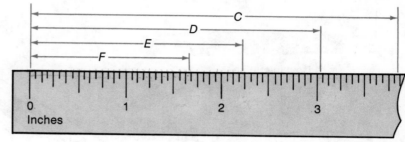

Metric System

Interpolate each length in centimeters.

5. $M \, 9.85$ **6.** $N \, 7.3$

7. $O \, 3.45$ **8.** $P \, .95$

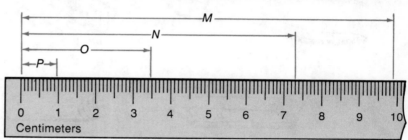

B When working with sheet metal, you often need several pieces that are exactly alike. You make your pieces from a pattern called a *template*, as shown below.

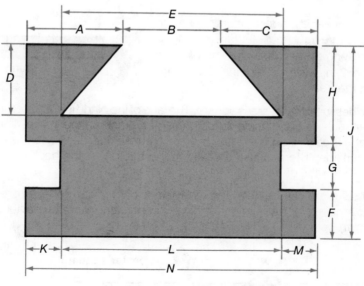

U.S. System

9. Interpolate each dimension in the template above using a rule showing tenths of an inch.

Metric System

10. Interpolate each dimension in the template above using a rule showing centimeters and millimeters. Write each dimension in centimeters.

11

11. Interpolate each dimension in the sketches of tools below in centimeters.

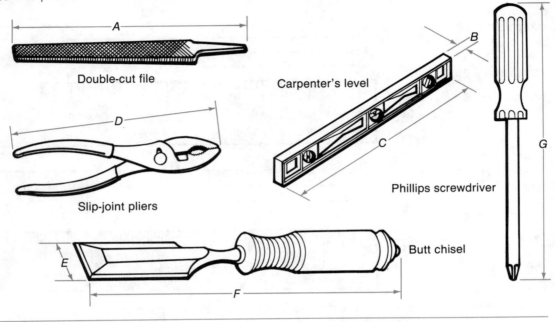

Double-cut file

Carpenter's level

Slip-joint pliers

Phillips screwdriver

Butt chisel

Tricks of the Trade

The Carpenter's Square

The carpenter's square is made from a single piece of steel bent to form a right angle (90°). The *tongue* is usually 16 in. long and $1\frac{1}{2}$ in. wide. The *blade* is 24 in. long and 2 in. wide. The point of the right angle is called the *heel*.

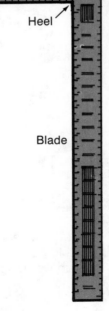

Tongue   Heel

Blade

There are two sides to the carpenter's square. The *face* of the blade is calibrated in sixteenths of an inch on the outside and eighths of an inch on the inside. The back of the blade is calibrated in twelfths of an inch on the outside and tenths of an inch on the inside. There is also a scale on the back near the heel that is calibrated in hundredths of an inch.

The carpenter's square can be helpful in many kinds of projects. For example, the angled cuts required for stair stringers are more easily done with a carpenter's square, as shown below.

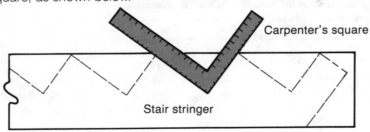

Carpenter's square

Stair stringer

Self-Analysis Test

1. Draw a line to the given length.

    a. $\frac{5}{8}$ in.   **b.** $\frac{14}{16}$ in.   **c.** $\frac{21}{32}$ in.   **d.** 1.5 in.   **e.** 23 mm   **f.** 1.4 cm

2. Measure the lengths shown in the diagram below:

    a. to the nearest millimeter;
    b. to the nearest sixteenth of an inch.

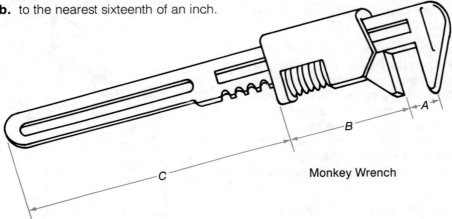

Monkey Wrench

3. A sewing machine has a bobbin hole that measures $\frac{29}{32}$ in. across. Will a bobbin with a $\frac{7}{8}$ in. diameter fit into this bobbin hole?

4. A computer keyboard measures 410 mm long and 180 mm wide. What are the dimensions in centimeters?

5. Interpolate each dimension in the template below:

    a. to the nearest thirty-second of an inch;
    b. in centimeters.

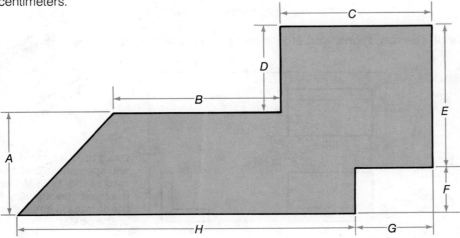

Reading Technical Drawings

1-6 Points of View

For every manufactured object there are drawings that describe its physical shape completely and accurately. The drawing is the means by which the design draftsperson communicates to the fabricator.

One problem a draftsperson must deal with when making drawings is to represent an object as three-dimensional to the eye on the flat plane of the drawing paper. To overcome this problem, different views of the object can be arranged on the drawing paper to convey the necessary information to the reader, as shown. The method of showing the top, front, and side view is known as **orthographic projection**.

Since some objects drawn are quite simple, only one or two views may be needed to describe them completely. More complex objects may require more than three views.

To make the side view easier to draw, it has been rotated and moved beside the top view in the drawing at the right.

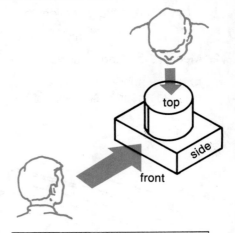

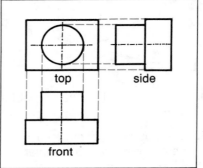

Example

Sketch the top, front, and side views of the object shown.

Solution

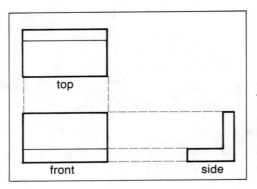

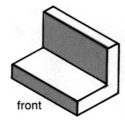

←Drawings of most objects will appear as three views arranged like this. Notice how the front view usually shows the dominant characteristics of the object.

EXERCISES

A Match the three-dimensional drawing A, B, or C to its set of top-, front-, and side-view drawings.

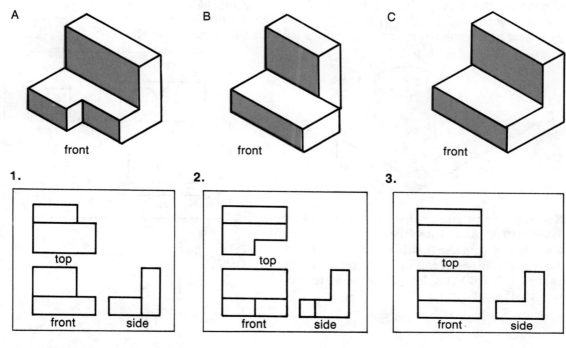

A front   B front   C front

1.   2.   3.

B Sketch the front view for each object.

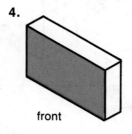

4. front

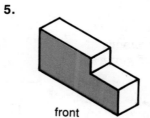

5. front

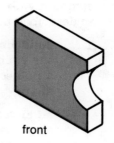

6. front

C Sketch the top, front, and side view of each object.

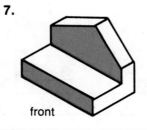

7. front

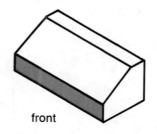

8. front

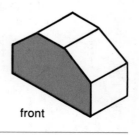

9. front

1-7 Lines

To help you read technical drawings more precisely, you should know about the different kinds of lines used. Look at the object below. Then notice how the technical drawings of the top, front, and side of the object show hidden features and center lines with the use of special dashed lines.

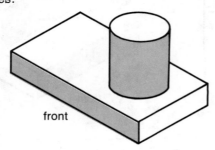

front

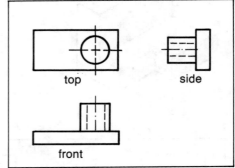

top   side

front

Notice how the center of the cylinder is indicated with dashed lines that are interrupted with dots. Since the inside walls of the cylinder cannot be seen from the side or the front, dashed lines of equal length are used to indicate their position.

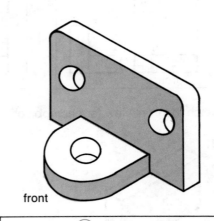

front

Three commonly used lines in technical drawings are described below.

(S) ——— **Solid lines** are used to indicate the visible edges and corners of an object and to give its general shape.

(H) - - - - - **Hidden feature lines** show those surfaces, edges, or corners of an object that are hidden from view.

(C) —..—..— **Center lines** represent axes that divide objects into two symmetrical parts.

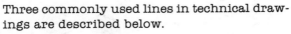
For circular objects and features, two perpendicular center lines are usually shown.

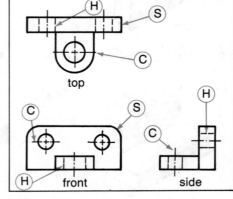

top

front   side

Example 1

Draw the front, the top, and the side view of the object at the right using solid lines, hidden lines, and center lines.

Solution

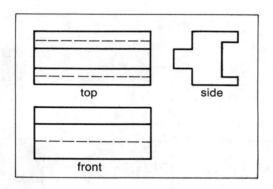

top

side

front

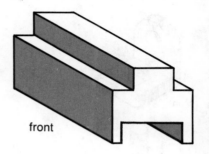

front

Example 2

Sketch three views of the object at the right. Show solid, hidden, and center lines.

Solution

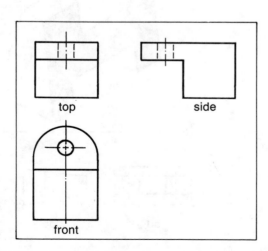

top

side

front

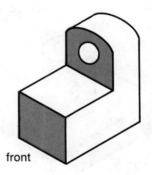

front

EXERCISES

Which technical drawing A, B, or C is correct? Identify the view of the object shown in the correct drawing.

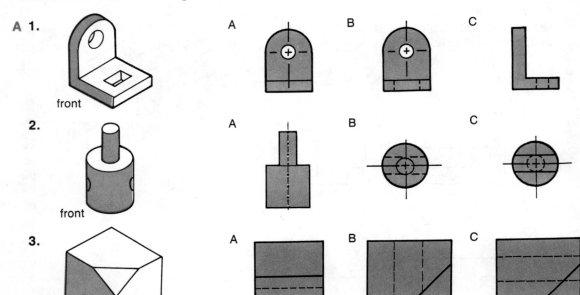

A 1. front   A   B   C

2. front   A   B   C

3. front   A   B   C

The top and side views of each object given are drawn for you. Draw the front view.

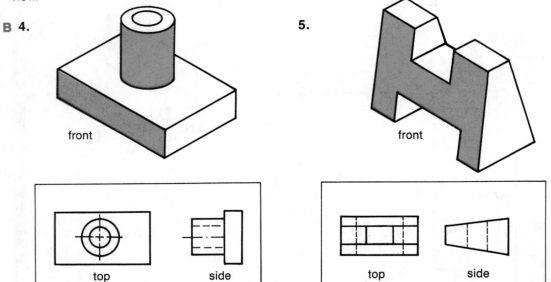

B 4. front

top   side

5. front

top   side

18

Sketch the top, the front, and the side views of each object. Show hidden lines, center lines, and so on.

6.

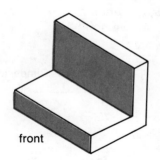

front

7.

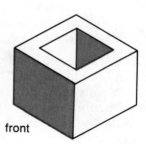

front

8.

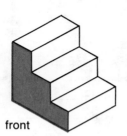

front

9.

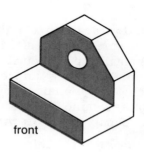

front

Check Your Skills

1. Round to the nearest tenth.

    a. 104.72   **b.** 3.245   **c.** 45.991   (Skill 1)

2. Round to the nearest foot.

    a. 18.5 ft   **b.** 13.48 ft   **c.** 4.76 ft

Add, subtract, multiply, or divide.

3. 65,432 − 5543   **4.** 54,324 − 3542   **5.** 45,098 + 2305   (Skill 2)

6. 296 × 86   **7.** 47 × 803   **8.** (984,233)(12)   (Skill 3)

9. 52.13 + 6.21   **10.** 5.44 − 1.987   **11.** 19.833 − (4.98 + 3.7)   (Skill 6)

Find the numerator to make the pair of fractions equivalent.

12. $\frac{3}{4} = \frac{?}{8}$ **13.** $\frac{16}{20} = \frac{?}{5}$ **14.** $\frac{2}{3} = \frac{?}{18}$ (Skill 10)

Multiply by powers of ten to convert each measurement to millimeters, decimeters, and kilometers.

15. 8496 m   **16.** 235 cm   **17.** 21.3 m   (Skill 31)

18. 2.1 cm   **19.** 456.123 m   **20.** 9987.2 cm

1-8 Dimensioning

Dimensions are indicated on drawings by extension lines, dimension lines, leaders, arrowheads, and other symbols to define the various characteristics of the object drawn.

Dimension lines are used to indicate the extent and direction of dimensions. They are terminated by neatly made arrowheads, as shown at the right. Dimension lines should be placed outside the view where possible and should extend to extension lines rather than thicker object lines.

Extension lines are used to indicate the point or line on the drawing to which the dimension applies. A small gap of about 1 mm is left between the extension line and the outline of the object so that direction lines are not crowded into the space around the object. This adds to the overall clarity of the drawing.

A **leader** is used to direct information and symbols to a place in the drawing.

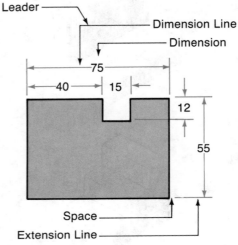

(Dimensions are in millimeters.)

Dimensions may be shown in U.S. units as fractional inches and decimal inches, or in metric units. The commonly used metric linear unit on engineering drawings is the millimeter. Usually the units used are specified at the bottom of a drawing. Thus the unit abbreviation is not needed beside the numerical dimension.

Dimensions are positioned in one of two ways in technical drawings. With **unidirectional dimensioning**, all dimensions can be read from the bottom of the page. For **aligned dimensioning**, the dimensions are readable from either the bottom or the right side of the drawing.

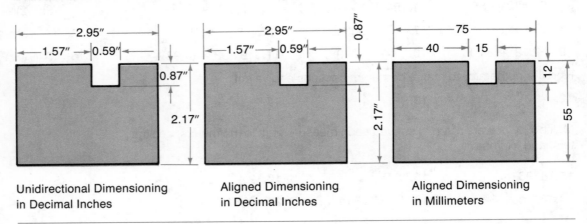

Unidirectional Dimensioning
in Decimal Inches

Aligned Dimensioning
in Decimal Inches

Aligned Dimensioning
in Millimeters

Example

Trace the figure. Then indicate the dimensions in decimal inches using extension lines, dimension lines, and leaders as needed.

Solution

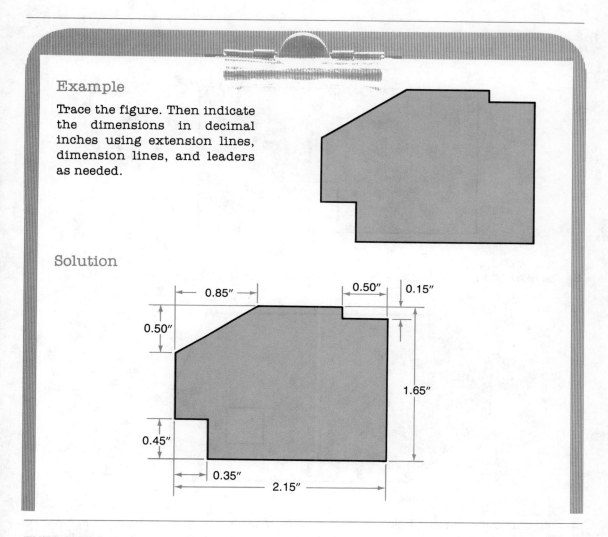

EXERCISES

A 1. Redraw the figure, using aligned dimensions.

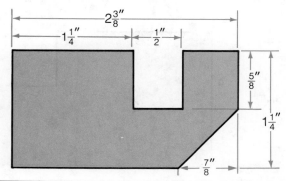

B **2.** Trace the figure. Then indicate the dimensions in fractional inches using extension lines, dimension lines, and leaders.

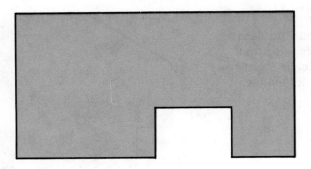

3. Trace the figure. Then indicate the dimensions in decimal inches.

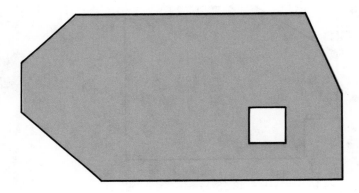

4. Reproduce the drawing below in *full size.* Place extension lines, dimension lines, and dimensions on the drawing.

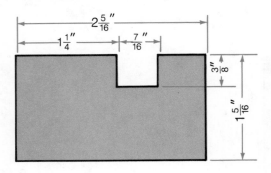

5. Measure the drawing below in millimeters. Then reproduce it to *twice its size.* Place extension lines, dimension lines, and the original dimensions on the drawing.

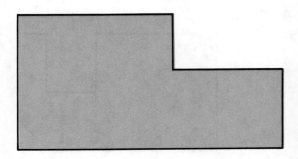

Reproduce each drawing below in *full size*. Place extension lines, dimension lines, and dimensions on the drawing.

6.

7.

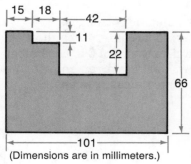

(Dimensions are in millimeters.)

Measure the drawing below in millimeters. Then reproduce it to *twice its size*. Place extension lines, dimension lines, and the original dimensions on the drawing.

8.

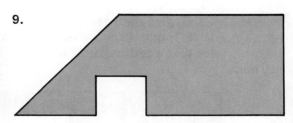

9.

Technology Update

The Laser Beam

The discovery of the laser has changed many industries. A laser is a high-energy beam that can be focused on points smaller than a millimeter that are kilometers away. Clothing manufacturers are now cutting cloth with computer-directed laser beams. In precision work with microscopic circuits and other electronic equipment, lasers are now a necessary tool. Holes can be drilled in a wire as small as a paper clip and wires as small as 0.08 mm can be welded.

Many uses for lasers have also been found in medicine, including eye surgery and improvements in contact lenses. A number of holes about 0.1 mm across are drilled in contact lenses to allow better circulation of the fluids around the eye and to prevent itching. Common gum diseases are now treated with lasers. Lasers have greatly reduced excessive bleeding and post-operative discomfort when a tooth is pulled.

Research Project: Find three other uses of laser beams today.

1-9 Dimensioning Curves

Dimensions for circles are given in terms of the diameter, as shown below. The center lines show the location of the center of the circle. The diameter symbol Ø is used most commonly today and must precede the diameter measurement. (If the abbreviation DIA is used, it follows the diameter measurement.)

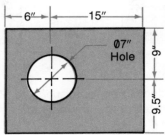

To dimension an arc, show the center of the curvature and the radius of the curve. We indicate the radius with the abbreviation R, as shown below. Notice that a radius dimension line uses one arrowhead at the arc end.

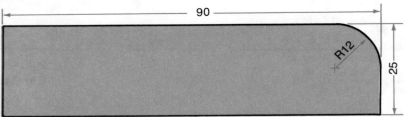

Example

Trace the diagram at the right. Then add the dimensions in millimeters to your drawing.

Solution

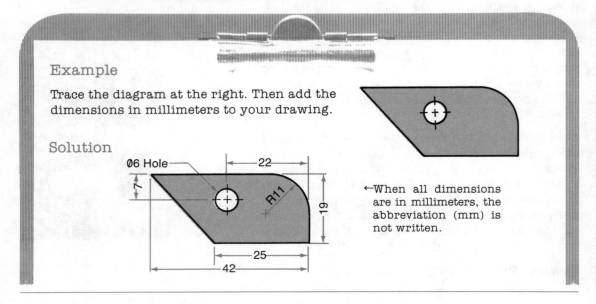

←When all dimensions are in millimeters, the abbreviation (mm) is not written.

EXERCISES

Trace the diagram. Then add the dimensions in fractional inches.

A 1.

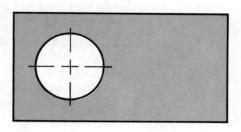

2.

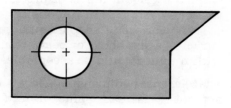

Trace the diagram. Then add the dimensions in decimal inches.

B 3.

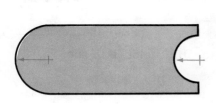

4.

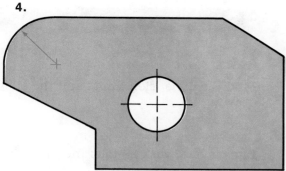

Trace the diagram. Then add the dimensions in millimeters.

C 5.

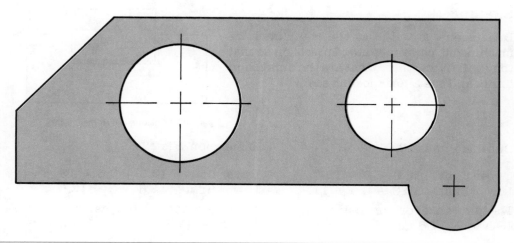

1-10 Dimension Tolerances

The bushing drawn at the right must be machined to certain limits of precision. Notice that the **basic size** of dimension A must be 1.25 in. However, due to manufacturing inaccuracies, a size above or below this basic size within limits would be acceptable.

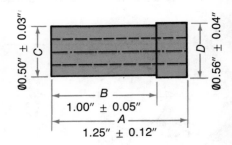

The **tolerance** of dimension A is ±0.12 in. (The symbol ± is read "plus or minus.") This means the length of A can be over 1.25 in. by 0.12 in. (1.37 in.) or below 1.25 in. by 0.12 in. (1.13 in.) and still be within the **dimension limits**. The difference between the dimension limits is called the **allowance**. The allowance of dimension A is 1.37 in. − 1.13 in., or 0.24 in.

Example 1   Skill(s) _____ 6

Find the dimension limits and the allowance for dimension B in the drawing above.

Solution

Dimension B with its tolerance is 1.00 in. ± 0.05 in.

Upper limit: 1.00 in. + 0.05 in., or 1.05 in.
Lower limit: 1.00 in. − 0.05 in., or 0.95 in.

The allowance is 1.05 in. − 0.95 in., or 0.10 in.

Example 2

Dimensions A and B at the right must be machined to certain limits of precision so that they will fit together. What are the dimension limits and the allowance for each?

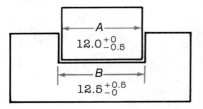

(Dimensions are in millimeters.)

Solution

Dimension A is $12.0^{+0}_{-0.5}$.

Upper limit: 12.0 + 0 = 12
Lower limit: 12.0 − 0.5 = 11.5

The allowance for dimension A is 12 mm − 11.5 mm, or 0.5 mm.

Dimension B is $12.5^{+0.5}_{-0}$.

Upper limit: 12.5 + 0.5 = 13
Lower limit: 12.5 − 0 = 12.5

The allowance for dimension B is 13 mm − 12.5 mm, or 0.5 mm.

EXERCISES

Copy and complete the tables.

U.S. System

	Basic Size	Tolerance	Maximum Size	Minimum Size
A 1.	$2\frac{1}{2}$ in	$\pm \frac{1}{8}$ in.	?	?
2.	$4\frac{3}{8}$ in.	$\begin{matrix}+0\\ -\frac{1}{16}\text{ in.}\end{matrix}$	?	?
3.	2.445 in.	± 0.002 in.	?	?
4.	0.828 in.	± 0.005 in.	?	?
5.	1.875 in.	$\begin{matrix}+0\\ -0.100\text{ in.}\end{matrix}$	?	?

Metric System

	Basic Size	Tolerance	Maximum Size	Minimum Size
6.	322 mm	± 1 mm	?	?
7.	45.0 mm	$\begin{matrix}+0.5\text{ mm}\\ -0\end{matrix}$	?	?
8.	625 mm	± 2 mm	?	?
9.	83.0 mm	$\begin{matrix}+0.5\text{ mm}\\ -0\end{matrix}$	?	?
10.	39.0 mm	$\begin{matrix}+0.2\text{ mm}\\ -0\end{matrix}$	?	?

B **11.** Draw two tables like those shown above. Complete the tables for dimensions
A, *B*, *C*, *D*, and *E* in each figure below.

U.S. System

a.

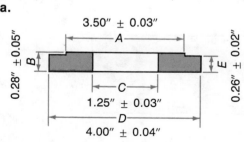

3.50″ ± 0.03″

1.25″ ± 0.03″

4.00″ ± 0.04″

0.28″ ± 0.05″

0.26″ ± 0.02″

Metric System

b.

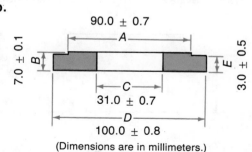

90.0 ± 0.7

31.0 ± 0.7

100.0 ± 0.8

7.0 ± 0.1

3.0 ± 0.5

(Dimensions are in millimeters.)

12. The cylinder bore on a certain engine is 4.062 in. $^{+0}_{-0.005}$. Find the dimension limits and the allowance.

13. The specifications for an economy automobile engine show the bore as 856 mm ± 2 mm. Find the dimension limits and the allowance.

Self-Analysis Test

1. Sketch the top view of each object.

a.

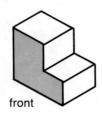

front

b.

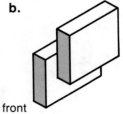

front

c.

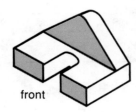

front

2. Which technical drawing A, B, or C is correct? Identify the view of the object shown by the correct drawing.

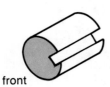

front

A   B   C

3. Trace the figure. Then indicate the dimensions in decimal inches using extension lines, dimension lines, and leaders.

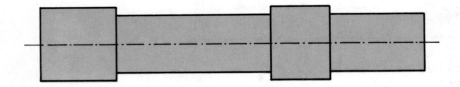

4. Trace the drawing. Then show all dimensions in millimeters.

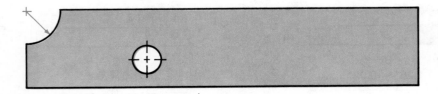

Find the dimension limits and the allowance of each aligning pin.

5.

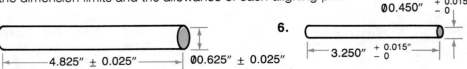

4.825″ ± 0.025″   ⌀0.625″ ± 0.025″

6.

⌀0.450″   + 0.015″   − 0

3.250″ + 0.015″ − 0

28

COMPUTER CONNECTION

In recent years, design firms have experienced a revolutionary change in drafting techniques that before long could make the drafting board obsolete. This is CAD (Computer-Aided Drafting or Design). This is also referred to as CADD, when drafting and design software is available on the same system.

In its simplest form, a CAD drawing is a two-dimensional representation of an object. In its more sophisticated forms, CAD may be a three-dimensional dynamic representation, like a moving crankshaft. A CAD drawing can also model a solid or show an assembly of layered drawings, which permit a peel-off of layers, exposing the inner-most core of the assembly, as in the dis-assembly of a car.

Since these drawings are mathematically based, important data can be automatically accessed, such as coordinate locations, volume, mass, and area. Stress and strain analysis, fluid friction, and wind tunnel tests are some of the more specialized capabilities of CAD, not to mention such routine tasks as parts lists, bills of material, and process plans.

Some of the many industries using CAD systems include: engineering, electronics, mining, manufacturing, home design, and interior design. The following examples show the usefulness of a CAD system.

1. Home Design

    A company offering semi-custom houses has placed several basic house designs in a CAD system. After the buyer has chosen a basic design, he or she can customize the basic design by changing the positions of doors, windows, and interior walls. These changes are quickly made on a CAD system and plotted out for the buyer. The builder automatically sets a list of the new materials, if any, required for this custom design.

2. Manufacturing

    A small company manufactures a line of four machines. Each machine requires 1500 drawings of simple two-dimensional parts. All of the drawings are in a CAD system so that updates and minor changes can be easily made.

CAREER PROFILE

The builders of homes, bridges, and factories and the manufacturers of tools, appliances, and cars are just some of the workers who are guided in their projects by the technical drawings made by draftspersons.

Job Description

What do draftspersons do?

1. They use various drafting instruments such as drafting machines, T-squares, set squares, scales, compasses, and templates to make technical drawings.

2. They make drawings of items to be built or manufactured based on the sketches and calculations of engineers, architects, and designers.

3. Draftspersons know how to make various kinds of drawings, such as schematic, preliminary, general arrangement, sub-assembly, and detail drawings. They can also draw different projections, such as orthographic, oblique and perspective.

4. They must be able to portray maximum information with a minimum of lines, dimensions, and views that are in conformity with architectural and/or engineering standards.

5. Draftspersons also know how to add additional information to their drawings, such as a detailed parts list, which tells the quantity, size, material, and mass of all items drawn.

6. They visit work sites to take measurements and gain first-hand information about a project.

Qualifications

Draftspersons must be proficient in using drafting instruments with skill, accuracy, neatness, and speed. They should have a good ability to picture objects in their minds and to portray them on drawings. Good draftspersons should also be able to work as part of a team and be able to follow directions.

A good background in mathematics, physical sciences, mechanical drawing, and drafting is required. Both high-school and post-high-school programs offer courses in drafting.

Cotton, wool, and silk were once the basic raw materials for textile manufacturing. Today, synthetic fibers such as nylon, acetate, rayon, and polyester are leading the raw materials.

The materials of the textile industry must pass through many processes before they become finished fabrics. All textiles are produced by spinning a fiber into yarn, weaving or knitting the yarn into fabric, and then applying dye and other treatments that might prevent shrinkage, give added strength, or add a silky luster.

Spinning and weaving generally take place in the same plant. Fabric is produced on looms, which weave the yarn. Skilled workers operate as many as 200 looms at a time. Knitted cloth is produced by machines that intermesh yarn loops. Then the textile designers create the patterns or designs that are woven into or printed onto the fabrics.

Once the fabric has been made, the clothing design stage takes place. After designs are deemed acceptable, patterns for each article of clothing to be mass produced and sold are made. Then textile workers follow the patterns as they sew the clothing we wear.

---

1. Create and draw a design that is at least 8 in. long and 4 in. wide. How many times could you repeat this design on a piece of fabric that is 45 in. wide and 2 yd long?

2. A textile worker cuts out several pieces of material using the pattern below.
   a. Measure each dimension in the pattern to the nearest inch.
   b. How many times could this pattern be cut out of a piece of fabric that is 60 in. wide and 3 yd long?

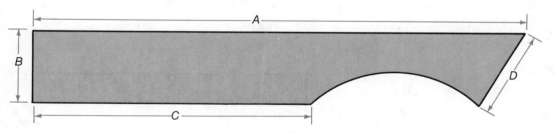

Chapter 1 Test

April 3

1. If the March 3 reading was 7783 for the meter at the right, how many kilowatt hours were used for the month? (1-1)

December 5

2. If the November 5 reading was 1002 for the meter at the right, how many kilowatt hours were used for the month? (1-1)

3. Read the tachometer below to the nearest hundred revolutions per minute. (1-2)

4. Measure the lengths shown to the nearest thirty-second of an inch. (1-3)

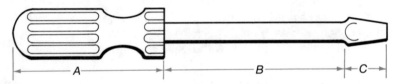

5. Measure the lengths shown to the nearest millimeter. (1-3)

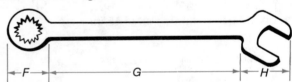

6. **a.** Change $\frac{7}{8}$ in. to sixteenths. (1-4)   **b.** Reduce $\frac{24}{32}$ in. to eighths. (1-4)

   c. Change 5 cm to millimeters. (1-4)   **d.** Change 0.35 m to millimeters. (1-4)

7. Interpolate each length in thirty-seconds of an inch. (1-5)

   a. E   **b.** F

   Interpolate each length in millimeters. (1-5)

   c. G   **d.** H

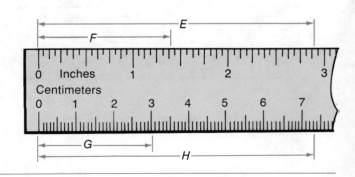

8. Sketch the front view of the object below. (1-6)

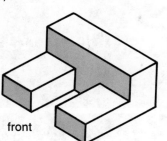

front

9. Sketch the top view of the object. Show solid, hidden, and center lines. (1-7)

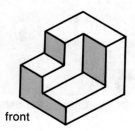

front

10. Trace the figure. Then indicate the dimensions in millimeters. (1-8)

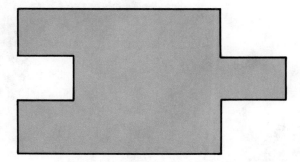

11. Trace the figure. Then indicate the dimensions in decimal inches. (1-9)

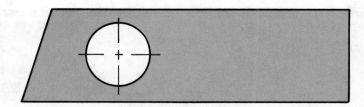

12. Find the dimension limits and allowance for the cylinder below. (1-10)

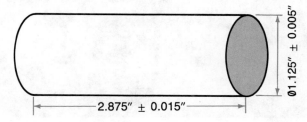

Ø1.125" ± 0.005"

2.875" ± 0.015"

Chapter 2
Measurements, Fractions, Decimals, and Percent

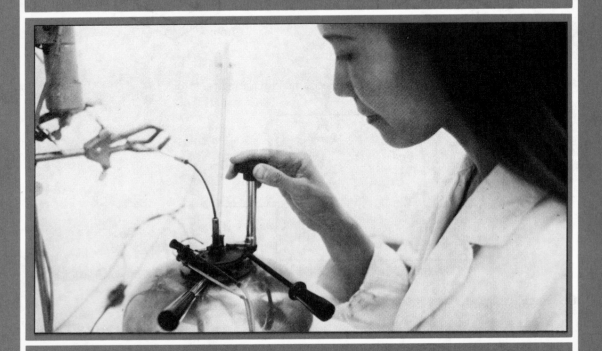

After completing this chapter, you should be able to:

1. Add, subtract, multiply, and divide measurements in the U.S. and metric system.

2. Work with dimensions expressed in inches or feet and inches.

3. Use a micrometer to measure length, width, or thickness to the nearest thousandth of an inch or hundredth of a millimeter.

4. Estimate and calculate the solutions to problems involving fractions and decimals.

5. Use percents to solve various work-related problems.

Measurement Calculations

2-1 Approximate Numbers and Significant Digits

Numbers arrived at by definition or by the counting process are *exact*. For example, if a computer counts the pages it has printed as 986, this number is exact. When we say there are seven days in one week, the seven is exact, since this is a definition.

When numbers are arrived at through measurement, they are *approximated*. Thus, we use the number of **significant digits** in a measurement to determine its accuracy. The significant digits are the non-zero digits or any zero whose purpose is not just to place the decimal point. In other words, significant digits include:

- all non-zero digits;
- zeros separating two non-zero digits;
- zeros that appear to the right of a non-zero digit in a decimal place.

Note the significant digits in the following measurements.

43.6 in. has *three* significant digits.	(All digits are non-zero.)
1700 mi has *two* significant digits.	(We assume that the two zeros are place holders.)
0.027 in. has *two* significant digits.	(The zeros are for proper location of the decimal point.)
604.3 kg has *four* significant digits.	(The zero is not used for the location of the decimal point. It shows specifically the number of tens in the number.)
3.50 m has *three* significant digits.	(The zero is not necessary as a place holder, and should not be written unless it is significant.)

In calculations with measurements, it is important that the result does not give a false indication of its accuracy. Consequently, the following guidelines are used.

- When measurements are *added* or *subtracted*, the result is precise to the place to which the least accurate measurement involved in the calculation is given.

  5.2 in. (precise to tenths)   + 6 in. (precise to ones)   = 11 in. (precise to ones)

- When measurements are *multiplied* or *divided*, the result is expressed with the number of significant digits of the least accurate measurement.

  64 in. (2 significant digits)   ÷ 0.2 s (1 significant digit)   = 300 in./s (1 significant digit)

Example 1

A cord 7.2 m long and a cord 0.525 m long are spliced together. What is the combined length of cord?

Solution

At first, it appears from the calculator key presses below that the total cord length is 7.725 m.

7 **·** **2** **+** **·** **5** **2** **5** **=**   7.725

However, the first length is precise only to the tenths place and the digit in this position was obtained by rounding off. It could have ranged from 7.15 m to 7.24 m. Therefore, the result must be rounded to the nearest tenth. The combined length of cord is 7.7 m, to two significant digits.

Example 2

What is the area of a garden plot having a length of 18.56 ft and a width of 25.4 ft?

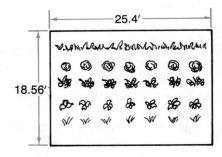

Solution

The measurement 18.56 ft has four significant digits, while 25.4 ft has three. The multiplication result should then be rounded to three significant digits.

1 **8** **·** **5** **6** **x** **2** **5** **·** **4** **=**   471.424

The area of the garden plot is 471 ft², to three significant digits.

EXERCISES

A Are the measurements given *exact* or *approximate*?

1. There are 12 months in one year.

2. The velocity of light is 298,000 km/s.

3. A 5 kg package of nails was bought on sale.

4. A building lot is 48 m by 130 m.

5. There are 31 days in December.

B How many significant digits are in the given measurements?

6. 19.5 ft

7. 3907 mi

8. 420 L

9. 7.0 in.

10. 0.0375 in.

11. 0.0032 km

12. 6.08 gal

13. 3.004 m

14. 0.01 in.

Round off each of the given measurements to:

a. three significant digits;
b. two significant digits.

15. 3.944 in.

16. 50.38 yd

17. 75,398 mi

18. 40,320 mi

19. 0.8250 in.

20. 0.20604 in.

Add or *subtract* the given pair of measurements with a calculator. Round off the result to the correct number of significant digits.

21. 3.5 in. + 9.95 in.

22. 6.7 ft − 0.032 ft

23. 0.1 lb + 78.0 lb

24. 20.9 cm − 0.02 cm

25. 5 m − 0.5 m

26. 1.7 cm + 0.75 cm

Multiply or *divide* the given pair of measurements with a calculator. Round off the result to the correct number of significant digits. (Assume numbers without units to be exact.)

27. 6 ft × 2.8

28. 7.5 mi ÷ 2.375

29. 0.25 × 0.001 in.

30. 9.2 m ÷ 6.575

31. 0.002 km × 12

32. 700.5 m ÷ 100.0

Tricks of the Trade

Thickness Gauge

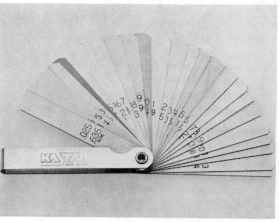

A thickness or feeler gauge can be used to set the proper gap on spark plugs for all kinds of cars. The gauge can also be used to check the spark plug gap on farm and home power equipment such as tractors, boats, snowmobiles, lawn mowers, outboard engines, motorcycles, and so on.

If the spark plug gap is not properly set, you will use excess fuel and cause more pollution. Spark plugs should be cleaned, checked, and adjusted every 5000 mi (for cars) and 12.5 mi (for small engines).

Tempered steel leaves, which have the thickness marked in the U.S. and metric systems, are nested in a steel frame. The thickness gauge at the right allows measurements ranging from 0.0015 in. to 0.025 in. or from 0.38 mm to 0.635 mm.

2-2 Adding and Subtracting Measurements

There are a wide variety of jobs that require skill in adding and subtracting measurements. Remember that when decimal measurements are added or subtracted, the result should be precise to the place to which the least accurate measurement involved in the calculation is given.

Example 1

Find the length of the handle of the carpenter's brace shown at the right in centimeters.

Skill(s) 1, 6, 10, 13, 14, 15

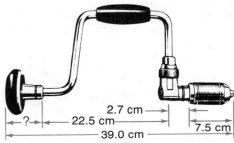

2.7 cm
?
22.5 cm
7.5 cm
39.0 cm

Solution

Estimate the solution first: 39.0 − (22.5 + 2.7 + 7.5) is about 40 − (23 + 3 + 8), or 6.

A calculator can be used to solve the problem.

| 2 | 2 | · | 5 | + | 2 | · | 7 | + | 7 | · | 5 | = | 32.7

| 3 | 9 | − | 3 | 2 | · | 7 | = | 6.3

The length of the handle of the carpenter's brace is 6.3 cm.
The result is precise to the tenths place and the estimate shows the solution is reasonable.

Example 2

If a fashion designer worked overtime $3\frac{1}{2}$ h, $2\frac{1}{4}$ h, and 5 h on three different days, what is the total number of overtime hours worked? (State the result as a mixed fraction.)

Solution

Estimate the solution first: $3\frac{1}{2} + 2\frac{1}{4} + 5$ is about $4 + 2 + 5$, or 11.

$$3\frac{1}{2} + 2\frac{1}{4} + 5 = 3\frac{2}{4} + 2\frac{1}{4} + 5$$

$$= 10\frac{3}{4}$$

| 3 | · | 5 | + | 2 | · | 2 | 5 | + | 5 | = | 10.75

The total amount of overtime worked is $10\frac{3}{4}$ h.

EXERCISES

A Use a calculator where possible.
Find the sum.

1. $\frac{5}{8}$ in. $+ \frac{1}{16}$ in. $+ \frac{1}{8}$ in.

2. $6\frac{1}{4}$ oz $+ \frac{3}{4}$ oz $+ 12\frac{3}{8}$ oz

Find the difference.

3. 45.5 lb $-$ 3.75 lb

4. 2.5 gal $-$ 0.625 gal

U.S. System

B 5. How long is the hammer head in inches?

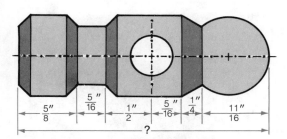

Metric System

6. What is the length of the base of the platform in meters?

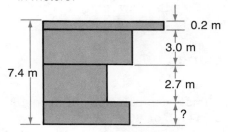

7. What is the total length of the rivet in inches?

8. What are lengths *A* and *B* in the switch plate? (Dimensions are in millimeters.)

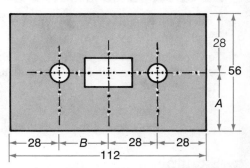

9. What is the height of the base of the pedestrian bridge in feet?

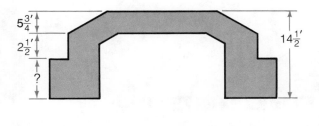

10. What is the total length of the kitchen chopping block in centimeters?

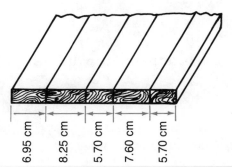

2-3 Multiplying Measurements

Skilled technicians often need to multiply measurements to solve problems that normally come with their jobs. Remember that when measurements are multiplied or divided, the result is expressed with the number of significant digits of the least accurate measurement.

Example 1

Putting caps on bottles.

An automatic bottler has a capacity of 90.5 L/min. How many 1 L bottles can it fill in one $7\frac{1}{2}$ h workday?

Skill(s) 1,7,9,10,11,16

Solution

Change $7\frac{1}{2}$ h to 450 min.

Estimate the solution: 90.5 × 450 is about 90 × 450, or 40,500.

A calculator can be used to solve the problem.

| 9 | 0 | . | 5 | × | 4 | 5 | 0 | = | 40725

The time, $7\frac{1}{2}$ h or 450 min, has three significant digits.

To three significant digits, 40,700 bottles can be filled in one workday. The result is close to the estimate.

> Note: When time is given in this text in hours, or hours and minutes, assume the measurement accurate to the nearest minute.

Example 2

If a printing press uses $4\frac{3}{4}$ oz of ink in 1 h, how much will it use during a $3\frac{1}{2}$ h run?

Solution

Estimate the solution first: $4\frac{3}{4} \times 3\frac{1}{2}$ is about 5 × 4, or 20.

$$4\frac{3}{4} \times 3\frac{1}{2} = \frac{19}{4} \times \frac{7}{2}$$
$$= 16\frac{5}{8}$$

| 4 | . | 7 | 5 | × | 3 | . | 5 | = | 16.625

The printing press will use $16\frac{5}{8}$ oz of ink in $3\frac{1}{2}$ h.
The solution is reasonable.

EXERCISES

Use a calculator where possible. (For this text, round all calculations with money to the nearest cent where needed.)

A　**1.** $6\frac{1}{2}$ pages in 1 h, ___?___ pages in $5\frac{1}{4}$ h

2. 1 lb costs \$3.95, $5\frac{3}{4}$ lb cost ___?___

3. $8\frac{1}{2}$ mi in 1 h, ___?___ mi in $7\frac{1}{3}$ h

4. 1 load costs \$49.50, $3\frac{1}{2}$ loads cost ___?___

5. A 16.5 mm rod weighs 1.05 kg. A rod 2.5 times as long weighs ___?___ .

6. There is an average of 36.2 posts along 1 mi of road. In 3.125 mi there are ___?___ *whole* posts.

B　**7.** If one kilowatt hour of electricity costs 7.75¢, what is the cost for 8.5 kW·h?

8. The thickness of a sheet of No. 1 gauge sheet metal is 0.28 in. How high is a stack containing 48 sheets?

9. A furniture refinisher is calculating the amount of material needed to repair the four legs of a stool. It is found that a piece of lumber $13\frac{3}{8}$ in. long for each leg is needed and $\frac{1}{4}$ in. is allowed for waste at each end.

a. How many inches of lumber are needed for four legs?
b. What is the answer in feet and inches?

10. A surveyor marks off 12 adjacent lots. The frontage of each lot measures $\frac{1}{10}$ mi. What is the total frontage in miles?

11. The distance, F, across the top of a hexbolt is about 0.866 times D. Find F when D is 4.40 cm.

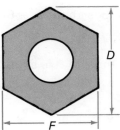

12. An automobile gasoline tank can hold $15\frac{1}{2}$ gal. How many gallons are in the tank when it is $\frac{3}{4}$ full?

13. A tile mason uses $\frac{7}{8}$ in. square tiles to form a wall mosaic. How wide is the wall if 184 tiles are used in each row?

C　**14.** Each pound of brass alloy used for stampings contains 0.58 lb of copper, 0.405 lb of zinc, and 0.015 lb of lead. A bar of this alloy weighs 8.5 lb. Find the amounts of copper, zinc, and lead in the bar.

15. Suppose a man breathes 18 times per minute. When resting, he takes in 0.75 L of air per breath. When doing light work, he takes in 1.62 L. For heavy work, he takes in 2.44 L. How many litres of air does the man use if he rests for 15 min, works lightly for 15 min, and then works hard for 15 min?

2-4 Dividing Measurements

Often a measurement needs to be divided in order to solve an on-the-job problem. Recall that when dividing measurements, results are expressed with the number of significant digits of the least accurate number.

Example 1

Skill(s) 1,8,9,10,11,17

The cross section of a laminated helicopter rotor blade is shown below. Find the thickness of each sheet of laminating material if the five layers of the blade measure 6.4850 in.

6.4850 in.

Solution

Estimate the solution first: 6.4850 ÷ 5 is about 6 ÷ 5, or $1\frac{1}{5}$.

6 · 4 8 5 ÷ 5 =   1.297

The thickness of each sheet is 1.2970 in., to five significant digits. The result is close to the estimate.

Example 2

A wood screw advances 1.8 mm for each complete turn. How many *complete* turns will advance it at least 12.4 mm?

Solution

Estimate the solution first: 12.4 ÷ 1.8 is about 12 ÷ 2, or 6.

1 2 · 4 ÷ 1 · 8 =   6.8888889

Seven *complete* turns will advance it at least 12.4 mm, since 7 × 1.8 = 12.6.

EXERCISES

Use a calculator where possible.

A **1.** $6\frac{1}{2}$ gal for 5 batches, _?_ gal for 1 batch

2. $17\frac{1}{2}$ loads in $3\frac{1}{2}$ h, _?_ loads in 1 h

3. 24.5 ft for 8.75 h, _?_ ft for 1 h

4. $3\frac{3}{8}$ lb for $12.50, 1 lb costs _?_

B **5.** A recipe calls for $2\frac{1}{3}$ packages of frozen spinach. How many *whole* recipes can be made from 20 packages?

6. A certain task takes $1\frac{1}{2}$ h to complete. How many times could the *complete* task be repeated in a 40 h workweek?

7. How many *whole* roofing boards, that are each $\frac{13}{16}$ in. thick, are there in a stack 2 ft 2 in. high?

8. How many *whole* pieces of solid copper wire $2\frac{1}{2}$ in. long can be cut from a piece 15 ft 6 in. long?

U.S. System

9. Eight *complete turns* of a nut advance it one inch. How far does it advance on each full turn?

Metric System

10. How many *whole* rectangular insulators 2.5 cm long can be cut from an 8.75 cm nylon strip?

11. A 250 in. roll of TV antenna lead costs $8.75. What is the cost per foot?

12. An earthmover removed 64.5 m³ of gravel in 4 h. What is its rate per hour?

13. A sheet of vinyl plastic is 25.5 in. wide. How many strips, each 2.625 in. wide, can be cut from the sheet?

14. A strip of aluminum 25.2 cm long is to be sheared into *exactly* six pieces of equal length. What is the length of each piece?

15. Weight bars are used to press tablet sheets together before the glue is applied. How many *whole* bars $\frac{9}{16}$ in. wide can be placed side by side in a space $8\frac{1}{2}$ in. wide?

16. A street 106.5 m long is to be curbed with blocks of granite 0.75 m long. If 0.50 cm between the blocks is allowed for expansion, how many *whole* blocks would have to be bought to completely make the curb?

17. A piece of rubber weather stripping 26.25 in. long is required along the bottom of a storm window. How many *whole* strips can be cut from a roll 43.5 ft long?

18. An airplane propeller 10.75 cm thick is made of *exactly* five layers of uniformly thick laminating material. Find the thickness of one layer of this material.

19. Five and one half yards of ribbon cost $7.10. How much does one yard cost?

20. Six and one half meters of drapery fringe cost $21.13. What is the cost per meter?

21. Sixteen and a half feet of elastic cord cost $3.47. What is the cost per foot?

C 22. Pieces of bias tape, each 0.7 m long, are to be cut from a piece 3.0 m long. One extra centimeter of tape is to be folded back at each end of each 0.7 m piece.

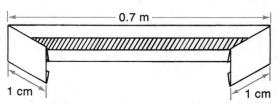

0.7 m

1 cm   1 cm

    a. How many *whole* 0.7 m pieces can be cut?

    b. How much bias tape will be left over?

23. Eight T-rails 0.705 m long are to be cut from a piece of stock 6.00 m long. In cutting each piece, 0.01 m of stock is wasted.

    a. How much stock is used in cutting the eight pieces?

    b. What is the length of the piece of unused stock?

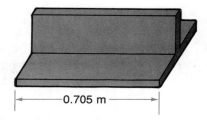

0.705 m

Technology Update

Railroads have turned to the computer to solve the problem of sorting cars in switching yards. A central computer keeps a record of every car's cargo, destination, location, and position on a train. This information, in coded labels on each car, is "read" by electronic scanners as the car enters the yard. The scanners send the data to the central computer. The computer then decides on the proper destination for that car. It relays this information to a computer in each yard, which automatically switches the car to the proper track.

To safely guide the car, track sensors pick up the weight and speed of the car, wind velocity, and other data needed for correct braking. The car is then controlled by the computer until it is stopped and coupled to its new train. This new computerized car sorting has resulted in fewer errors and faster, safer service.

Research Project:

Find two other examples of using computers for sorting.

Self-Analysis Test

Use a calculator where possible.

1. How many significant digits are in each measurement?

   a. 37.2 in.   **b.** 0.0075 ft   **c.** 107 mm   **d.** 730 cm

2. Find lengths G and H on the door hinge below.

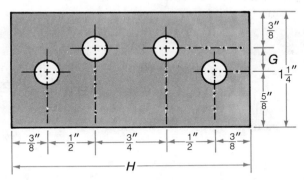

3. A machinist uses a grinder to remove $\frac{5}{16}$ in. from the end of a steel bar that is $3\frac{7}{8}$ in. long. How long will the bar be after the grinding?

4. An electronics kit contains a spool of insulated wire 18 ft long. The following lengths are to be cut: $12\frac{5}{8}$ in., $33\frac{3}{16}$ in., $4\frac{7}{8}$ in., and $25\frac{1}{2}$ in. The instruction manual suggests allowing $\frac{1}{16}$ in. for each cut.

   a. Draw a diagram to show how to cut the wire.
   b. What length will be cut from the spool?
   c. How much wire will remain on the spool after cutting?

5. Bricks 20 cm long and 7 cm high are used to build a brick wall 3.70 m long and 23.51 m high. The mortar between the bricks is 1 cm thick.

   a. How many bricks are needed for one layer? (Count a fraction of a brick as a whole brick.)
   b. How many layers of brick are in the completed wall?

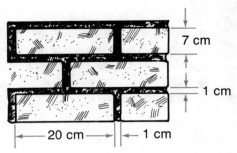

6. A computer printer can print two lines in one second. At this rate, how many *complete* lines can be printed in $3\frac{1}{2}$ min?

7. How many *whole* $10\frac{5}{8}$ in. pieces can be sawed from a board $5\frac{5}{8}$ ft wide?

8. In a square, the side is always about 0.707 times the length of the diagonal. Find the length of the side when the diagonal is 3.55 cm long.

Measurement Applications

2-5 Changing Inches to Feet and Inches

In industry, most lengths up to 72 in. are stated in inches only. Lengths greater than 72 in. are usually expressed in feet and inches. For example, a strip of plastic may measure 21 in. wide (*not* $1\frac{3}{4}$ ft) but a door frame may be 7 ft 2 in. high (*not* 86 in.). However, we should be able to make calculations with dimensions in either form.

Example 1

Skill(s) *11, 13, 14, 15*

A section of upholstery fabric $17\frac{1}{2}$ in. long is cut from a piece 6 ft $4\frac{3}{4}$ in. long. What is the length of the remaining piece of fabric?

Solution

Change 6 ft $4\frac{3}{4}$ in. to $76\frac{3}{4}$ in. and then calculate.

$$76\frac{3}{4} - 17\frac{1}{2} = 76\frac{3}{4} - 17\frac{2}{4}$$
$$= 59\frac{1}{4}$$

| 7 | 6 | . | 7 | 5 | − | 1 | 7 | . | 5 | = | 59.25 |

A piece of upholstery fabric $59\frac{1}{4}$ in. long is left.

Example 2

When installing an automatic car wash, a plumber joins three sections of pipe. The pipes are 6 ft 8 in., 11 ft 6 in., and 8 ft 4 in. in length. What is the total length of the pipes?

Solution

Step 1: Add the three lengths.

    6 ft 8 in.
  11 ft 6 in.
+ 8 ft 4 in.
  25 ft 18 in.

Step 2: Simplify the number of feet and inches.

25 ft 18 in. = 25 ft + 18 in.
             = 25 ft + (1 ft + 6 in.)
             = 26 ft 6 in.

The total length of the pipes is 26 ft 6 in.

EXERCISES

Use a calculator where possible.

A Add or subtract. Express answers in inches, or feet and inches.

1. 14 in. + $3\frac{3}{4}$ in. + 6 in.

2. 19 in. + $27\frac{1}{2}$ in.

3. $38\frac{3}{8}$ in. + $46\frac{3}{4}$ in.

4. 6 ft 4 in. − 26 in.

5. 9 ft $8\frac{3}{8}$ in. − $24\frac{1}{4}$ in.

6. 7 ft 2 in. − $10\frac{1}{2}$ in.

7. 13 ft + 6 ft $3\frac{3}{4}$ in. + $5\frac{1}{4}$ in.

8. 9 ft + 2 ft $4\frac{1}{2}$ in. − $5\frac{1}{2}$ in.

9. 9 ft 4 in. − (32 in. + $6\frac{1}{2}$ in.)

10. 13 ft $\frac{1}{2}$ in. − (36 in. + $12\frac{1}{4}$ in.)

11. 8 ft + (42 in. − $10\frac{1}{8}$ in.)

12. 6 ft 5 in. − (3 ft − $2\frac{1}{2}$ in.)

B 13. A plumber needs pieces of copper tubing of the following lengths: 8 ft 6 in., 12 ft 4 in., 9 ft 8 in., and 7 ft 3 in. How much pipe is needed in all?

14. A book binding machine used 237 ft 3 in. of thread from a new spool to finish a job. If a spool holds 1000 ft of thread, how much thread is left on the spool?

15. Three hundred fifty feet of pipe must be laid in a new industrial center.

    a. So far, 95 ft 8 in. and 80 ft 6 in. sections of pipe have been laid. How much pipe has been put in place?

    b. How much more pipe must be laid?

16. A computer printer uses $14\frac{7}{8}$ in. wide paper. If the left- and right-hand margins are each $\frac{7}{8}$ in., how much space remains for printing?

17. Five laboratory desks are placed along a 25 ft wall, side by side. If each desk is 54 in., how much space is there at the end of the wall?

18. A pattern maker cuts a 36 in. piece and a $12\frac{1}{4}$ in. piece of vinyl from a 14 ft $\frac{1}{2}$ in. piece of stock vinyl. How much vinyl is left on the stock piece?

C 19. From a 4 yd piece of broadcloth, a 45 in., a 30 in., and a 2 ft 8 in. length are cut. How much fabric is left over?

2-6 Micrometers

A **micrometer** is an instrument used to make precise measurements of length, width, or thickness. Depending on whether you are using a U.S. or a metric micrometer, measurements can be made to the thousandth of an inch or hundredth of a millimeter.

A micrometer is calibrated according to two scales: one on the *sleeve*, the other on the *thimble*, as shown at the right.

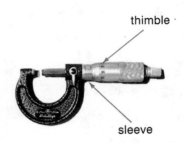

thimble

sleeve

Example 1

Skill(s) **6.7**

The U.S. system micrometer reading at the right is a measure of the thickness of a sheet of steel. What is the thickness to the nearest thousandth of an inch?

0.001"

0.025"   0.100"

Solution

1. Multiply the largest number on the sleeve by 0.100 in.
2. Multiply the number of additional subdivisions by 0.025 in.
3. Multiply the number on the thimble by 0.001 in.
4. Total the three products.

$5 \times 0.1 = 0.5$
$2 \times 0.025 = 0.05$
$24.5 \times 0.001 = 0.0245$
_____
0.5745

The steel sheet is 0.575 in. thick.

Example 2

A measure of the outside diameter of a copper tube is shown by the micrometer reading at the right. What is the thickness to the nearest hundredth of a millimeter?

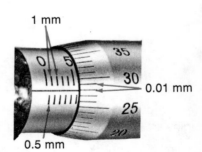

1 mm

0.01 mm

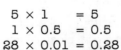

0.5 mm

Solution

1. Multiply the largest number on the sleeve by 1 mm.
2. Multiply the number of additional subdivisions by 0.5 mm.
3. Multiply the number on the thimble by 0.01 mm.
4. Total the three products.

$5 \times 1 = 5$
$1 \times 0.5 = 0.5$
$28 \times 0.01 = 0.28$
_____
5.78

The outside diameter is 5.78 mm.

EXERCISES

U.S. System

B Read each micrometer to the nearest thousandth of an inch.

Metric System

Read each micrometer to the nearest hundredth of a millimeter.

1.

2.

3.

4.

5.

6.

7.

8.

9.

10.

2-7 Solving Measurement Problems

Skill(s) ___5___

Example 1

A cutter in a textile factory needs to cut four pieces of fabric $13\frac{3}{4}$ in. wide from a 60 in. width. Is the fabric wide enough if a $\frac{5}{8}$ in. seam allowance is needed on each side of each piece?

Solution

Make a sketch of the problem; then calculate.

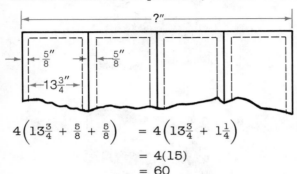

$$4\left(13\frac{3}{4} + \frac{5}{8} + \frac{5}{8}\right) = 4\left(13\frac{3}{4} + 1\frac{1}{4}\right)$$
$$= 4(15)$$
$$= 60$$

[1][3][·][7][5][+][·][6][2][5][+][·][6][2][5][=][×][4][=]   60

Yes, the fabric is wide enough.

Example 2

The renovation of the basement of a house required $85\frac{1}{2}$ h of a carpenter's services at \$42/h and $15\frac{3}{5}$ h of a plumber's services at \$50/h. What was the total cost of labor for the job to the nearest dollar?

Solution

$$\left(85\frac{1}{2} \times 42\right) + \left(15\frac{3}{5} \times 50\right) = \left(\frac{171}{2} \times 42\right) + \left(\frac{78}{5} \times 50\right)$$
$$= 3591 + 780$$
$$= 4371$$

[8][5][·][5][×][4][2][=][+][1][5][·][6][×][5][0][=]   4371*

The total cost of labor for the job was \$4371.

* For calculators with built-in order of operations

EXERCISES

Use a calculator where possible.

U.S. System

A 1. $4\frac{3}{4}$ h available in the office
$1\frac{1}{3}$ h used to input data
$2\frac{3}{4}$ h used to make corrections
How much unused time in hours is there?

3. four loads of 16 baskets delivered
contents of $1\frac{1}{4}$ loads damaged
How many undamaged baskets are there?

Metric System

2. 3.50 m long
45 cm cut off
1.70 m also cut off
How much is left in meters?

4. 44 plates of 0.75 kg each
32 kg per crate
Can the crate hold all 44 plates?

B 5. Elastic cord is selling for $0.21/ft. What is the price of 12 ft 6 in. of elastic cord to the nearest cent?

6. A technical drawing 15.9 cm wide by 20.6 cm high is to be centered on a sheet of paper 297 mm by 420 mm. How much of a horizontal and a vertical margin should be left on each side of the drawing?

7. A service technician finds that three pieces of copper tubing $8\frac{1}{4}$ in., $9\frac{7}{8}$ in., and $14\frac{1}{16}$ in. long are needed to install the cooling unit in an air conditioner. Allowing $\frac{1}{16}$ in. for each cut, how much tubing is needed to assemble 16 such units? Give the answer in feet and inches.

8. A tie rack is to be manufactured from a piece of clear plastic according to the plan below. Equally spaced holes must be drilled so that plastic pegs can be inserted. How many holes can be drilled?

Drill

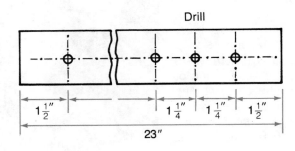

9. A computer lab uses $3\frac{3}{4}$ boxes of printer paper each month. At $26.75 per box, will $500 cover the cost of paper for one year?

10. Two pieces of inlaid asphalt flooring each 10 ft $4\frac{3}{4}$ in. long must be cut from a 24 ft roll. A cutting allowance of 2 in. is made for each piece. How much flooring will remain after cutting?

11. Twenty-five brackets with the dimensions shown must be stamped from a strip of sheet metal. **(All dimensions are in centimeters.)**

    a. How long is one bracket?

    b. What is the width of the piece of stock needed for the bracket?

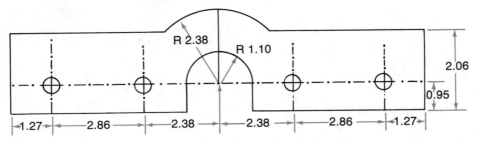

12. One wire has a diameter of 1.48 cm and a second wire has a diameter of 0.05 cm.

    a. How many times greater is the diameter of the first wire than that of the second?

    b. A **micrometer** (μm) is 0.001 mm. What is the diameter of each wire in micrometers?

Technology Update

Industry today has turned to the use of robotics in order to maintain a competitive edge in national and international markets. The introduction of automation in the workplace has led to increased productivity and quality while reducing costs.

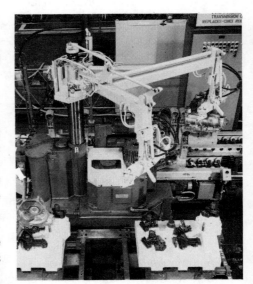

Essentially, industrial robots are machines with a manipulator arm that can perform motions repetitively and tirelessly. Robots can do a surprising range of tasks, but most are simple pick-and-place maneuvers such as loading or unloading machines. There are also sophisticated welding or painting robots in which the manipulator arm can be programmed to follow a continuous path through space instead of simply going to various predetermined points. Robots have already established their value in the performance of dangerous, dull, and dirty functions. Some robotic functions include: 1. picking up and moving parts and/or hazardous materials; 2. welding; 3. spray painting; 4. mixing chemicals at a safe distance; 5. performing inspection functions of manufactured products by vision and tactile sensing systems; and 6. assembly operations.

Research Project:

Describe how two companies in your area use robots.

Self-Analysis Test

Use a calculator where possible.
For questions 1 and 2, give the answers in inches, or feet and inches.

1. Find the length of woven wire needed to enclose a rectangular plot of ground with the dimensions shown below.

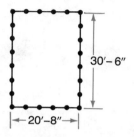

2. A piece of plywood 6 ft 8 in. long must be trimmed by $10\frac{3}{4}$ in. to fit between shelf supports. What will be the finished length?

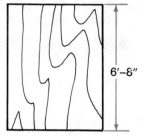

U.S. System

3. Find the cost of 44.5 gal of asphalt sealer at $2.13/gal to the nearest cent.

5. Give the micrometer reading to the nearest thousandth of an inch.

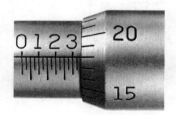

Metric System

4. Surgical tubing costs $3.27/m. What is the cost of 6.6 m of tubing to the nearest cent?

6. Give the micrometer reading to the nearest hundredth of a millimeter.

7. The weight of 7.5 in. of aluminum wire is 1.575 oz. What is the weight of 1 in. of the wire?

8. Yellow multi-strand polypropylene rope with a 1 cm diameter costs $1.14/m. What is the cost of 42.5 m of the rope to the nearest cent?

9. Orchard growers use 100 ft rolls of netting to cover certain trees. If a smaller tree needs 37 ft 6 in. of netting and a larger one needs 42 ft 1 in., is there enough netting to cover another smaller tree?

10. A 12 m roll of wall covering must be cut into strips to cover a wall 2.4 m high. An allowance of 0.1 m is made at the end of each piece for matching the pattern.

  a. How many *whole* 2.4 m strips can be cut from the roll?
  b. How much material will be wasted?

Using Percent

2-8 Percent of Enlargement or Reduction

Machines that reproduce copies of print or pictorial material have the ability to make either an enlargement or a reduction of the material. Some of these machines use a specified set of percents for enlargements and reductions, like a photocopy machine. Others, such as those used by phototypesetters, are capable of making enlargements and reductions of any reasonable percent.

Example 1

Skill(s) _22, 25_

The dimensions of a photograph are 5.0 in. by 7.0 in. What are its dimensions if each is reduced to 70%?

Solution

The dimensions of the photograph will be 70% or 0.7 times their original lengths. Use a calculator to solve the problem.

| 5 | × | · | 7 | = | 3.5

| 7 | × | · | 7 | = | 4.9

After being reduced to 70%, the photograph will be 3.5 in. by 4.9 in., to two significant digits.

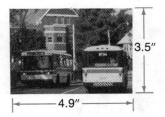

Example 2

The dimensions of a 4.0 in. by 6.0 in. photograph are to be enlarged to 250%. What will the dimensions of the enlarged photograph be?

Solution

The dimensions of the photograph will be 250% or $2\frac{1}{2}$ times their original lengths. Use a calculator to solve the problem.

| 4 | × | 2 | · | 5 | = | 10

| 6 | × | 2 | · | 5 | = | 15

After a 250% enlargement, the photograph will be 10 in. by 15 in., to two significant digits.

EXERCISES

Use a calculator where possible.

A Will the measurement be reduced or enlarged after the multiplication? What will the new measurement be?

1. 75% × 9 in.

2. 30% × 15 lb

3. 150% × 42 L

4. 20% × 840 m

5. $12\frac{1}{2}$% × 8 km

6. 110% × 90 lb

7. 118% × 12 ft

8. 150% × 16 in.

9. $37\frac{1}{2}$% × 16 m

B **10.** A 16 in. by 28 in. picture frame needs to be reduced to 75%. What will be the new dimensions?

11. A carpenter decides to enlarge the dimensions of a planned deck to 130%. What will the dimensions be if the original plans were for a 15.0 ft by 32.0 ft deck?

12. A 2.0 mm by 3.0 mm slide preparation will appear to be enlarged to 400% when the laboratory technician places it under a microscope. What will be its dimensions as seen under the microscope?

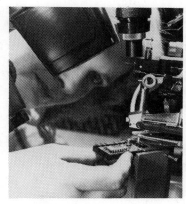

13. A plumbing-supply house wants to reduce its inventory of 65 laundry faucets to 40%. How many faucets does it need to sell?

14. An appliance store had 15 air-conditioning units before the start of summer. Before summer begins, the number of units is expected to increase to 180%. How many more units must be obtained?

15. A 15.0 in. by 24.0 in. opening in a furnace needs to be enlarged to 120% in order to receive a filtering unit. What will be the new dimensions?

16. A roofer needs to increase the dimensions of a vent pipe opening to 150%. What will be the new dimensions if the present opening is 8.0 in. by 12.0 in.?

17. An electrician needs to enlarge a 2.5 cm diameter hole in order to run a conduit through a wall. What will be the diameter of the hole if it is increased to 115%?

C **18.** The amount of salt stored for use on icy roads is reduced to 95% each time a salt truck is loaded. If 5000 t of salt is available in storage the first time a truck is loaded, how much will be left after four salt trucks have been loaded?

2-9 Quality Control

In order to sell a product, a manufacturer must be sure that it is free from major defects. It also should meet the requirements for which it was designed. The testing of a product before, during, and after its development is called **quality control**.

Inspection is one way of testing the quality of a product. Measuring various parts and even x-raying internal parts can determine the quality without destroying the product.

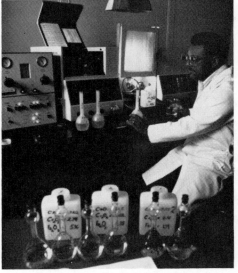

Some products are tested to see how long it will take for them to wear out. This can be done by using the product excessively over a short period of time. For example, a car might be driven 10,000 mi, but in the period of only a week. Other types of wear testing use machines to wear out the product.

During the production of some items it may not be practical to test every item. Of course, this is true if destructive testing is involved. If it is impractical to test every item, a **representative sample** is tested.

Example

Skill(s) _20, 22, 24, 25, 26_

During the testing of a representative sample of transistors it was found that 5, or 0.80%, were defective. How many transistors were tested?

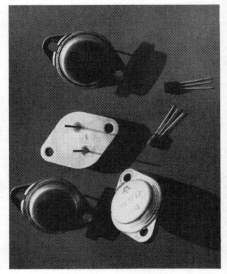

Solution

In this problem, you are looking for the *whole*. Find it by dividing the *part*, 5, by the *percent*, 0.80%.

Change 0.80% to 0.0080 and use a calculator to solve the problem.

| 5 | ÷ | · | 0 | 0 | 8 | = | 625

There were 625 transistors tested.

EXERCISES

Use a calculator where possible.

A Find the *percent* by dividing the *part* by the *whole*.

1. 4 defective out of 200
What percent are defective?

2. 6 defective out of 1200
What percent are defective?

Find the *part* by multiplying the *whole* by the *percent*.

3. 16% of 450 jars were sampled.
How many were sampled?

4. 12% of 350 boxes were sampled.
How many boxes were sampled?

Find the *whole* by dividing the *part* by the *percent*.

5. 0.6% of the movies produced is 30.
How many movies were produced?

6. 0.5% of the cars built is 40.
How many cars were built?

B Solve each problem by finding the *part*, *percent*, or the *whole*.

7. In a visual inspection on an assembly line, 18 defective bottles were found in one hour. If the bottles move through the line at the rate of 500/min, what percent were defective?

8. The quality control of a bottling company requires that no more than 0.12% of the bottles filled per month be defective.

   a. If 4,800,000 bottles are filled in one month, how many are expected to be defective?
   b. How many bottles are expected to be defective in one year?

9. Standards set by a certain company require that the maximum percent of defective parts is 0.15%. How many defective parts are allowed for 10,000 parts?

10. A representative sample of razor blades manufactured is considered to be five, or 2%, of the blades produced. How many razor blades were produced?

11. The quality control division of a company requires that no more than five out of every 1000 packages produced can be defective. What percent of the packages produced can be defective?

12. A company tested one out of every 150 batteries produced. They found 0.03% to be defective. How many defective batteries would they expect to find if they tested 10,000?

C 13. Fourteen pens in a sample of 280 were found to be defective. If more than 3% of the sample is defective, the process is said to be "out of control." Was the process out of control?

2-10 Marketing the Product

The **marketing** of a product involves advertising, pricing, and selling. The price of the product is determined by many factors. The cost of making the product, plus all the expenses of packaging, shipping, storing, advertising, and selling it determine the manufacturer's cost. The manufacturer also includes an amount for profit in the selling price. The profit cannot be too large, or the product will not sell. Also, many industries have government regulations about the amount of profit they can have. The following formula relates the selling price to cost, expenses, and profit.

Selling price = cost + expenses + profit

Example 1

Skill(s) *20, 22, 24, 25, 26*

A shoe store has set the goal of selling shoes to 25% of the local market in the coming year. If 24,520 pairs of shoes were sold in the local area last year, how many pairs of shoes can the shop expect to sell in the coming year?

Solution

$\boxed{2}\ \boxed{4}\ \boxed{5}\ \boxed{2}\ \boxed{0}\ \boxed{\times}\ \boxed{\cdot}\ \boxed{2}\ \boxed{5}\ \boxed{=}$ 6130

The shoe store can expect to sell 6130 pairs of shoes.

Example 2

What percent of the selling price is the profit on a sweater that sells for $49.95 having expenses of $6 and a cost of $39.95?

Solution

Step 1: Find the amount of profit, $49.95 − ($6 + $39.95).

$\boxed{6}\ \boxed{+}\ \boxed{3}\ \boxed{9}\ \boxed{\cdot}\ \boxed{9}\ \boxed{5}\ \boxed{=}$ 45.95

$\boxed{4}\ \boxed{9}\ \boxed{\cdot}\ \boxed{9}\ \boxed{5}\ \boxed{-}\ \boxed{4}\ \boxed{5}\ \boxed{\cdot}\ \boxed{9}\ \boxed{5}\ \boxed{=}$ 4

The amount of profit is $4.

Step 2: Find the percent of profit, $4 ÷ $49.95.

$\boxed{4}\ \boxed{\div}\ \boxed{4}\ \boxed{9}\ \boxed{\cdot}\ \boxed{9}\ \boxed{5}\ \boxed{=}$ 0.08008008

The profit is about 8%.

EXERCISES

Use a calculator where possible.

A 1. The selling price of a chair is $78.
The amount of profit is $9.36.
What is the percent of profit?

2. The selling price of a video is $59.95.
The percent of profit is 8%.
What is the amount of profit?

3. The amount of profit on a coat is $11.25.
The percent of profit for the coat is 9%.
What is the selling price?

4. A book's cost to a merchant is $3.50.
The merchant's expenses are $0.25.
The cost and expenses are 30% of the selling
price. What is the selling price of the book?

B 5. A belt sells for $12.98. The merchant knows that 65% of the selling price
must be used to pay expenses. How much were the expenses on this item?

6. The cost and expenses to a furniture dealer on a lamp is $150. This is 75%
of the selling price. What is the selling price of the lamp?

7. Marketing analysis indicates that 80,000 appliances of a certain type can
be sold each year. One company believes that it can sell its model to 43%
of the potential market. How many appliances should the company expect
to sell?

8. One company sells 45,000 books each year. This is 24% of the market.
What is the total market for this product?

9. Research indicates that a company must sell 25,000 pairs of shoes a year
to break even. Last year, the company sold 15% more than was necessary
to break even. How many pair of shoes did the company sell?

10. A dealer's sales increased from one year to the next by 12%. In one year,
$185,000 worth of merchandise was sold. What were the sales for the
following year?

11. During one year, 2,150,000 electric ranges were sold. The following year,
2,408,000 electric ranges were sold. Find the percent of increase in sales
from the first year to the second.

12. In order to make a profit, a merchant *marks up* the price of the merchandise by a certain percent of the total expenses paid. Calculate the missing
items in the table below.

	Total Expenses	Mark Up	Selling Price (Expenses + Mark Up)
a.	$ 42	50%	?
b.	$ 56	100%	?
c.	$360	200%	?
d.	$ 2	?	$10

2-11 Discounts

In order to attract more customers, store owners and suppliers often sell their products at sale prices. The **sale price** is the regular price less a discount. **Discount rates** are often stated as percents.

Example

A bookstore is offering a $12.95 atlas at 20% off the regular price. What is the sale price of the atlas?

Solution

Use a calculator to solve the problem.

Method 1: a. Find the amount of discount.

$$\boxed{1}\,\boxed{2}\,\boxed{\cdot}\,\boxed{9}\,\boxed{5}\,\boxed{\times}\,\boxed{\cdot}\,\boxed{2}\,\boxed{=} \quad 2.59$$

The discount is $2.59.

b. Find the sale price.

$$\boxed{1}\,\boxed{2}\,\boxed{\cdot}\,\boxed{9}\,\boxed{5}\,\boxed{-}\,\boxed{2}\,\boxed{\cdot}\,\boxed{5}\,\boxed{9}\,\boxed{=} \quad 10.36$$

The sale price is $10.36.

Method 2: a. Find the percent of the sale price.

$$100\% - 20\% = 80\%$$

b. Find the sale price.

$$\boxed{1}\,\boxed{2}\,\boxed{\cdot}\,\boxed{9}\,\boxed{5}\,\boxed{\times}\,\boxed{\cdot}\,\boxed{8}\,\boxed{=} \quad 10.36$$

The sale price is $10.36.

EXERCISES

Use a calculator where possible.

A 1. 20% off the regular price of $840
What is the sale price?

2. $21 discount on an $84 item
What is the percent of discount?

3. $12\frac{1}{2}$ % off the regular price of $80
What is the sale price?

4. $24.12 discount on an $402 item
What is the percent of discount?

B 5. A restaurant chef estimates that the amount of food purchased with discount coupons in one day is about 12.5%. If 584 lb of food is served on a given day, how much can be expected to be purchased on discount?

6. Copy and complete the following purchase order.

Purchase Order No. 4279

HOUSE OF PLUMBING CO.
Date: 4/5

For: Ortega's Plumbing   Phone: 555-2921
Address: 1434 Ridge Blvd.   City: Waco, Texas

QUANTITY	ITEM	UNIT PRICE	COST	DISCOUNT	SALE PRICE
5	$\frac{3}{4}$ in. unions	$0.85	$4.25	20%	$3.40
8	$\frac{3}{4}$ in. 45° ells	$0.65	?	30%	?
4	$\frac{3}{4}$ in. valves	$2.50	?	25%	?
18	$\frac{3}{4}$ in. × 6 in. nozzles	$0.95	?	15%	?
5	$\frac{3}{4}$ in. faucets	$2.00	?	20%	?
12	$\frac{3}{4}$ in. 90° ells	$0.75	?	30%	?
18	$\frac{3}{4}$ in. fittings	$0.35	?	35%	?
8	$\frac{3}{4}$ in. T-joints	$0.55	?	20%	?

Total	?
Less 2% Cash Discount	?
Net Total	?

7. Gold chains that regularly sell for $55 are on sale for $38.50. What is the discount rate?

8. A hardware supply house is offering a carton of hinges at $5.88. This is 20% off the regular price. What is the regular price?

9. A farm co-op purchases a new truck at an 18% discount. The truck originally cost $23,500. How much did the co-op pay for the truck?

C 10. A paint sale offers 10 gal of latex paint at $5.34/gal, and 18% off for each additional gallon over 10. How much will 14 gal of paint cost?

11. Which is less expensive?

    a. a radio on sale for $79.95
    b. a radio regularly priced at $89.95 but on sale at 15% off the regular price

2-12 Electronic Resistors

Electronic resistors like the one at the right are used in various items, such as radios and television sets, to control the flow of electric current. Resistors are rated in **ohms** (Ω), the standard unit of electrical resistance.

Because resistors are quite small in actual size, colored bands are used to indicate their resistance value and tolerance limits. The resistor color-code table shown below lists the meanings of various combinations of colored bands.

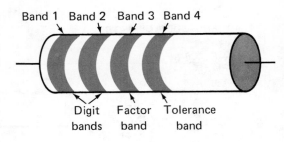

Band 1 Band 2 Band 3 Band 4

Digit bands   Factor band   Tolerance band

Resistor Color Code			
Color	Band 1	Band 2	Band 3	Band 4
Black (BK)	0	0	× 1	
Brown (BR)	1	1	× 10	
Red (R)	2	2	× 100	
Orange (O)	3	3	× 1000	
Yellow (Y)	4	4	× 10,000	
Green (GN)	5	5	× 100,000	
Blue (BL)	6	6	× 1,000,000	
Violet (V)	7	7	× 10,000,000	
Gray (GY)	8	8	× 100,000,000	
White(W)	9	9	× 1,000,000,000	
Gold (GD)			× 0.1	± 5%
Silver (S)			× 0.01	± 10%

← The *tolerance*, or the allowable error, is stated as a percent in the fourth band.

← No fourth band indicates a tolerance of ± 20%.

Study the example below to see how you use the table to find the resistance value and tolerance limits of a resistor.

Example

Skill(s) _22, 25, 31_

What is the resistance value and the tolerance limits of the resistor at the right?

Solution

The color code shows:

	Band 1	Band 2	Band 3	Band 4
	gray	green	gold	silver
	8	5	× 0.1	± 10%

The *resistance* is 85 × 0.1, or 8.5 Ω.
The *tolerance* is ± 10%.

The resistance value is 8.5 Ω ± 10%.
The tolerance limits are: 8.5 + (10% × 8.5) and 8.5 − (10% × 8.5);
8.5 + 0.85 and 8.5 − 0.85;
9.35 Ω and 7.65 Ω.

EXERCISES

Use a calculator where possible.
Indicate the tolerance limits in ohms for the given resistance values.

A   **1.** $10.5 \, \Omega \pm 20\%$ **2.** $6.2 \, \Omega \pm 5\%$ **3.** $11.4 \, \Omega \pm 10\%$

    4. $560 \, \Omega \pm 5\%$ **5.** $10{,}000 \, \Omega \pm 20\%$ **6.** $210 \, \Omega \pm 10\%$

    7. $(65 \times 1000) \, \Omega \pm 5\%$ **8.** $(39 \times 1{,}000{,}000) \, \Omega \pm 20\%$ **9.** $(57 \times 0.01) \, \Omega \pm 10\%$

    10. $(48 \times 0.1) \, \Omega \pm 20\%$ **11.** $(13 \times 10) \, \Omega \pm 10\%$ **12.** $(20 \times 1) \, \Omega \pm 5\%$

B   Copy and complete the table.

	Band 1	Band 2	Band 3	Band 4	Resistance Value
13.	Yellow	Brown	Green	Silver	?
14.	Violet	Yellow	Red	None	?
15.	Black	Blue	Gray	Gold	?
16.	White	Green	Orange	None	?

17. What are the tolerance limits for the resistor described in question 13 if there is no fourth band?

18. What are the tolerance limits for the resistor described in question 15 if band 4 is silver?

For each resistor find: **a.** the resistance value; **b.** the tolerance limits in ohms.

19.    **20.**

21.    **22.**

23.    **24.**

25.    **26.**

27.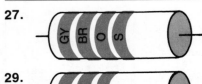

28.

29.

30.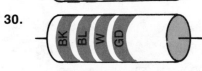

31.

32.

Check Your Skills

1. Round to the nearest hundred.

   a. 136   **b.** 851   **c.** 749   (Skill 1)

Simplify.

2. $8.2 + 4 \div 2 - 1$ **3.** $7.5 + 6 \div 3$ **4.** $9 + (8 - 4) \times 2$ (Skills 2, 5 and 6)

5. $(4.23)99$ **6.** 0.743×0.96 **7.** $(9.42)(0.02)$ (Skill 7)

8. $52 \div 0.416$ **9.** $4.1784 \div 0.5223$ **10.** $6.0027 \div 0.187$ (Skill 8)

Change to a fraction in simplest form.

11. $5\frac{6}{8}$ **12.** $\frac{24}{72}$ **13.** 0.35 (Skills 9-11)

Express the answer in simplest form.

14. $7\frac{1}{2} + 5\frac{1}{3}$ **15.** $12 - 6\frac{4}{9}$ **16.** $5\frac{1}{3} - 4\frac{1}{2}$ (Skills 13-17)

17. $\frac{4}{5} \times 5\frac{1}{3}$ **18.** $16 + \frac{3}{4}$ **19.** $\frac{4}{5} + \frac{1}{2}$

Write each decimal as a percent.

20. 0.43 **21.** 5.21 **22.** 0.0098 (Skill 20)

Write each percent as a decimal.

23. 9% **24.** 12.4% **25.** 0.62% (Skill 22)

Find the percent, part, or whole.

26. 4 is what percent of 5?   **27.** What is 7% of 25.32?   (Skills 24–26)

28. What is 97% of 534?   **29.** 12% of what number is 62.4?

Multiply by powers of ten to convert each measurement to centimeters, meters, and millimeters.

30. 23.543 km   **31.** 12 dm   **32.** 14.832 dm   (Skill 31)

Self-Analysis Test

Use a calculator where possible.

1. The dimensions of the photograph below are 1.75 in. by 4.00 in. What are its dimensions if it is reduced to 80%?

2. A 5.0 in. by 8.0 in. computer printout is to be enlarged to 150%. What will the dimensions of the enlarged printout be?

3. If 0.05% of the light bulbs tested are defective, how many defective bulbs should you expect to find if 10,000 are tested?

4. In a representative sample of cotton shirts, 21 were found to be defective. If this is 2% of the shirts in the sample, how many shirts were sampled?

5. If the selling price of a book is $4.50 and the cost to the dealer is $3.56 plus $0.40 in expenses, what is the percent of profit?

6. A laboratory technician purchases a new electronic balance for $2550. Originally the balance cost $3000. What was the percent of discount?

7. If a 15% cash discount is allowed on a $49 drill, what is the sale price?

8. There is a $1.74 discount on a pair of pliers that is priced at $14.50. What is the percent of discount?

9. What are the tolerance limits for each resistance value?

    a. $(54 \times 0.1)\ \Omega \pm 20\%$ **b.** $(37 \times 100{,}000)\ \Omega \pm 5\%$

What is the resistance value and the tolerance limit of each resistor below?

10.

11.

12.

13.

Medical laboratory technicians perform a wide variety of tests to help physicians diagnose and treat disease. Their principal activity involves the analysis and identification of substances in biological specimens.

Job Description
What do medical lab technicians do?

1. Medical lab technicians perform chemical tests to determine cholesterol levels, white and red blood cell counts, or carry out microscopic examinations of blood to detect diseases such as leukemia.

2. Medical lab technicians make cultures of body fluid and tissue samples in order to detect bacteria, parasites, or other microorganisms.

3. They also type and cross-match blood samples for transfusions.

4. Medical lab technicians use radioactive materials to help detect disease.

5. They store and label blood plasma.

6. They clean and sterilize laboratory equipment, glassware, and instruments.

7. Medical lab technicians prepare solutions following standard laboratory formulas and procedures.

8. They keep records of tests and identify specimens.

9. Medical lab technicians also do research, develop laboratory techniques, teach, or perform administrative duties.

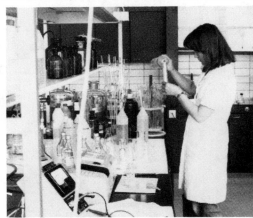

Qualifications

Medical lab technicians generally have an associate degree or certificate from a private or post-secondary trade or technical school. Accuracy, dependability, and the ability to work under pressure are important personal characteristics. Mechanical and electronic skills are becoming increasingly important.

Food Service Management

In recent years, the business of providing and managing kitchen and food services to hospitals, schools, factories, large and small offices, and other institutions has become increasingly important. There is a definite market for cohesive food service programs that provide the following kinds of services:

- prudent purchasing of well-balanced and appealing meals;
- competitive pricing of meals;
- setting up of kitchens with all necessary items from china to ovens;
- hiring site managers, head cooks, bakers, pot washers, and all other personnel required to effectively serve the customer;
- preparing cash control analyses, inventories, cost accounting studies, payroll, and the like.

1. When an 8 oz portion of lasagna is to be served, the cost of each ingredient in that portion is calculated before the cost of serving lasagna to 500 people is projected. Use the table below showing the cost of the ingredients of an 8 oz portion of lasagna to project the cost for serving lasagna to:

   a. 75 people;

   b. 180 people;

   c. 500 people.

Lasagna — 8 oz serving	
pasta	$0.49
sauce	$0.37
cheese	$0.74
meat	$0.82

   d. If the customer pays $3.95 per serving, how much money is charged to cover profit and expenses other than the cost of the food?

2. Complete the following table for each kitchen staff worker below.
   Gross pay = hours worked × hourly rate
   net pay = gross pay − deductions

	Employee	Hours Worked	Hourly Rate	Gross Pay	Deductions	Net Pay
a.	Burke, J.	35	$12.75	?	$41.35	?
b.	Garcia, A.	42	$10.50	?	$38.52	?
c.	Wood, B.	24	$8.75	?	$34.28	?
d.	Ponti, L.	38	$9.75	?	?	$328.00
e.	Lynn, V.	37.5	$11.50	?	?	$398.49

Chapter 2 Test

Use a calculator where possible.

1. Round off each measurement to two significant digits. (2-1)

    a. 0.6150 in.
    b. 729,450 mi
    c. 30.8 m
    d. 1.836 km

U. S. System

2. What are dimensions *C* and *D* in inches? (2-2)

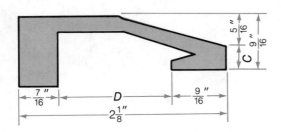

Metric System

3. What are dimensions *G* and *H*? (Dimensions are in millimeters.) (2-2)

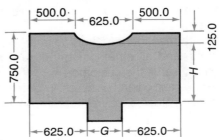

4. Gold leaf has a thickness of 0.03 in. What is the thickness of seven sheets of gold leaf? (2-3)

5. A strip of copper used in making printed circuits is 0.4 mm thick. Find the thickness of nine layers of this material. (2-3)

6. How many *whole* chain links $2\frac{1}{4}$ in. long can be cut from a piece of chain 14 ft 3 in. long? (2-4)

7. It costs $74.25 to cover a roof with $5\frac{1}{2}$ rolls of roofing paper. What is the cost per roll? (2-4)

8. Three sections of insulated telephone cable are cut from a 1000 ft reel. The pieces measure 37 ft 4 in., 66 ft 2 in., and 92 ft 8 in. How much cable is left on the reel? (2-5)

9. Read the micrometer to the nearest thousandth of an inch. (2-6)

10. Read the micrometer to the nearest hundredth of a millimeter. (2-6)

11. A panel of an airplane wing measures 9 ft $3\frac{1}{2}$ in. by $54\frac{1}{2}$ in. The rivet holes are drilled $1\frac{3}{16}$ in. apart, and the end holes are located $1\frac{1}{8}$ in. from the ends of the section. How many holes are needed:

   a. along the length of the panel?
   b. along the width of the panel?
   c. around the perimeter of the panel? (2-7)

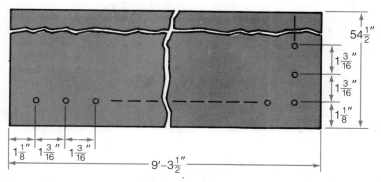

12. Plans have been prepared for a 3.00 ft by 6.00 ft workbench top. What will the dimensions of the top be if the plans are enlarged to 125%? (2-8)

13. During the testing of a representative sample of fasteners, it was found that 0.67% were defective. If you were to test 1200 fasteners, how many should you expect to be defective? (2-9)

14. A mechanical pencil sells for $6.25. If the cost to the merchant is $5.23 plus $0.52 in expenses, what is the percent of profit? (2-10)

15. A computer printer retails for $7395. Find the sale price if a 5% discount is allowed for full payment within 90 days. (2-11)

Give the resistance value and the tolerance limits of each resistor. (2-12)

16.

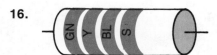

17.

Cumulative Review

Use a calculator where possible.

1. If the October 6 reading was 6958 for the meter below, how many kilowatt hours of electricity were used for the month?

November 6

2. Read the tachometer below to the nearest hundred revolutions per minute.

3. Measure the length of each line below to the nearest tenth of an inch.

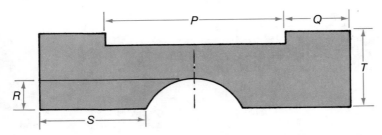

4. Copy the figure at the right. Then indicate the dimensions in decimal inches using extension lines, dimension lines, and leaders as needed.

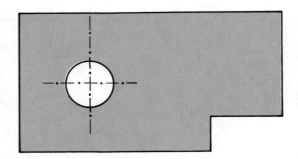

5. Find the dimension limits and the allowances of dimensions A and B below.

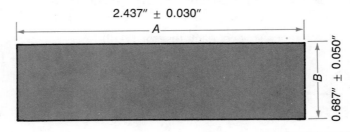

6. Find dimensions *A* and *B* on the C-clamp below.

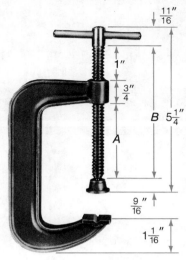

7. What are the dimensions of an 8.0 in. by 10.0 in. photograph after a reduction to 60%?

8. During a sale, a set of high-speed drills are marked down 20%. If the regular price is $16.95, how much is the sale price?

9. What is the selling price of a grease gun if the cost to the merchant is $5.23, the expenses are $0.52, and the profit is $0.48?

10. A tree-pruning pole is put together in three sections. The sections are 5 ft 6 in., 4 ft 4 in., and 2 ft 10 in. in length. What is the total length of the pole?

11. To fit between two fence posts, a split rail 8 ft long must be shortened by 15.3 in. What will be the length of the rail after sawing?

12. Galvanized carbon steel aircraft cable sells for $0.33/ft. What is the cost of 20.5 ft of the cable?

13. During the testing of computer chips it was found that three out of 1500 tested were defective. What percent were defective?

Give the resistance value and the tolerance limits of each resistor.

14.

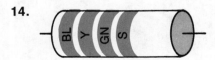

15.

Chapter 3
Expressions and Equations

After completing this chapter, you should be able to:

1. Find solutions to problems by simplifying expressions and solving equations.
2. Use Ohm's law and wattage formulas to solve problems relating to electricity.
3. Write and use formulas to solve work-related problems.

Using Expressions and Equations

3-1 Using Expressions

A strip of stainless-steel molding must go around the edge of the table at the right. To cut the strip to the required length, we need to find the perimeter of the table top. (The **perimeter** is the distance around a closed figure.) For the table top, the perimeter can be represented with the numerical expression 24.0 + 30.0 + 36.0.

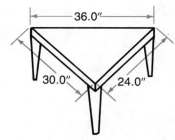

Simplify this expression to find the perimeter of the table top.
24.0 + 30.0 + 36.0 = 90.0 (in.)

The perimeter of the table top is 90.0 in.

Often, a problem is clarified by simplifying an expression.

Example   Skill(s) __5__

The dimensions of the steps shown at the right indicate that the second step is twice as high as the first and the third step is three times as high as the first. What is the overall height of the three steps?

Solution

Write the overall height as a variable expression. Then combine *like terms*. Remember that x means $1x$.

$$\text{height} = 3x + 2x + x$$
$$= (3 + 2 + 1)x$$
$$= 6x$$

Thus, the overall height of the steps is six times as large as the height of the first step.

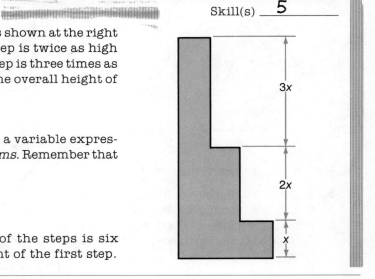

EXERCISES

A The variable expressions below represent lengths of objects. Write a simplified expression of the overall length for each.

1. $4a + a$

2. $7c - 2c$

3. $x + x$

4. $5m + 3m + 2m$

5. $2t - t$

6. $5y + 2y + 9y$

7. $3x + 4x + 1$

8. $9y - 2y + 6$

9. $4t + 8 + 3t$

10. $3x + 2x + 4x + 2x$

11. $5a - (3a + a)$

12. $12p - (7p + 2p)$

B Write a variable expression in simplest form for each measure.

13. Express the perimeter of the triangle below.

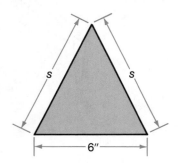

14. The square shaft below is milled from a round steel bar. Write an expression for the diameter of the bar.

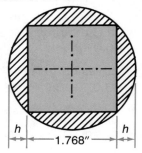

15. Express the perimeter of the hexagon below.

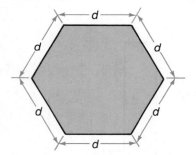

16. Express the overall length of the template below.

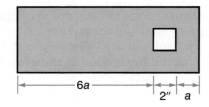

17. Refer to the diagram at the right.

   a. Express the diameter of the smallest, the middle-size, and the largest pulley.
   b. Express the thickness of the three pulleys. Assume all pulleys to have the same thickness.

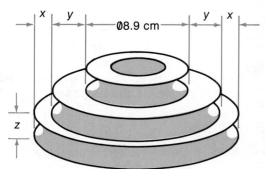

18. Three production lines at the EZ Company produce circuit boards at the hourly rates of $8x + 20$, $9x + 16$, and $10x - 14$. Write an expression that represents the total hourly production.

19. Express the overall length of the inside calipers.

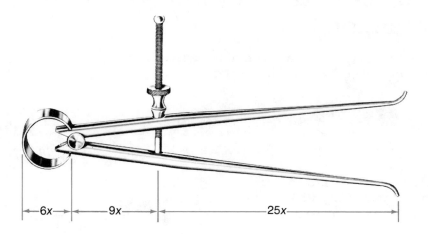

|←—6x—→|←——9x——→|←————25x————→|

C **20.** Two companies are digging a tunnel under a mountain starting at opposite ends. Company A completes 60x feet per day on weekdays and 50y feet on Saturdays. Company B completes 65x feet per day on weekdays and 45y feet on Saturdays. In a particular month, there are twenty-one weekdays and five Saturdays. Write an expression for the number of feet completed during the month.

Technology Update

Solar Power

As our natural resources of fuel are being used up, scientists are turning to the sun as a source of energy. Solar power has been used in satellites, which use the sun as their continuing power source. Their photo power cells convert solar energy directly into electrical energy. It has been suggested that a great number of these cells be put in a desert to supply electrical power. Another idea is to have satellites with photo power cells beam electrical power back to the earth.

Solar power can also be converted to electrical power by heating water. The steam produced could drive turbines in existing power plants. These ideas have not yet been applied on a large scale. However, scientists estimate that solar power could provide up to 20% of the world's electrical power needs.

Research Project:
Identify three ways solar power is used today.

3-2 Solving Equations by Addition and Subtraction

Problems often arise that require an equation to be written that represents a situation. Then the equation can be solved to find the solution to the problem.

Example 1

Skill(s) 6,13,14,27

After some money was withdrawn from a bank and $165.87 was spent, $84.13 is left. How much money was withdrawn?

Solution

Let m be the amount of money withdrawn. Write an equation and solve for m using a calculator.

$$m - 165.87 = 84.13$$
$$m - 165.87 + 165.87 = 84.13 + 165.87$$
$$m = 250$$

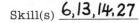

250

The amount of money withdrawn was $250.00.

Example 2

A 2.7 cm hole must be enlarged to allow a cable connector to pass through it. The cable is 3.6 cm in diameter. By how much must the hole be enlarged?

Solution

Write an equation representing the situation. Let y be the amount by which the hole must be enlarged. Solve for y using a calculator.

$$y + 2.7 = 3.6$$
$$y + 2.7 - 2.7 = 3.6 - 2.7$$
$$y = 0.9$$

The hole must be enlarged by 0.9 cm.
The result is precise to the tenths place.

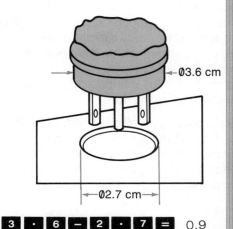

EXERCISES

A Write an equation for each situation. Let *x* represent all unknown values.

1. A number of invoices added to 37 invoices is 198.

2. The cost of an electric cord less a $1.60 discount is $19.

3. The overall length less 6.33 m is 27.96 m.

4. The total number of days decreased by $3\frac{1}{4}$ d is $7\frac{1}{2}$ d.

5. The cost of $89.95 added to the cost of a second item is $295.44.

6. A number of meters plus 3.4 m is 21.4 m.

7. The total number of thermostats less 1784 thermostats already tested leaves 14,693 items still to be tested.

8. $1\frac{3}{4}$ h added to a certain number of hours is 8 h.

B Write an equation and use a calculator to solve the problem.

U.S. System

9. A walnut panel needed for a stereo cabinet must be $22\frac{1}{2}$ in. long. The piece is made by cutting off $1\frac{3}{4}$ in. from a longer piece of stock. What is the length of the stock before cutting?

Metric System

10. Find the outside diameter of a pipe having an inside diameter of 24 mm and a thickness of 5.0 mm.

11. Find dimension *Y* on the plate.

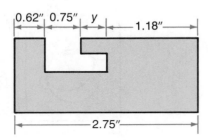

12. Find dimension *Z* on the plate.

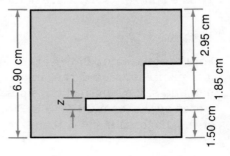

13. The inside diameter of a plastic pipe is 0.4375 in. This is 0.1250 in. less than the outside diameter. Find the outside diameter.

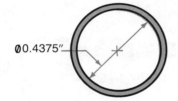

14. An air duct cover has an inside length of 34.9 cm. This is 4.6 cm less than the outside length. Find the outside length.

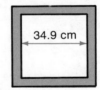

3-3 Solving Equations by Multiplication and Division

The equations representing some on-the-job problems require the use of multiplication and division for their solutions.

Example 1

A number of bolts of fabric were purchased by a dressmaker at $65 each. Altogether, $780 was spent. How many bolts of fabric were bought?

Skill(s) 7, 8, 16, 17, 28

Solution

Write an equation representing the situation. Let x be the number of bolts of fabric.

$$65x = 780$$
$$\frac{65x}{65} = \frac{780}{65}$$
$$x = 12$$

| 7 | 8 | 0 | ÷ | 6 | 5 | = | 12

Twelve bolts of fabric were bought.

Example 2

For the square nut shown at the right, the relationship between d, the length of the diagonal, and s, the length of the sides, is approximated by the formula $\frac{d}{1.4} = s$. Find the length of the diagonal if $s = 1.9$ cm.

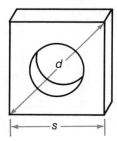

Solution

Substitute $s = 1.9$ into the equation $\frac{d}{1.4} = s$. Then solve for d.

$$\frac{d}{1.4} = 1.9$$
$$(1.4)\frac{d}{1.4} = 1.9\,(1.4)$$
$$d = 2.66$$

| 1 | . | 9 | × | 1 | . | 4 | 2.66

If each side of the square nut is 1.9 cm long, the length of the diagonal is about 2.7 cm, to two significant digits.

EXERCISES

A Write an equation for each situation. Let x represent all unknown values.

1. Six times a number of newspaper ads is 24.

2. Twenty-four times the number of boxes in two truckloads is 672 bottles.

3. Seven percent of a population is 5600 people.

4. An amount of money spent divided by 14 persons is \$1.90/person.

5. A length of plastic tubing cut into five equal parts leaves each part 0.83 m long.

6. A dozen hard hats at a certain price per unit costs \$719.40.

B Write an equation and use a calculator to solve the problem.

7. A temporary spacer made of the three washers on a bolt measures 1.2 in. in length. How thick is each washer?

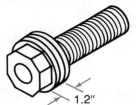

1.2"

8. Four table legs can be turned from a piece of pine 4.22 m long. How long is each leg?

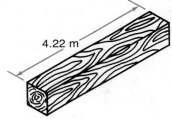

4.22 m

9. A company has received a shipment of parts from England. The charges for the shipment are listed in British pounds. On that date, the conversion formula for changing pounds, p, to U.S. dollars, d, was $d = 1.42\,p$. Find the U.S. dollar cost of the parts having the following shipment charges.

  a. 3.72 British pounds   **b.** 14.70 British pounds   **c.** 168.75 British pounds

10. The relationship between dimensions d and F of the hex-nut below is approximated by the equation $d = \frac{F}{0.866}$. Find dimension F for each size of hex-nut.

  a. $d = 0.34$ in.
  b. $d = 5.5$ in.
  c. $d = 15$ mm
  d. $d = 1.7$ cm

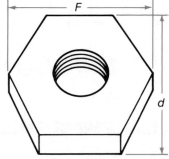

11. When a load is hung on a spring, the spring stretches. The amount it stretches, or the extension, e, is related to the mass, m, of the load by the formula $e = 0.75m$. Find the mass attached to the spring for each extension.

  a. 0.3 cm
  b. 2.7 cm
  c. 1.2 cm
  d. 2.1 cm

3-4 Solving Equations by Several Methods

Some problem situations found in the workplace require more complex equations to represent them.

Example 1

Skill(s) __29__

A store cashier took some five-dollar bills and $38 in one-dollar bills out of the till to exchange for more coins. Altogether $103 was taken out of the till. How many five-dollar bills were exchanged for coins?

Solution

Let x be the number of five-dollar bills. Write an equation and solve for x using a calculator.

$$5x + 38 = 103$$
$$5x + 38 - 38 = 103 - 38$$
$$5x = 65$$
$$\frac{5x}{5} = \frac{65}{5}$$
$$x = 13$$

$\boxed{1}\,\boxed{0}\,\boxed{3}\,\boxed{-}\,\boxed{3}\,\boxed{8}\,\boxed{=}\,\boxed{\div}\,\boxed{5}\,\boxed{=}$ 13

Thirteen five-dollar bills were exchanged for coins.

Example 2

Half of a shipment of books had broken bindings while 17 books of the same shipment had good bindings but blank pages. If 2517 of the books in the shipment had either broken bindings or blank pages, how many books in all were in the shipment?

Solution

Let b be the total number of books in the shipment. Write an equation and solve for b using a calculator.

$$\frac{b}{2} + 17 = 2517$$
$$\frac{b}{2} + 17 - 17 = 2517 - 17$$
$$\frac{b}{2} = 2500$$
$$(2)\frac{b}{2} = 2500\,(2)$$
$$b = 5000$$

$\boxed{2}\,\boxed{5}\,\boxed{1}\,\boxed{7}\,\boxed{-}\,\boxed{1}\,\boxed{7}\,\boxed{=}\,\boxed{\times}\,\boxed{=}$ 5000

Altogether, there were 5000 books in the shipment.

EXERCISES

A Write an equation for each situation. Let *n* represent all unknown values.

1. Twenty-five hours worked at a regular hourly rate plus $78 in tips is $378.

2. A flat rental fee of $8 plus an hourly rate of $3 per hour gives a total rental cost of $23.

3. Number of hours worked times $5.50 less taxes of $13.50 equals $166.50.

4. Total cost of a truck tire is $472.50, which includes the list price plus 5% in sales tax.

B Write an equation and use a calculator to solve each problem.

U.S. System

5. A square shaft is milled from a round bar, as shown below. Find the depth of the cut *h*.

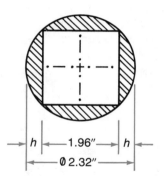

Metric System

6. Find the length *z* on the computer disk shown below.

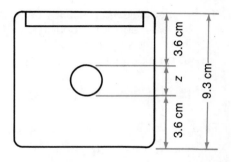

7. Find length *x* on the switch plate.

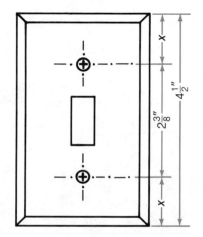

8. Find the missing dimension, *y*.

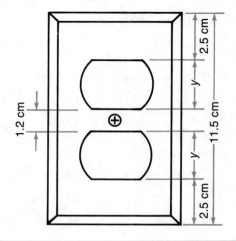

Find the unknown dimension on each object below.

9.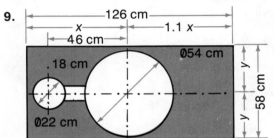

10.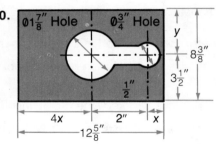

C **11.** A landscaper is planting a hedge along a 150 ft driveway. The plan is to use one plant for each $1\frac{1}{2}$ ft of driveway. There is a sidewalk that uses up 6 ft of space at the end of the driveway. How many plants are needed?

12. A fishing boat is 45 mi from its home port. After traveling 12 mi, the remainder of the trip is completed at a rate of 7.5 mi/h. How long does the rest of the trip take?

13. A cable television service uses an average of 15 ft of cable for each customer. In addition, about 100 ft of cable from each roll is wasted or is used in making connections. How many customers can they expect to service using a 2500 ft roll?

Check Your Skills

Find the value of the numerical expression.

1. $1 + 6 \times (5 + 3)$ **2.** $2 \div (4 - 2) + 1$ **3.** $3 \times 4 \div 2 - 1$ (Skill 5)

4. $14.23 + 9.78$ **5.** $6.123 - 5.234$ **6.** $143.272 - 2.311$ (Skill 6)

7. 1.354×6 **8.** $(11.1)(2.3)$ **9.** $15.64(1.25)$ (Skill 7)

Round to the nearest hundredth.

10. $6.56 \div 4$ **11.** $98.1 \div 2.12$ **12.** $7.7208 \div 0.12$ (Skill 8)

Express the answers in simplest form.

13. $7\frac{1}{2} + 3\frac{3}{4}$ **14.** $6\frac{3}{10} - 1\frac{1}{10}$ **15.** $10\frac{5}{8} - 2\frac{3}{4}$ (Skills 13–15)

16. $6 \times 5\frac{1}{2}$ **17.** $2\frac{1}{4} \times 3\frac{1}{8}$ **18.** $10 \div 2\frac{1}{2}$ (Skills 16 and 17)

Solve the equations.

19. $x + 7.23 = 25.3$ **20.** $x - 4.62 = 8.913$ **21.** $x + 2.1 = 3.6$ (Skill 27)

22. $3.2x = 36.6$ **23.** $12x = 7.938$ **24.** $9.12x = 43.32$ (Skill 28)

25. $12x - 5 = 7$ **26.** $5x - 1.6 = 7.4$ **27.** $4x + 3.21 = 5.6$ (Skill 29)

Self-Analysis Test

The variable expressions below represent lengths of objects. Write a simplified expression of the overall length for each.

1. $4n + 7n$

2. $8b - 3b$

3. $5x - 3x + x$

4. $8y + 7y - 5$

5. $5x - 0.5x$

6. $(8y - 4y) - y$

7. What is the overall length of the pipe wrench?

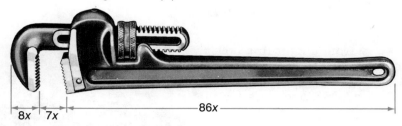

8x   7x   86x

Use a calculator to solve the problems where possible.

8. Write an equation for the unknown dimension in the figure below. Then solve the equation.

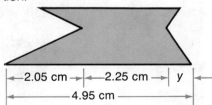

←2.05 cm→←2.25 cm→ y ←
←——— 4.95 cm ———→

9. Find the length x on the bracket below.

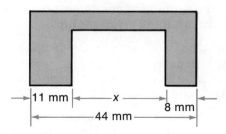

→11 mm←———x———→ 8 mm
←——— 44 mm ———→

10. The number of bushels per acre times 35 A yields 3934 bu. How many bushels are yielded per acre?

11. The height of a tree less 20.0 ft is 36.3 ft. What is the height of the tree?

12. Eighteen pounds of nails times the cost per pound plus $0.50 gives a total cost of $15.26. What is the cost per pound?

13. The number of hours worked times $22.50/h less $68 in deductions is $787. How many hours were worked?

14. A car-rental firm charges $22 per day plus 20¢ per kilometer.

    a. Write a formula for the total car-rental cost, C, if k represents the number of kilometers.

    b. What is the total cost if a car is rented for seven days and is driven 650 km?

    c. What is the total cost if a car is rented for four days and is driven 1200 km?

Using Formulas

3-5 Writing Formulas from Rules

Many common relationships and processes used in the workplace can be stated as rules. In the example below, notice how a rule can be written as an equation, or **formula**.

Example

The distance away from a lightning bolt, in kilometers, is estimated by dividing the number of seconds between the flash and the thunder by three. Write a formula to represent this relationship.

Solution

Define the variables in the situation in terms of the units involved. Then write an equation describing the relationship.

Let d be the distance away from the lightning, in kilometers.
Let n be the number of seconds between the flash and the thunder.

The formula for estimating the distance away from lightning is the following.

$$d = \frac{n}{3}$$

EXERCISES

Write a formula for each rule.

A **1.** The number of inches (i) is equal to 12 times the number of feet (f).

  2. The number of meters (m) is 0.01 times the number of centimeters (c).

  3. The number of kilograms (k) is 0.454 times the number of pounds (p).

B **4.** The net cost (c) is equal to the marked price (p) less the discount (d).

  5. The number of dollars (d) is the number of cents (c) divided by 100.

6. The diagonal length (x) of a square is about 1.4 times the side length (y) of the square.

7. The distance traveled (d) by an automobile equals the speed (s) multiplied by the time in hours (t).

8. The force (F) acting on an object is the product of the mass (m) of the object and its acceleration (a).

9. The average (x) of three numbers is the sum (y) of the numbers divided by three.

10. The earned-run average (ERA) of a pitcher is nine times the number of runs (r) divided by the number of innings (i) pitched.

11. The total wages (w) equals the time (t) in hours worked multiplied by the hourly rate (r).

12. The rent (x) on a power posthole digger is a flat fee of $12.50 and an hourly charge of $3.25. (Let y be the number of hours it was rented.)

13. The number of screw threads (t) per inch in the diagram below is equal to one divided by the pitch (p) of the screw.

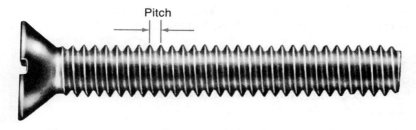

Pitch

Define each variable. Then write a formula for each rule.

14. The amount of take-home pay per week equals $1650 divided by four, minus the amount deducted for taxes.

15. The efficiency of a machine in percent form is 100 times the input divided by the output.

16. The number of cards left to be played is equal to 52 minus the number of hands dealt times five.

C 17. The flow rate for an intravenous transfusion is calculated by multiplying the total infusion volume by the drop factor and dividing the result by the total time of the infusion in minutes.

3-6 Electrical Formulas

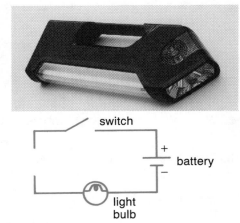

A flashlight contains one or more batteries, a bulb, and a switch, all connected together by materials that conduct electricity. A flashlight circuit is drawn at the right. The diagram illustrates a *series circuit*.

Voltage (*V*), which acts like water pressure in a hose, is the electrical pressure in a circuit. Electricity cannot flow if the circuit is not complete. A break in the circuit, such as an open switch, prevents the flow but the pressure is still there. Voltage is measured in **volts** (V).

Current (*I*) is the rate of flow of electricity when the circuit is complete. It can be compared to the rate of flow of water through a hose. Current is measured in **amperes** (A).

Resistance (*R*) is any device in an electrical circuit that determines the amount of flow. Resistance can be given by a bulb, motor, wire, radio, heater, or anything that runs by electricity. In the flashlight diagram, the light bulb provided resistance to the circuit. Resistance is measured in **ohms** (Ω).

The voltage, current, and resistance in electrical circuits are related by a formula first stated by Georg Simon Ohm, a German physicist who lived from 1787 to 1854. According to Ohm's law, the voltage is equal to the product of the current and the resistance. $V = IR$

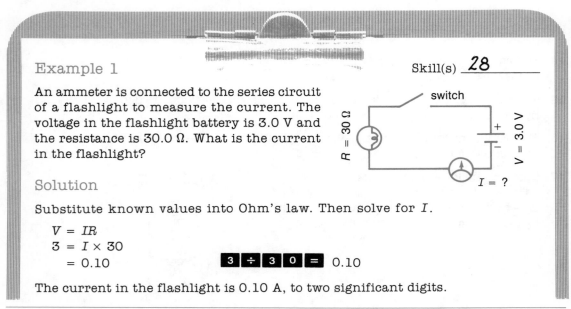

Example 1

Skill(s) __28__

An ammeter is connected to the series circuit of a flashlight to measure the current. The voltage in the flashlight battery is 3.0 V and the resistance is 30.0 Ω. What is the current in the flashlight?

Solution

Substitute known values into Ohm's law. Then solve for *I*.

$V = IR$
$3 = I \times 30$
$ = 0.10$ `3 ÷ 3 0 =` 0.10

The current in the flashlight is 0.10 A, to two significant digits.

Electrical **power** is the work performed by an electrical circuit. The unit of power used for electricity is the **watt**. To calculate the power, P, you simply multiply the voltage, V, by the current, I (in amperes).

$$P = VI$$

As the watt is such a small unit, a larger unit, the kilowatt (kW) is sometimes used for large amounts of power. To find the number of kilowatts, divide the number of watts by 1000.

$$1000 \text{ W} = 1 \text{ kW}$$

Most electrical devices have a *power rating*. This indicates the rate at which the device converts electrical energy. The power rating of an electrical device can be read directly from its nameplate or can be calculated from other information given.

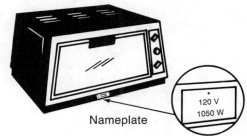

Nameplate

120 V
1050 W

Example 2

A portable electrical appliance draws a current of 10.0 A and is connected to a 120 V power source.

a. What is the power rating of the appliance in watts? in kilowatts?

b. Find the amount of resistance in the circuit.

Solution

a. To find the power rating, substitute known values into the power formula.

$P = VI$
  $= \text{Voltage} \times \text{Current}$
  $= 120 \times 10$
  $= 1200$

`1 2 0 ÷ 1 0 =` 12

The power rating would be 1200 W or 1.2 kW, to two significant digits.

b. Find the resistance by substituting known values into Ohm's law.

$V = IR$
$120 = 10 \times R$
  $R = 12$

`1 2 0 × 1 0 =` 1200

The resistance is 12 Ω, to two significant digits.

EXERCISES

A Use Ohm's law for each problem.

1. How much voltage is needed to send a current of 2.0 A through a resistance of 60.0 Ω?

2. How many amperes will 120 V send through a resistance of 6 Ω?

Two of the three quantities in an electrical circuit are given.
Calculate the unknown quantity.

3.

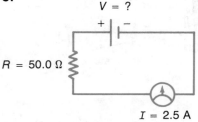

$V = ?$

$R = 50.0\ \Omega$

$I = 2.5\ A$

4.

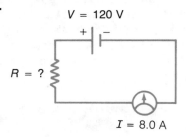

$V = 120\ V$

$R = ?$

$I = 8.0\ A$

Use the electrical power formula for each problem.

5. What is the power rating in watts of an electric range that draws 5.0 A of current on a 240 V line?

6. In a 12 V auto electrical system, it takes 8.0 A to light the headlights. How much power in watts is used?

7. A 12 V battery has an output of 75 A. How much power in watts is available from this battery?

8. An industrial machine uses 50.0 A of current on a 240 V line. How many kilowatts does it use?

B You may need more than one formula to solve the following problems.

9. It takes 110 V to send a current of 5.0 A through an electric toaster. Find the resistance of the toaster in ohms.

10. What is the power in watts produced by 12.0 A when the resistance is 14.0 Ω?

11. An iron has an electrical power rating of 1100 W. If the iron is connected to a 110 V source, what is the current drawn?

12. Find the resistance of a 60.0 W light bulb on a 120 V line.

13. A 5.0 A electric drill requires 110 V. What is the resistance?

14. Find the power rating of an electrical device that has a resistance of 240 Ω on a 120 V line.

15. A 750 W electric motor is connected to a 120 V line. How many amperes of current does it require?

16. A 3300 W motor is connected to a 220 V line. How much resistance does this motor have?

c 17. The electrical circuit below shows two resistors connected in series.

    a. What is the total resistance, R_T, if $R_T = R_1 + R_2$?

    b. What is the voltage supplied by the battery if the power of the circuit is 1342 W?

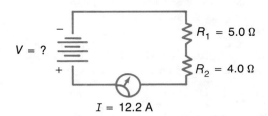

$$V = ?$$

$$R_1 = 5.0 \ \Omega$$

$$R_2 = 4.0 \ \Omega$$

$$I = 12.2 \ A$$

Tricks of the Trade

Voltmeter / Ammeter / Ohmmeter

The three basic instruments for making electrical measurements are the voltmeter, ammeter, and ohmmeter.

A **voltmeter** measures the voltage of a power source. For example, when testing a car battery, a voltmeter reading of 8 V would indicate that the battery is nearly dead and does not have enough voltage to start the car.

An **ammeter** measures the electrical current flow in a circuit. If an ammeter were connected to a faulty car starter motor, a reading of 500 A would show that the motor is drawing too much current.

Some voltmeters and ammeters use single-range scales as shown at the right.

Reading: 7.2 V or 7.2 A

An **ohmmeter** is used to measure the amount of resistance. For example, an ohmmeter can be used to test a fuse. A good fuse should indicate a resistance near 0 Ω. A blown fuse will show an infinite value of resistance. The scale on an ohmmeter is not uniformly divided, as shown at the right. Some ohmmeters can be set at four different ranges.

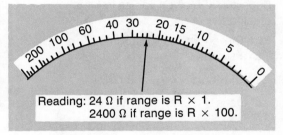

Reading: 24 Ω if range is R × 1.
2400 Ω if range is R × 100.

3-7 Solving Problems with Formulas

Many jobs use formulas to represent problem situations.

Example

For young adults, a blood-pressure reading of between 110/70 and 120/80 is considered normal. The upper number in the blood-pressure reading measures the force of the blood pushing against the walls of the arteries and is called the *systolic pressure*.

A person's systolic pressure, P, can be found by adding 100 to half the person's age, a.

$$P = 100 + \frac{a}{2}$$

What is the possible age of someone whose systolic pressure is 120?

Solution

Substitute the known value into the systolic pressure formula and solve for a using a calculator.

$$P = 100 + \frac{a}{2}$$
$$120 = 100 + \frac{a}{2}$$
$$120 - 100 = 100 + \frac{a}{2} - 100$$
$$20 = \frac{a}{2}$$
$$2 \times 20 = \frac{a}{2} \times 2$$
$$40 = a$$

| 1 | 2 | 0 | – | 1 | 0 | 0 | = | × | 2 | = | 40

A person with a systolic pressure of 120 is likely to be 40 years old.

EXERCISES

A Substitute the given values into the formula. Solve the resulting equation using a calculator.

1. The length of the diagonal of a square d: $d = 1.4\,s$, where s is the length of a side.

    a. Find d, if $s = 8.2$ cm.   **b.** Find s, if $d = 12.831$ cm.

2. Density d: $d = \frac{m}{v}$, where m is the mass of an object and v is the volume.

    a. Find d if $m = 15$ kg and $v = 300$ L.   **b.** Find m if $d = 0.05$ kg/L and $v = 15.4$ L.

3. Gasoline mileage M: $M = \frac{R_2 - R_1}{G}$, where G is the number of gallons used, and R_1 and R_2 are the odometer readings in miles.

   a. Find M if R_1 is 78,563 mi, R_2 is 78,905.2 mi, G is 14.5 gal.
   b. Find R_2 if M is 18.2 mi/gal, R_1 is 9763.4 mi, and G is 12.5 gal.
   c. Find G if M is 28 mi/gal, R_1 is 12,680 mi, and R_2 is 13,430 mi.

B Use formulas previously stated in this lesson to solve the problem.

4. a. What is a normal systolic pressure for an eighteen-year-old?
   b. What is the possible age of a person whose systolic pressure is expected to be 112?
   c. A 30 year old has a systolic pressure of 148. How much above normal is this?

U. S. System

5. Find the length of the diagonal of a square table whose side is 32 in.

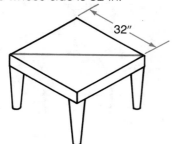

Metric System

6. Find the length of the crossbeam of a square shed whose side is 2.74 m long.

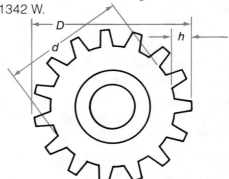

2.74 m

7. A delivery truck averages 16.7 mi/gal. If the truck is driven 8500 mi this month, how many gallons of gasoline will be needed?

8. Find the voltage of a battery in an electrical circuit if the ammeter reading indicates 12.2 A and the power in the circuit is 1342 W.

C 9. The depth, h, of the gear tooth at the right is found by dividing the difference between the major diameter, D, and the minor diameter, d, by 2, or $h = \frac{D - d}{2}$.

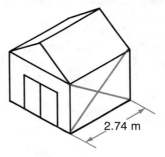

   a. Find the height of the gear tooth if the major diameter is 3.4 in. and the minor diameter is 2.85 in.
   b. Find the major diameter if the minor diameter is 5.25 cm and the depth of the gear tooth is 1.25 cm.

10. The minor diameter of a clock gear is 6 in. and the depth of the gear tooth is 0.3 in. Find the major diameter of the gear.

11. The minor diameter of a watch gear is 0.7 cm and the major diameter is 0.8 cm. What is the depth of the gear tooth?

Self-Analysis Test

Write a formula for each rule.

1. The density is found by dividing the mass by the volume. Let d be the density, m be the mass, and v be the volume.

2. The reaction distance to stop a car is 0.22 times the speed of the moving car. Let d be the distance and s be the speed.

3. The amount of simple interest (I) received on a loan or investment is the product of the principal (p), the rate (r), and the time (t).

4. The total cost (C) of producing computer disks in a factory is a fixed cost (f) plus a variable cost (v) times the number of computer disks produced (n).

Use a formula and a calculator to solve each problem.

5. a. Write a formula that relates the quantities x, R, and r in the diagram below.
    b. Find r if x is 8.93 cm and R is 5.25 cm.
    c. Find R if x is 11.325 cm and r is 2.175 cm.
    d. Find x if R is 8.45 cm and r is 6.7 cm.
    e. Find r if x is 24.8 cm and R is 18.95 cm.

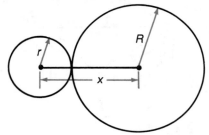

Use one of the given formulas to solve the following problems.

$$M = \frac{R_2 - R_1}{G} \qquad P = VI \qquad d = \frac{m}{v} \qquad P = 100 + \frac{n}{2} \qquad V = IR$$

6. How much resistance does an electric mixer offer if it is rated at 5 A with 110 V?

7. How many watts are used by a 3 A electric dishwasher when it is connected to a 240 V line?

8. A car's odometer showed 45,678 mi at the beginning of a business trip and 47,198 mi at the end. If 32 gal of gas were used for the trip, what was the gasoline mileage?

9. Find a normal systolic pressure of a person who is:

    a. 18 years old.   **b.** 25 years old.
    c. 40 years old.   **d.** 65 years old.

10. Find the density in pounds per cubic foot of a three-dimensional object with a mass of 28 lb and a volume of 0.50 ft³.

Integrated circuits (IC's) perform a variety of functions in many items we use every day. These functions include:

- counting oscillations of the quartz crystal in digital watches and clocks to count the time;
- calculating and displaying results in electronic calculators;
- amplifying the sound in radio and tape recorders;
- controlling color and drawing screen images in television sets;
- controlling an automobile's fuel system or providing a car's digital display dashboard;
- holding information in a computer's memory.

The picture at the right shows someone checking a section of a 10 ft by 14 ft blowup of an integrated circuit. When applied to the inside of a computer, the circuitry will fit on a microchip.

Probably the most significant use of integrated circuits today is in computers. If you were to look at the inside of a microcomputer, you would see several integrated circuits connected to a printed circuit board. You might be able to locate the integrated circuit for the Central Processing Unit (CPU), the Random Access Memory (RAM), and the Read Only Memory (ROM).

The first integrated circuit was made in 1958. At this time, the integrated circuit was designed as an improvement over the transistor whose main function was to amplify electrical current. The constant upgrading of integrated circuits has paved the way for continuous improvements in today's computers and other electronic items.

An integrated circuit consists of several components within a single piece of silicon. The early IC's contained fewer than ten components; today an IC can contain thousands. Integrated circuits are grouped into one of five classifications:

- Small Scale Integration (SSI) containing 1 to 10 components;
- Medium Scale Integration (MSI) containing 10 to 500 components;
- Large Scale Integration (LSI) containing 500 to 20,000 components;
- Very Large Scale Integration (VLSI) containing 20,000 to 100,000 components;
- Super Large Scale Integration (SLSI) containing over 100,000 components.

Today's microcomputers can contain the SSI, MSI, LSI, and VLSI integrated circuits. It will not be long before they will contain a SLSI chip as well.

Heating, lighting, power, air-conditioning, and refrigeration components all operate using electrical systems that are assembled, installed, and maintained by electricians. Electricians generally specialize in either construction or maintenance, although some electricians do both.

Job Description

What do electricians do?

1. They read and use technical drawings to install electrical systems in factories, office buildings, homes, and other structures.

2. Electricians install conduit to enclose electrical wires and fasten metal or plastic boxes to surround electrical switches and outlets.

3. They make the required connections to circuit breakers and transformers to complete the electrical circuits.

4. Electricians trouble shoot and repair all kinds of electrical equipment. This often involves referring to the technical drawings.

5. They inspect electrical systems and correct defects before breakdowns occur.

Qualifications

An electrician should have a good knowledge of the basic principles and laws of electricity. A good electrician is familiar with AC (alternating current) and DC (direct current) series, parallel, and complex circuits. It is also important for them to be acquainted with the basic concepts of AC and DC measuring instruments as well as electrical computations.

Because of these requirements, it is necessary for electricians to have taken courses in the fundamentals of electricity and algebra. In most places, apprenticeship programs for electricians, lasting four or five years, provide both classroom instruction and on-the-job training. Areas covered in these programs include blueprint reading, electrical theory, electronics, mathematics, electrical code requirements, and safety. Under the supervision of experienced electricians, apprentices must demonstrate mastery of the trade. Beginning by drilling holes, setting anchors, and setting up conduit, the trainee progresses to measuring, bending, and installing conduits well as installing, connecting, and testing wiring. Trainees also learn to set up and draw diagrams for entire electrical systems.

We use electricity every day for light, heat, appliances, as well as for radio, television and other communication. Industries depend on electricity to run tools and machines, provide heat for processes such as steelmaking, and to serve many other purposes. The source of electrical power is sometimes water power from dams.

Grand Coulee Dam on the Colorado River, for example, has the largest hydroelectric-production capacity in the United States and the second largest potential capacity in the world. In a recent year, the capacity was almost seven million kilowatts. The ultimate capacity is expected to be over ten million kilowatts.

Grand Coulee Dam

The advantages of producing electricity using water power include the following.

- Natural resources, like coal, oil, and gas are not depleted. No expensive fuels are required for the generation of hydroelectricity.
- The atmosphere is not polluted with waste products.
- The operation of hydroelectric facilities is relatively low in cost, even though the initial building of the dam requires a large capital investment.

---

1. The percentages of electricity produced by various sources in the United States in a recent year are approximated in the table below.

   a. How much greater is the usage of coal than water for electrical power?
   b. Is all electricity production in the United States represented in the table?

Source	Percentage
Coal	54.5
Oil	6.3
Nuclear Fuel	12.7
Hydroelectric	14.4
Natural Gas	11.9

2. The Grand Coulee Dam has the potential of producing 10,080 MW (megawatts) of electricity. How many light bulbs would that amount of power be able to light at 100.0 W each? (1 MW = 1,000,000 W)

3. The Brumley Gap hydroelectric plant in Virginia has the potential of producing 3300 MW of electricity. How many kilowatts is this?

Chapter 3 Test

The expressions below represent lengths of objects. Write a simplified expression of the overall length for each. (3-1)

1. $5m + 8m - 2m$

2. $50x + 30x + 20x - 14y$

Rental costs for three power tools are shown in the table below. Write a separate expression for the cost of renting each of the three items for n hours. (3-1)

	Item	Flat Fee	Additional Hourly Rate
3.	Saw	$4.50	$1.25
4.	Plane	$5.50	$2.00
5.	Sander	$2.50	$0.75

Write an equation and use a calculator to solve each problem.

6. Find dimension A and B on the metal plate below. (3-2)

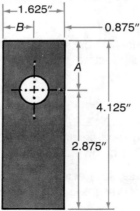

7. Find length Y in the diagram below. (3-2)

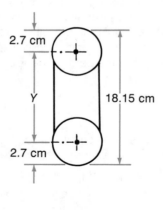

8. What is the size of a rod if 0.230 in. must be ground off the end in order to have the proper length of 1.375 in.? (3-2)

9. Find the outside diameter of a piece of tubing that has an inside diameter of 4.31 cm and a thickness of 0.51 cm. (3-2)

10. A post-office worker picked up 940 lb of mail from 50 drop boxes. What is the average amount picked up per box? (3-3)

11. A construction supervisor earns 1.4 times as much as a laborer. If the supervisor earns $20.70/h, how much does the laborer earn per hour? (3-3)

12. A mechanic charged $83.25 for a repair job. If the mechanic worked for $4\frac{1}{2}$ h on the job, what is the rate per hour? (3-3)

13. A television technician charges a flat fee of $24 plus $22.50/h for labor. Find the number of hours of labor included in a repair bill of $71.25. (3-4)

14. A taxi driver charges a flat fee of $2 plus $1.50/mi. The total charge for a trip was $11.60. How long was the trip? (3-4)

15. If the temperature is 67°F and is increasing at the steady rate of 2.6°F/h, how long will it take the temperature to reach 80.0°F? (3-4)

Write a formula for the statements.

16. The amount of work done is equal to the product of the force and the distance. Let W represent the amount of work done, F represent the force, and d represent the distance. (3-5)

17. A baseball player's batting average is the quotient of the number of hits and the number of official times at bat. Let A represent the batting average, h represent the number of hits, and b represent the number of official times at bat.(3-5)

Use a formula and a calculator to solve each problem.

18. A 5 A electric drill requires 110 V. What is its resistance? (3-6)

19. How much electrical power is used for a 6.0 A electric lawn mower connected to a 240 V line? (3-6)

20. Find the depth of the screw thread below using the formula $H = \frac{P}{2} + 0.01$, where H is the depth of the screw thread and P is the pitch. (3-7)

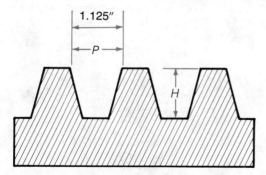

21. The total cost, T, of the manufacture of a certain number of appliances, n, is represented by the formula: $T = 10,000 + 30n$.

   a. Find the total cost if the number of appliances manufactured is 200.
   b. Find the number of appliances manufactured if the total cost is $19,000. (3-7)

Chapter 4
Working with Measurement Formulas

After completing this chapter, you should be able to:

1. Apply equation-solving skills to measurement formulas.

2. Use perimeter and circumference formulas to solve work-related problems.

3. Apply area of parallelogram, triangle, trapezoid, and circle formulas to find the solutions to various industrial problems.

Perimeter and Circumference Formulas

4-1 Using Perimeter Formulas

A **rectangle** has two pairs of sides of equal length. Hence, the sum of the sides of the rectangle, or the **perimeter**, P, is equal to the sum of twice the length, l, and twice the width, w. This relationship can be stated in the formula, $P = 2l + 2w$.

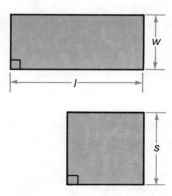

All four sides of a **square** have the same length. Thus, the perimeter of a square is equal to four times the length of one side, s. The formula for the perimeter of a square is $P = 4s$.

The perimeter formulas above are often used to solve problems, as shown below.

Example

A building is to have a perimeter of 100.0 ft. The two sides that will have common walls with adjoining buildings are 18.5 ft each. The other two sides are to be made of glass. What will be the length of each side of glass?

Skill(s) _29_

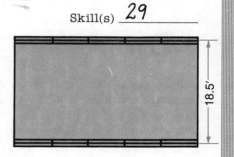

Solution

Substitute the known values into the perimeter formula and solve for the unknown length, l, using a calculator.

$$2l + 2w = P$$
$$2l + 2(18.5) = 100.0$$
$$2l + 37.0 = 100.0$$
$$2l + 37.0 - 37.0 = 100.0 - 37.0$$
$$2l = 63.0$$
$$\frac{2l}{2} = \frac{63.0}{2}$$
$$l = 31.5$$

| 1 | 0 | 0 | − | 2 | × | 1 | 8 | · | 5 | = | 63

| 6 | 3 | ÷ | 2 | = | 31.5

The length of each glass side of the building is 31.5 ft. This result is accurate to three significant digits.

EXERCISES

A Substitute the given values into a perimeter formula. Then solve the resulting equations for the unknown quantity using a calculator.

U.S. System

1. Find the perimeter of a square if one side is $5\frac{5}{8}$ in.

3. The length of a rectangle is 16 in., and its width is 12 in. Find the perimeter.

5. Find the width of a rectangle if its perimeter is $36\frac{1}{3}$ in. and its length is $12\frac{3}{4}$ in.

Metric System

2. Find the length of a side of a square if the perimeter is 58.8 m.

4. Find the length of a rectangle if its perimeter is 48.0 cm and its width is 9.5 cm.

6. Find the width of a rectangle with perimeter 84.6 cm, and length 315.0 mm.

B **7.** Find the perimeter of a rectangular panel with a length of 5.0 ft and a width of 39.0 in.

9. Find the perimeter of a triangular plate with sides of length 11.30 in., 12.60 in., and 15.75 in.

11. A rectangular ceiling panel has a perimeter of 16.0 ft. If the width of the panel is 2.0 ft, find the length.

13. A landscaper is planting a hedge around a rectangular plot that is 42.0 ft by 20.0 ft. If one plant is used for each 2.0 ft of perimeter, how many plants are needed?

8. Find the perimeter of a rectangular field with a width of 230 m and a length of 0.560 km.

10. Find the perimeter of a triangular garden with sides of length 15.25 m, 17.30 m, and 21.40 m.

12. A rectangular table has a perimeter of 4.0 m. If the width of the table is 0.80 m, find the length.

14. A fence is being built around an 18.0 m by 30.0 m plot. If one fence post is used for each 2.0 m, how many fence posts are needed?

C **15.** The perimeter of a triangular template is 26.0 in. The sides marked *s* are equal in length. Find *s*.

16. The lengths of the sides of a triangular wall are given below. The perimeter of the triangle is 4660 mm. Find the lengths of sides *AB* and *BC*.

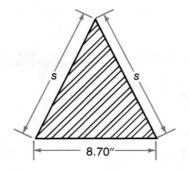

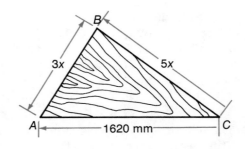

Tricks of the Trade

The dimensions of the room at the right are in meters. To replace the dimensions with measurements in the U.S. system, what approximate measurements would you use?

This problem is typical of the kind of problem found in industry today. To solve this problem, a conversion table like the one below, or a metric conversion calculator can be used.

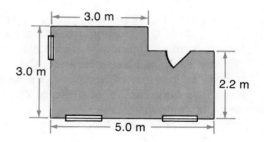

Linear Metric Conversion Table (approximations)		
When You Know	Multiply By	To Find
millimeters	0.039	inches
centimeters	0.39	inches
meters	3.3	feet
meters	1.1	yards
kilometers	0.62	miles
inches	25	millimeters
inches	2.5	centimeters
feet	31	centimeters
feet	0.31	meters
yards	0.91	meters
miles	1.6	kilometers

1. Convert the dimensions in the above diagram to feet.

2. What is the perimeter of the room below in meters?

3. What is the perimeter of the parking lot below in meters? (Start by changing each dimension to meters.)

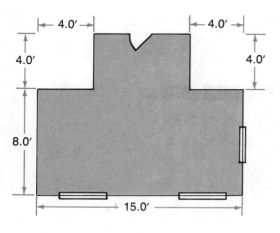

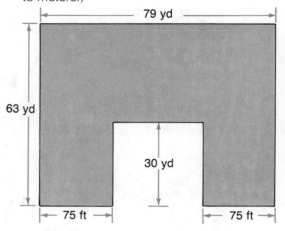

4. Convert the dimensions in question 2 to meters.

4-2 Using Circumference Formulas

Many parts of machines and tools are circular. Wheels, saw blades, rings, gears, and pulleys are examples of some of the many circular objects commonly used in industry and the trades. Thus, it is important to know that the **circumference**, C, is the distance around a circle. The circumference and **pi** (π, approximately 3.14) are related to the **radius**, r, and the **diameter**, d, of a circle. These relationships are stated in the formulas below.

The diameter is exactly twice the radius.

$d = 2r$

The circumference is about 3.14 times as long as the diameter.

$C = \pi d$

The circumference is about 3.14 times as long as twice the radius.

$C = 2\pi r$

The value of pi cannot be expressed exactly. Rounded to six significant digits, π is 3.14159. For calculations in this text, use $\pi = 3.14$.

Example

Skill(s) ___28___

A tree has a circumference of 35 in. What is its diameter?

Solution

Substitute the known values into the formula relating the circumference and the diameter. Then solve for d using a calculator.

$$C = \pi d$$
$$35 = (3.14)d$$
$$\frac{35}{3.14} = \frac{3.14d}{3.14}$$
$$11 = d$$

 3 5 ÷ 3 . 1 4 = 11.146497

The diameter of the tree is 11 in., to two significant digits.

EXERCISES

Use a calculator where possible.

A Use the formula $C = \pi d$ to find the unknown values.

1. Find the circumference of a circle if the diameter is 5.00 in.

2. The circumference of a circle is 13.74 m. Find its diameter.

Use the formula $C = 2\pi r$ to find the unknown values.

3. Find the circumference of a circle if the radius is 12.0 ft.

4. The radius of a circle is 9.20 cm. Find the circumference.

B **5.** A bicycle has tires with radii of 13.0 in. each. How far does the bicycle move forward with each revolution of the wheels?

6. A circular table top has a diameter of 110.0 cm. Find the circumference.

7. The circumference of the base of a grain storage silo is 81.64 ft. Find the radius of the silo.

8. A computer disk has a circumference of 39.9 cm. Find the diameter of the disk.

9. An engine piston has a circumference of 9.42 in. Find the diameter of the piston.

10. A flat washer has an outside diameter of 2.0000 in. and an inside diameter of 1.0625 in.

    a. What is the circumference around the outside rim of the washer?
    b. What is the circumference around the hole of the washer?

11. The square piece of wood shown at the right has been rounded using a quarter of a circle having a radius of 7.25 in. Find the perimeter of the piece of wood with one rounded corner.

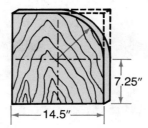

7.25″

—— 14.5″ ——

C **12.** Refer to the diagram of the race track below.

    a. Find the shortest distance around the track.
    b. Why could the track be referred to as a one-mile track?

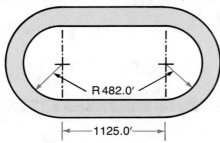

R 482.0′

——1125.0′——

4-3 A Brake Horsepower Formula

The horsepower output of an engine can be found by measuring the force required to brake the engine. This is also known as finding the brake horsepower (bhp) of an engine. The prony brake is a device used to measure engine horsepower output. A diagram of a prony brake is shown below.

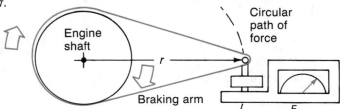

Engine shaft

r

Circular path of force

Braking arm

F

Notice that the arm of the prony brake travels in a circular path. The circumference of the circular path, $2\pi r$, is part of the simplified brake power formula. Other quantities in the formula are listed below.

- The force, F, needed to brake a running engine is measured in pounds.
- The length of the braking arm, r, is measured in feet. Notice that this length is the radius of the circular path on which the braking arm travels.
- The engine speed, n, is written as the number of revolutions per second.

Brake horsepower (bhp) in foot pounds per second $= \dfrac{F \times 2\pi r \times n}{550}$

Example

Skill(s) __29__

The force needed to brake a 3.0 ft prony brake arm is 100.0 lb. If the engine speed is 16 r/s, what is the brake horsepower in foot pounds per second?

Solution

Substitute the known values into the brake horsepower formula. Use a calculator to solve the problem.

Brake horsepower (bhp) $= \dfrac{F \times 2\pi r \times n}{550}$ or $\dfrac{100.0 \times 2 \times 3.14 \times 3.0 \times 16}{550}$

`1 0 0 × 2 × 3 . 1 4 × 3 × 1 6 ÷ 5 5 0 =` 54.807273

The brake horsepower is 55 ft lb/s, to two significant digits.

EXERCISES

Use a calculator where possible. (For this text, consider all digits in revolutions per second, including zeros, to be significant.)

A Substitute the given values into the brake horsepower formula and solve for the brake horsepower in foot pounds per second.

1. The force, F, is 250 lb.
The length of the prony brake arm, r, is 2.8 ft.
The engine speed, n, is 30 r/s.

2. The force, F, is 250 lb.
The length of the prony brake arm, r, is 3.1 ft.
The engine speed, n, is 25 r/s.

3. The force, F, is 250 lb.
The length of the prony brake arm, r, is 3.4 ft.
The engine speed, n, is 20 r/s.

4. The force, F, is 250 lb.
The length of the prony brake arm, r, is 3.7 ft.
The engine speed, n, is 15 r/s.

B Find the brake horsepower in foot pounds per second.

5. What is the brake horsepower of a gasoline engine if the prony brake arm is 3.0 ft long, the force is 300.0 lb, and the engine speed is 50.0 r/s?

6. Find the brake horsepower of an engine that registers 135 lb at the end of a prony brake arm 2.5 ft long. The engine is running at 76 r/s.

7. What is the brake horsepower of an engine if the length of the prony brake arm is 10 ft, the force is 130 lb, and the engine speed is 50 r/s?

8. Find the brake horsepower of an engine where the load on the scale is 125 lb at the end of a 30.0 in. prony brake arm. The engine is running at 92 r/s.

9. What is the brake horsepower of an engine that registers 200 lb on the end of a prony brake arm 2.0 ft long when the engine speed is 48 r/s?

C 10. How much force does an engine register at the end of a prony brake arm 4 ft long when the engine speed is 32 r/s and the brake horsepower is 9 ft lb/s?

4-4 A Speed Formula for Circular Tools

Many different processes are involved in the manufacturing of a product. One of these involves the removal of excess material by using a tool like a saw, drill, grinding wheel, milling cutter, or lathe.

The **speed** of the cutting edge of any circular tool used for the removal of excess material is important to know. The distance that the cutting edge of the tool moves in one revolution is the circumference of the tool. Therefore, the speed of the cutting edge of the tool is the circumference of the tool times the number of revolutions per second (r/s).

Notice that the simplified formula for the speed of the cutting edge for circular tools uses the πd formula for circumference.

Speed of the cutting edge in feet per second = circumference × revolutions per second
$$= \pi d \times n$$

The variable, d, is the diameter in feet of the circular tool. Since the diameters of drills and grinding wheels are usually given in inches, we must change the inches to feet so the speed of the cutting edge is in feet per second. The variable, n, is the number of revolutions per second.

Example

Skill(s) ___28___

A grinding wheel with a 9.000 in. diameter turns at 1860 r/min. What is the speed of the cutting edge in feet per second?

Solution

Change 9.000 in. to 0.7500 ft. Change 1860 r/min to 31 r/s. Then substitute the known values into the speed of the cutting edge formula. Solve for the speed using a calculator.

Speed of the cutting edge = $\pi d \times n$
$$= (3.14 \times 0.7500) \times 31$$

`3 . 1 4 x . 7 5 x 3 1 =` 73.005

The speed of the cutting edge is 73.0 ft/s.
The result is accurate to three significant digits.

EXERCISES

Use a calculator where possible.

A Find the circumference of each tool in feet.

1. circular saw blade: 12.0 in. diameter

2. grinding wheel: 6.00 in. diameter

3. surface planer: 4.50 in. diameter

B Find the speed of the cutting edge in each of the following tools in feet per second. (For this text, consider all digits in revolutions per minute, including zeros, to be significant.)

4. a circular saw blade, 12.0 in. diameter, rotating at 3450 r/min

5. a grinding wheel, 6.00 in. diameter, rotating at 1740 r/min

6. a surface planer with blades 4.50 in. diameter, rotating at 3600 r/min

7. a drill bit with a 0.624 in. diameter, rotating at 420 r/min

8. a grinding wheel with an 8.04 in. diameter, rotating at 2400 r/min

9. The maximum number of safe revolutions per minute for a circular carbon steel milling cutter while working on aluminum is 150. If the cutter has a diameter of 4.50 in., what is the maximum speed of its cutting edge in feet per second?

10. A high-speed steel drill bit rotates at 660 r/min and has a diameter of 0.372 in. What is the speed of its cutting edge in feet per second?

11. A router with a 0.372 in. diameter cutting bit rotates at 25,500 r/min. What is the speed of the cutting edge in feet per second?

12. A wood lathe has a four-step pulley that allows speeds of 840, 1250, 2250, and 3450 r/min. If a job requires a cutting-edge speed of 15 ft/s, and the wood being turned has a diameter of 2.76 in., which speed setting should be used?

Hand router

Wood lathe

C 13. A buffing wheel has a cutting-edge speed of 6000.0 ft/min. If the diameter of the wheel is 6.24 in., at how many rotations per minute does the wheel run?

14. A cast-iron flywheel must have a cutting-edge speed of not more than 4800.0 ft/min. Find the maximum safe rotations per minute for a flywheel that is 4.50 ft in diameter.

15. The recommended cutting-edge speed for brass is 120.0 ft/min. If a piece of stock with a 3.00 in. diameter is being turned in a lathe, what should be the number of revolutions per minute of the lathe?

16. A piece of steel with a diameter of 3.48 in. is being cut in a metal lathe at 210 r/min. The recommended cutting-edge speed is between 100.0 ft/min and 150.0 ft/min. Is the speed too fast or too slow?

CNC lathe

Technology Update

CNC Machining Center

Computer numerical control (CNC) of machine tools is important in modern technology. Numerical control is a system of controlling a machine or process by instructions in the form of coordinate locations and codes. A numerically controlled machine can produce any number of parts with a great deal of accuracy. A program of numerical and code instructions remains constant. The machine does not become bored or fatigued, and the machine function can be changed by inserting a new program. With computer control, programs can be stored in computer memory for permanent use, and the machine can be operated directly. A long production run can be stopped and a short run can be inserted. The machine can then be quickly returned to its long run.

The CNC machining center shown in the photo is multipurpose. Some of its functions include drilling, milling, threading, and tapping. While the initial costs of the CNC machining center are considerable, the increased production, better accuracy, uniformity of parts, and lessened step-up time more than offset the increased costs for many companies.

Question: How does the development of such technology affect you as a future participant in the job market?

Self-Analysis Test

Use a calculator where possible.

1. Find the perimeter of a rectangular garden with the given dimensions.

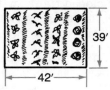

2. Find the length of one side of a square rug with a perimeter of 8480 mm.

3. A triangular sign has two equal sides. The perimeter of the sign is 42.6 in. Find the length of the third side.

4. Find the circumference of a circular table top with a diameter of 12.0 in.

5. Find the radius of a tree trunk having a circumference of 6.0 m.

6. A circular metal plate has a circumference of 33.0 cm. Find the diameter of the plate.

7. What is the brake horsepower of a gasoline engine in foot pounds per second if the prony brake arm is 2.6 ft long, the force is 350.0 lb, and the number of revolutions per second is 460?

8. Find the brake horsepower of an engine in foot pounds per second that registers 120.00 lb at the end of a prony brake arm 3.00 ft long. The engine is running at 42 r/s.

9. How much force does an engine register at the end of a prony brake arm 3.00 ft long when the engine speed is 45 r/s and the brake horsepower is 14 ft lb/s?

10. Find the speed of the cutting edge of a grinding wheel 6.00 in. in diameter turning at 5820 r/min. (Write your answer in feet per second.)

11. A circular saw has a diameter of 2.50 ft and rotates at 390 r/min. What is the speed of the cutting edge in feet per second?

12. A grinding wheel has a diameter of 6.00 in. and rotates at 768 r/min. What is the speed of the cutting edge in feet per second?

Area Formulas

4-5 Area of Parallelogram

A **parallelogram** is any four-sided closed figure with opposite pairs of sides parallel. A **rectangle** is a parallelogram with four right angles. A **square** is a rectangle with four sides of the same length.

To measure area, you always use *square units*. The formula for the area of any parallelogram is $A = bh$, where b is the base and h is the height. For the rectangle and square, the area formulas also involve multiplying the base by the height and are commonly written as shown below.

Parallelogram: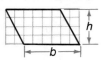
$A = bh$
$= 6 \times 4$
$= 24 \text{ units}^2$

Rectangle: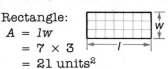
$A = lw$
$= 7 \times 3$
$= 21 \text{ units}^2$

Square: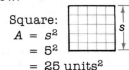
$A = s^2$
$= 5^2$
$= 25 \text{ units}^2$

Example

It took 5.76 m² of carpeting to cover the square floor of a hotel lobby. How long is each side of the square floor?

Skill(s) _28, 32_

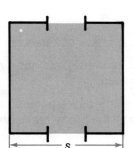

Solution

Substitute the known value into the area of a square formula. Then solve for the length of the side, s, using a calculator.

$$A = s^2$$
$$5.76 = s^2$$
$$\sqrt{5.76} = \sqrt{s^2}$$ ← The opposite of squaring a
$$s = 2.4$$ number is to find the square root.

 2.4

Each side of the square floor is 2.40 m, to three significant digits.

EXERCISES

Use a calculator where possible.

A Make a sketch of the problem situation. Then find the unknown measurement.

1. Find the area of a rectangular carpet if the length is $338\frac{1}{2}$ ft and the width is $45\frac{3}{4}$ ft.

2. Find the height of a template in the shape of a parallelogram if the area is 51.00 in.^{2} and the base is 17.00 in.

3. Find the length of the side of a square carpet if the area is 21.6 cm^2.

B 4. The blueprint of a house includes a 4.0 m by 5.0 m patio. What would be the cost of laying a flagstone patio if the flagstone costs \$23.35/m^2?

5. A 5.0 ft by 19.0 ft thermal-pane storefront window needs to be replaced. What will the cost be if the glass costs \$10.50/ft^2?

C 6. The table at the right shows annual real estate rental fees for a large city. What is the yearly cost to rent 50.0 ft by 72.0 ft space for:

Purpose	Annual Fee
retail	\$25/sq ft
office	\$20/sq ft
industrial	\$5/sq ft

   a. a retail business?
   b. an office?
   c. industrial purposes?

7. The cost of new windows is \$18/ft^2 plus \$7/ft^2 for installation. What is the total cost for installing all of the new windows with dimensions given at the right to the nearest dollar?

Number of Windows	Dimensions of Windows
2	3.0 ft by 5.0 ft
6	4.0 ft by 5.5 ft
3	2.0 ft by 6.0 ft

Tricks of the Trade

Square Measure Conversions

To convert an area measurement to a different unit of measure, multiply by the appropriate conversion factor, shown in the table below.

When You Know	Multiply By	To Find
square inches	6.5	square centimeters
square feet	0.093	square meters
square yards	0.84	square meters
square miles	2.6	square kilometers
square centimeters	0.16	square inches
square meters	1.2	square yards
square kilometers	0.39	square miles

When You Know	Multiply By	To Find
square millimeters	0.01	square centimeters
square centimeters	0.0001	square meters
square meters	0.000001	square kilometers
square kilometers	1,000,000	square meters
square meters	10,000	square centimeters
square centimeters	100	square millimeters

When You Know	Multiply By	To Find
square inches	0.007	square feet
square feet	0.1	square yards

When You Know	Multiply By	To Find
square yards	9	square feet
square feet	144	square inches

1. A house contains approximately 2040 ft^2 of living space. What is the area in square yards? in square meters?

2. A storage shed contains approximately 100.0 ft^2. How much would it cost to carpet the shed floor at a price of \$9.75/yd^2 to the nearest cent? at \$9.75/m^2?

4-6 Area Formulas for Triangles and Trapezoids

The area of a **triangle** is related to the area of a parallelogram. Each diagram below shows a triangle with base, b, at right angles with the height, h. The shaded region and the triangle form a parallelogram. You can see that the area of the triangle is *half* of the area of the parallelogram.

Area of a triangle:

$A = 0.5bh$
$= 0.5(10)5$
$= 25$ units^2

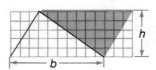

$A = 0.5bh$
$= 0.5(8)10$
$= 40$ units^2

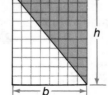

A **trapezoid** is a four-sided figure having exactly one pair of parallel sides. Two different trapezoids are shown below. All trapezoids have two different-length bases. They are refered to as b_1 and b_2. The area of a trapezoid can be found using the formula given below.

Area of a trapezoid:

$A = 0.5h(b_1 + b_2)$
$= 0.5(5)(8 + 12)$
$= 0.5(5)(10)$
$= 50$ units^2

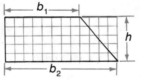

$A = 0.5h(b_1 + b_2)$
$= 0.5(5)(8 + 14)$
$= 0.5(5)(22)$
$= 55$ units^2

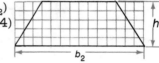

Example

Find the height of a triangular wooden panel having an area of 56 in.^{2} and a base of 14 in.

Skill(s) ___28, 29___

Solution

Substitute the known values into the formula for the area of a triangle. Then solve for the height, h, using a calculator.

$A = 0.5bh$
$56 = 0.5(14)h$ | 0 | · | 5 | × | 1 | 4 | = | 7
$56 = 7h$
$h = 8$ | 5 | 6 | ÷ | 7 | = | 8

The height of the wooden triangular panel is 8.0 in., to two significant digits.

EXERCISES

Use a calculator where possible.

A Find the unknown measurement for each wooden panel.

1. Find the area of the triangular panel below if its height is 10.0 in.

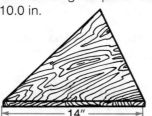

14"

2. Find the area of the trapezoidal panel below if its height is 61.0cm.

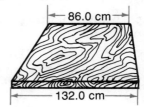

86.0 cm

132.0 cm

Make a sketch of the problem situation.

3. Find the height of a triangular panel if its area is 6765 mm^2 and its base is 5.000 cm.

4. Find the height of a trapezoidal panel with an area of 12 m^2 if the combined length of its two bases is 6 m.

B **5.** A triangular window contains 240 in.^{2} of glass. What is the height of the window if it measures 18 in. at its base?

6. A triangular sign has an area of 3.9 m^2. If the height is 2.6 m, find the length of the base.

7. A brick house has two wooden triangular peaks at opposite ends. If the bases of the triangles measure 24.0 ft and the heights measure 8.0 ft, how many quarts of paint are needed for the two peaks? A quart of paint covers 100 ft^2.

C **8.** A glass table top is to be built in the shape of a trapezoid with bases of 2.5 ft and 5.0 ft and a height of 2.2 ft.

  a. How many square feet of glass are needed for the table top?
  b. If the glass costs $5.50/ft^2, what will be the cost of the table top to the nearest cent?

9. A lumber yard sells treated wooden edging to be put around trees that is shaped like the figure at the right. The bases of each trapezoid in the edging are 18 in. and 11 in. The height of each trapezoid is 6 in. What is the total area covered by the wooden edging?

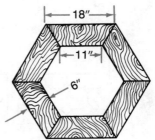

18"

11"

6"

4-7 Using Two or More Area Formulas

The figure at the right is composed of a triangular and a rectangular region. To find the total area of the figure, A_t, you would *add* the areas of its parts, using the formula for the area of a triangle as well as the area of a rectangle formula.

The example below illustrates another situation where more than one area formula is used.

$$A_t = A_1 + A_2$$

Example

Skill(s) *27, 28, 29*

What is the area of the flagstone patio and walkway at the right?

Solution

Find the area of the patio by calculating the *difference* between the area of the rectangle enclosing the patio and the combined areas of the two congruent trapezoids at each side of the patio.

$$
\begin{aligned}
A_{\text{patio}} &= A_{\text{rectangle}} - A_{\text{trapezoids}} \\
&= (60 \times 80) - 2[0.5 \times 20(50 + 40)] \\
&= 4800 - 1800 \\
&= 3000 \ (\text{ft}^2)
\end{aligned}
$$

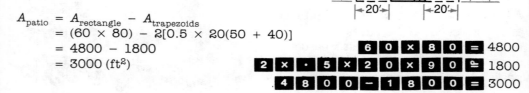

```
6 0 × 8 0 = 4800
2 × . 5 × 2 0 × 9 0 = 1800
4 8 0 0 − 1 8 0 0 = 3000
```

The area of the flagstone patio and walkway is 3000 ft², to one significant digit.

EXERCISES

Use a calculator where possible.

A Find the area of each patio.

1.

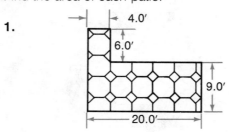

2.

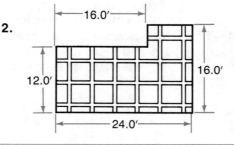

114

3.

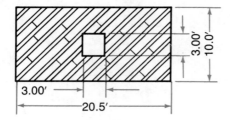

4.

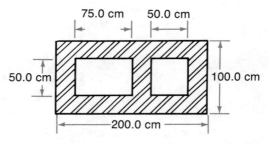

U.S. System

B 5. Find the area of the side of the building shown in the figure below.

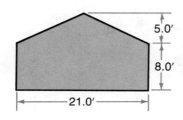

Metric System

6. Find the area of the template shown in the figure below.

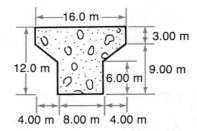

7. Find the area of the cross section of the concrete wall and footing shown below.

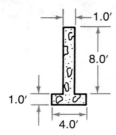

8. Find the area of the cross section of the concrete highway support shown below.

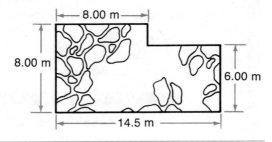

C 9. What is the cost to the nearest cent of fabricating the steel bracket below if the steel weighs 10.2 lb/ft^2 and the cost is $1.10/lb?

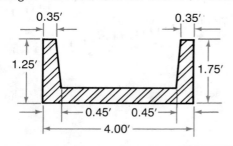

10. What is the cost to the nearest dollar of the bricks used in the patio below if the supplier charges $11.60/m^2?

4-8 The Area of a Circle Formula

The *approximate* area of a **circle** can be found by counting squares. In the figure at the right, there are about 24 whole squares and eight half squares. The area can be estimated as 28 square units.

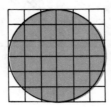

The area of a circle can be calculated using the formula $A = \pi r^2$. For the circle at the right, you can use the following calculation.

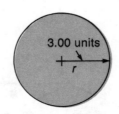

3.00 units

r

$$A = \pi r^2$$
$$= 3.14 \times 3.00^2$$
$$= 28.3 \text{ square units (to three significant digits)}$$

Example

Skill(s) ___28___

What is the cost of making a glass table top with the dimensions shown at the right if the cost of the glass is \$3.40/ft²?

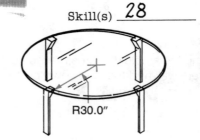

R30.0″

Solution

Step 1: Change 30.0 in. to 2.50 ft.

Step 2: Calculate the area of the circular piece of mirror.

$$A_{\text{table top}} = \pi r^2$$
$$= 3.14 \times 2.50^2$$
$$= 19.625 \ (\text{ft}^2)$$

`3 . 1 4 × 2 . 5 × 2 . 5 =` 19.625

The area of the mirror is 19.6 ft², to three significant digits.

Step 3: Calculate the cost of the mirror to the nearest cent.

$$\text{Cost} = 3.40 \times A_{\text{table top}}$$
$$= 3.40 \times 19.6$$
$$= 66.64 \ (\$)$$

`3 . 4 × 1 9 . 6 =` 66.64

The cost of making the mirrored table top is \$66.64.

EXERCISES

A *Estimate* the area of each circle.

1.

2.

3.

Use a calculator where possible.

B

4. Find the area of an end of a circular can if the end has a radius of 9.00 in.

5. A circular well cover has an area of 1450 in.². Find the radius.

6. Find the area of the top of a circular tank having a diameter of 12.0 ft.

7. Find the area of a camera lens having a diameter of 52 mm.

8. Find the radius of a circular pool with an area of 314 ft².

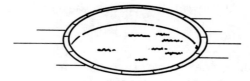

9. A cross section of a pipe is shown below. Find the area of the cross section using the following steps.

  a. Find the area of the outer circle.

  b. Find the area of the inner circle and subtract it from the area of the outer circle.

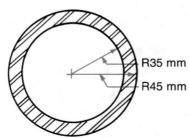

R35 mm
R45 mm

10. Find the area of the metal plate shown below using the following steps.

  a. Find the area of the outer square.

  b. Find the area of the inner circle and subtract it from the area of the square.

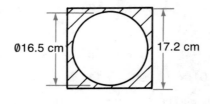

Ø16.5 cm   17.2 cm

11. What is the cost to the nearest dollar of fabricating the steel plate shown at the right if the steel weighs 20.4 lb/in.² and the manufacturer charges $1.10/lb?

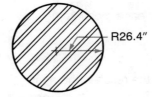

R26.4″

12. Which figure has the larger area, a circle with a diameter of 33.2 cm or a square with sides 33.2 cm long?

C 13. A steel plate weighing 10.2 lb/in.^{2} is shown at the right.

   a. What is the total weight of the plate?

   b. What would be the cost to the nearest dollar of fabricating the plate if the manufacturer charges $1.10/lb?

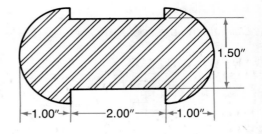

14. It is claimed that a lawn sprinkler waters an area of 250.0 ft^2. For this claim to be valid, what radius must the sprinkler throw the water?

15. A pizza shop sells 14 in. diameter pizzas for $7.70 and 12 in. diameter pizzas for $6.80.

   a. Which size costs more per square inch?

   b. How much more does the more expensive pizza cost per square inch?

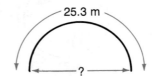

16. The arch length of a semicircular arch is 25.3 m. Find the diameter of the arch and the area of the semicircle.

Check Your Skills

Solve the equations. Round to the nearest tenth where needed.

1. $x + 2.1 = 17.34$	**2.** $x + \frac{1}{2} = 3$	**3.** $x - 4 = 3.5$	(Skill 27)
4. $3x = 5$	**5.** $4x = 2$	**6.** $7x = 42.7$	(Skill 28)
7. $2x + 11 = 13$	**8.** $3x - 2 = 13$	**9.** $6x + 3 = 4$	(Skill 29)
10. $0.01 \times 10{,}000$	**11.** $5 \div 10$	**12.** $1.356 \times 1{,}000{,}000$	(Skill 31)

Multiply each measurement by powers of ten to convert to meters.

13. 1230 mm   **14.** 18 cm   **15.** 16.4532 km

Multiply each measurement by powers of ten to convert to millimeters.

16. 5.1 km	**17.** 63 cm	**18.** 0.000453 m	
19. $\sqrt{169}$	**20.** $\sqrt{729}$	**21.** $\sqrt{625}$	(Skill 32)

Round the result to the nearest hundredth.

22. $\sqrt{45}$ **23.** $\sqrt{146}$ **24.** $\sqrt{983}$

25. $\sqrt{136}$ **26.** $\sqrt{401}$ **27.** $\sqrt{345}$

Self-Analysis Test

Make a sketch of the problem situation. Then use a calculator to solve the problem.

1. A rectangular airplane wing panel is 7.2 m² in area. The width of the panel is 0.60 m. What is the length?

2. A store sells a rectangular aquarium that has a base measuring 18.0 in. by 24.0 in. Each fish requires 7.5 in.² of surface area to survive. How many fish can safely be raised in the aquarium?

3. A rectangular driveway measures 51.0 yd by 4.00 yd. How much would it cost to the nearest dollar to blacktop the driveway at $12.78/yd²?

4. What is the height of a triangular tile having an area of 120 cm² and a base of 15 cm?

5. A "slow-moving-vehicle" sign on a tractor has an area of 72.8 in.². If the sign is a triangle with a height of 11.2 in., find the length of the base.

6. What is the cost to the nearest cent of fabricating a triangular thermal-pane window having a base of 2.4 ft and a height of 3.0 ft if the manufacturer charges $6.50/ft²?

7. What is the cost to the nearest dollar of the flagstone used in the patio shown below if the supplier charges $5.95/ft²?

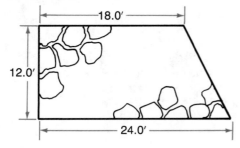

8. A parking lot has the shape shown below. Find the cost to the nearest hundred dollars of surfacing the lot with concrete at $8.40/yd² for material and $9.20/yd² for labor.

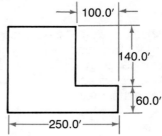

9. How many *whole* bags of fertilizer are needed for the lawn shown below if each bag of fertilizer covers 5000 ft² of area?

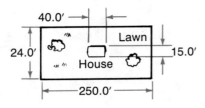

10. What is the cost to the nearest ten dollars of making the circular window shown below if the cost is $19/ft² plus $7/ft² for installation?

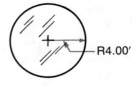

Machinists are skilled workers who use sophisticated metal-cutting machines like lathes, machining centers, punch presses, and grinders to manufacture precision parts for industrial machinery, aircraft, locomotives, automobiles, and other goods.

The increasing use of computer numerically controlled machines (CNC) has resulted in the evolution of a new type of machinist. This machinist has specialized and unique skills in the computer-aided techniques of machines. They are usually refered to as CNC machinists, as opposed to manual machinists, who operate conventional machines.

Job Description

What do machinists do?

1. They set up and operate many types of machines.

2. They review the blueprints and specifications for a job, select the tools, and plan the sequence of cutting operations required within the limits of the process plan given to them.

3. They load stock into the machine, set and make the cuts, and inspect the job with precision measuring instruments such as verniers and micrometers.

4. Maintenance machinists repair or make new parts for machines currently in use.

5. CNC machinists set up the computer control according to the program created for the job. They load the program into the memory, set the data, and manipulate tool length and wear offsets that compensate for the wear of the tool. CNC machinists also make minor changes in the program to solve any minor problems that may occur.

Qualifications

Machinists must be able to read blueprints and understand written specifications for jobs. They must have the mechanical and physical skills necessary to do high-precision work. Machinists must understand the operation of their machine tool and be totally familiar with the measuring instruments in order to check the parts made.

An apprenticeship is the usual way of becoming a machinist. A high-school or community-college education, including mathematics, blueprint reading, metal-working, physics, and drafting is desirable. A knowledge of computers and programming is helpful because of the increasing use of computer-controlled machines in industry.

Until the past century, there seemed to be an unending supply of timber and cutover areas were naturally reforested. Trees grew as a result of seeds being randomly scattered by the wind. Now, however, there is a concerted effort to save our forests and to plant trees to supply our future needs.

The use of tree products will sooner or later exceed the production of trees. One of the proposed solutions to this problem is to produce "super trees" by genetic selection. Forest scientists select trees that, for some reason, are faster growing, healthier, and have other desirable characteristics such as straight trunks and the ability to prune naturally. By selective crossbreeding, the scientists hope to produce seeds for a generation of "super trees" that will allow increased production to meet the increasing demands for tree products.

1. Using the table below, find the cost of each purchase.
   a. 3000 Norway spruce 2 year seedlings (3–6 in.)
   b. 850 Douglas fir 4 year transplants (15–20 in.)
   c. 12,000 Norway spruce 3 year seedlings (6–12 in.)
   d. 20,000 Douglas fir 4 year transplants (5–10 in.)

Norway Spruce	Price per Hundred	Price per Thousand
2 year seedlings (3–6 in.)	$ 19.00	$ 95.00
3 year seedlings (6–12 in.)	24.00	120.00
3 year seedlings (12–15 in.)	34.00	170.00
4 year transplants (6–12 in.)	78.00	390.00
5 year transplants (12–15 in.)	84.00	420.00
5 year transplants (15–20 in.)	89.00	445.00
Douglas Fir		
3 year seedlings (6–10 in.)	30.00	150.00
3 year seedlings (10–15 in.)	39.00	195.00
4 year transplants (5–10 in.)	84.00	420.00
4 year transplants (10–15 in.)	96.00	480.00
4 year transplants (15–20 in.)	110.00	550.00

2. A reforestation program requires 25 ft² of area for each seedling. How many seedlings, to the nearest hundred, should be planted in a 200 ft by 200 ft square region?

3. A fir-tree grower allows 56 ft² of space for each tree planted. How many trees, to the nearest hundred, should be ordered if 8 A of trees are to be planted? (1 A (acre) = 43,560 ft²)

4. A champion yellowwood tree has a circumference of 15 ft 4 in. What is its diameter? its radius?

Chapter 4 Test

Use a calculator to solve the problems.

1. A rectangular swimming pool has a length of 60 ft and a width of 40 ft. What is the perimeter of the pool? (4-1)

2. The perimeter of a rectangular feed lot is 190 m. If the length is 62 m, what is the width? (4-1)

3. The circular flower bed shown below has a radius of 48 in. How long is the wire fence that encloses the flower bed? (4-2)

4. The diameter of the bathroom floor in the diagram below is 11.8 ft. What is the circumference of the inside wall of bathroom? (4-2)

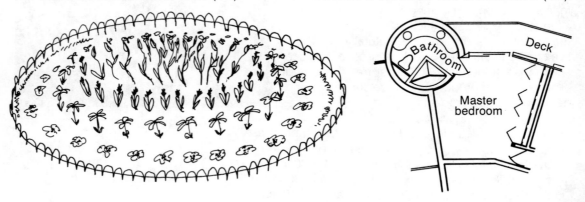

5. When the diameter of a circle is doubled, what happens to the circumference? (4-2)

6. What is the brake horsepower of an engine in foot pounds per second if the length of the prony brake arm is 8 ft, the force is 120 lb, and the engine speed is 50 r/s? (4-3)

7. Find the brake horsepower of an engine in foot pounds per second that registers 140.0 lb at the end of a prony break arm 5.0 ft long. The engine is running at 53 r/s. (4-3)

8. How long is the prony brake arm if a 235 ft lb/s engine running at 46 r/s registers a force of 95.0 lb? (4-3)

9. A hole with a 2.22 in. diameter is being drilled in an aluminum plate at 240 r/min. Find the speed of the cutting edge of the drill. (4-4)

10. The maximum recommended cutting-edge speed using a carbon-steel drill in a drill press on cast iron is 60.0 ft/min. Find the maximum safe revolutions per minute to drill a 1.50 in. diameter hole in a cast-iron block. (4-4)

11. A rectangular room is 8.00 yd by 5.00 yd. If a carpet for the room weighs 3.375 lb/yd^2, find the weight of the carpet. (4-5)

12. A pattern for a box has the dimensions shown below. How many square inches of material are needed to make each box? (4-5)

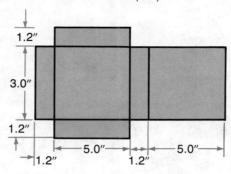

13. What is the area of one of the trapezoidal sides of the planter shown below? (4-6)

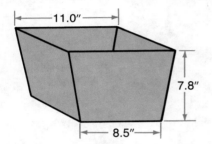

14. A shed roof is made up of four triangles. If each of the triangles has a base 20.0 ft long and a height of 12.2 ft, find the cost of roofing the shed at $0.32/ft^2. (4-6)

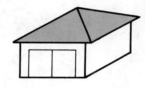

15. Find the area of the wooden panel. (Dimensions are in millimeters.) (4-7)

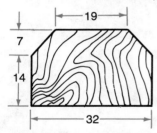

16. Find the area of the frame. (4-7)

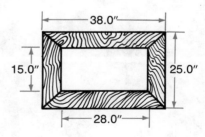

17. A circular pond has an area of 88.0 m^2. Find the diameter. (4-8)

18. A circular oak table top has a weight of 0.0355 lb/in.^{2}. Find the weight of the table top. (4-8)

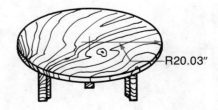

Chapter 5
Ratio, Scale, and Proportion

After completing this chapter, you should be able to:

1. Use ratios to compare quantities.

2. Read and make scale drawings.

3. Solve work-related problems involving rate, ratio, scale, and proportion.

4. Understand and use the mechanical advantage ratios.

5. Use ratio and proportion in problems involving gears and pulleys.

Ratios

5-1 Using Ratio

Ratios occur commonly when two or more numerical quantities are to be *compared*. Consider the following example.

Concrete mixtures are identified as ratios of two or more terms, depending on the number of ingredients. For example, a 1 : 2 : 4 mixture for concrete consists of one part cement, two parts sand, and four parts crushed rock. A 1 : 2 finish coat consists of one part cement and two parts sand.

The ratios 1 : 2 : 4 and 1 : 2 are expressed in *simplest* colon form. A two-term ratio, like 1 : 2, can also be expressed in fraction form as $\frac{1}{2}$.

Example 1

The *pitch*, or slope, of a roof is the ratio of the rise to the run. What is the pitch of a roof with an 8 ft rise and a 12 ft run, in simplest colon and fraction form?

Skill(s) ___10, 18___

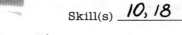

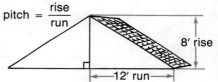

Solution

Write the rise to run ratio as a fraction. Then divide the numerator and denominator by 4 to simplify the ratio.

$$\frac{\text{rise}}{\text{run}} = \frac{8}{12} = \frac{2}{3}$$ The pitch is 2 : 3, or $\frac{2}{3}$.

Example 2

What is the pitch of a roof with a rise of 4.5 ft and a run of 18.0 ft in simplest colon and fraction form?

Solution

To find the simplest-form ratio, multiply the numerator and denominator by 10 and then divide both by 45.

$$\frac{\text{rise}}{\text{run}} = \frac{4.5}{18.0} = \frac{45}{180} = \frac{1}{4}$$ The pitch is 1 : 4, or $\frac{1}{4}$.

Note: A simplest-form ratio compares only whole number quantities.

EXERCISES

A Since a ratio is a comparison made in the same units, the units need not be written. Express each ratio in their simplest fraction form without units.

U.S. System

1. 18 in. : 24 in.

3. 3 ft : 30 ft

5. 3 in. : 12 in.

7. 2 in. : $3\frac{1}{2}$ in.

9. 4 in. : 18 in.

11. 2 yd : 1.5 yd

Metric System

2. 9 mm : 24 mm

4. 3 m : 6 m

6. 15 cm : 100 cm

8. 15 mm : 1 mm

10. 4 cm : 150 cm

12. 3 m : 1.5 m

Express the pitch of each roof as a ratio in simplest fraction form without units.

13.

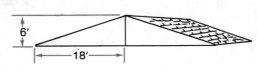

14.

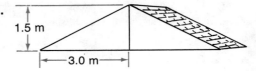

15.

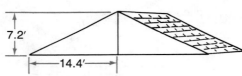

16.

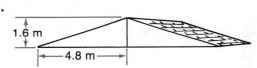

17.

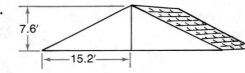

18.

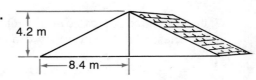

B Measure the line segments below in inches. Then write each ratio in its simplest colon form without units.

19. $a : c$ **20.** $d : a$ **21.** $c : d$

22. $a : e$ **23.** $a : b$ **24.** $e : a$

a _____

b _____

c _____

d _____

e _____

126

25. Write the ratio of 30 in. to 2 ft in simplest colon form.

26. Write the ratio of 750 cm to 3 m in simplest colon form.

C **27.** A surveyor's transit has a magnification scope. An object that is 0.125 in. long appears to be 2.5 in. long. What is the magnification of the scope?

28. A carpenter's level has a scope through which an object 2.0 mm long appears to be 3.9 cm long. What is the magnification of the scope?

29. The ratio of the density of an object to the density of water is called the *relative density* of the object.

    a. If the density of water is 62.4 lb/ft^3, and the density of a gold/copper alloy is 1176 lb/ft^3, find the relative density of the alloy in fraction form.

    b. If the density of limestone is 170 lb/ft^3, find the relative density of limestone in fraction form.

Technology Update

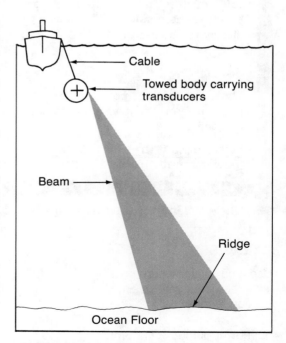

Towed-Array Sonar

Towed-array sonar is a sonar system that has many uses. It can aid the petroleum industry in locating offshore oil, help geologists determine what lies below the ocean floor, permit very accurate measurements of water depths for navigational maps, and supply the navy with a high-quality submarine-detection system.

A towed array can be over 2 km long. It consists of a collection of listening devices (hydrophones), which are spaced along a cable towed by a ship. Sound waves are bounced off the various layers of rock that make up the ocean floor. The hydrophones collect the returning sound waves and feed the data to a signal processor. A computer analyses the data, and a profile of the ocean floor is produced.

When used by the oil-exploration industry, the system makes use of the fact that the sound waves travel faster in rock than in gas or oil. Sub–sea-floor petroleum reserves often show up as strong reflections on the profiles.

Question: Why is exploration of the ocean floor profitable despite the large cost to get the oil?

5-2 Efficiency Ratio

Power-producing mechanisms, such as engines, convert the latent energy of gasoline to heat energy and then to mechanical energy. Much of the potential energy of gasoline is lost in the internal-combustion engine. This loss of energy within the engine itself detracts from the actual power delivered by an engine.

The efficiency of an engine is the ratio between the power output and the power input, measured in horsepower (hp). For example, in a car engine, the engine's power output compared to the power produced by the combustion of the gasoline entering the engine gives its efficiency. The engine-efficiency ratio is usually expressed as a percent.

Efficiency Ratio: $\dfrac{\text{Power output}}{\text{Power input}} \times 100 = \text{Percent efficiency}$

Example

Skill(s) *10, 20, 22*

A car engine takes in 120 hp of power in the form of gasoline and then delivers 100.0 hp to the transmission. What is the percent efficiency of the car engine?

Solution

Write the ratio of power output to power input and convert it to a percent using a calculator.

$$\frac{\text{power output (hp)}}{\text{power input (hp)}} \rightarrow \frac{100}{120} \times 100$$

`1 0 0 ÷ 1 2 0 × 1 0 0 =` 83.333333

The efficiency of the car engine is about 83% to two significant digits.

EXERCISES

Use a calculator where possible.

A Is the efficiency ratio in simplest fraction form? If not, write the ratio in simplest fraction form.

1. Power output: 12.5 hp
   Power input: 20.0 hp
   Efficiency ratio: $\frac{5}{8}$

2. Power output: 75,000 hp
   Power input: 90,000 hp
   Efficiency ratio: $\frac{15}{18}$

Write an efficiency ratio in simplest fraction form and as a percent for each.

	Power Output	Power Input	Efficiency Ratio	Efficiency (%)
3.	8 hp	10 hp	?	?
4.	6.3 kW	7.2 kW	?	?
5.	4.8 hp	6.4 hp	?	?
6.	3500 W	4500 W	?	?

B

7. The amount of power input to an automobile engine in the form of gasoline is 150 hp. Afterwards, the automobile engine delivers only 120 hp of power to the transmission. What is the percent efficiency of the automobile engine?

8. An automobile engine takes in 316 hp of power in the form of gasoline and converts it into mechanical power. If the transmission and rear-end drive receive 215 hp of power from the automobile engine, what is the percent efficiency of the automobile engine?

9. The power input to a generator in the form of steam is 15,000.0 hp. If the generator delivers 9450 hp of power, what is the percent efficiency of the generator?

10. A small generator receives an input of 13.5 kW of power in the form of gasoline. If the generator outputs 12 kW of power, what is the percent efficiency of the generator?

11. A motor has a percent efficiency of 89.0%. If 4200.0 W of power in the form of gasoline is input in the motor, what amount of power in watts is the motor expected to output?

12. A motor that outputs 3.5 hp of power has an efficiency of 90.0%. What amount of power is input to the motor to produce this outcome?

13. A turbine in a hydroelectric plant converts water power with an efficiency rate of 60.0%. If the power of the water that drives the turbine is rated at 750,000 hp, how much electrical power is generated by the turbine?

14. A power transformer has 75.0 kW of electrical power input to it. The transformer delivers 60.0 W of electrical power. What is the percent efficiency of the power transformer?

C **15.** A duplicating machine can make nine copies every 15 s. It is replaced by a newer model that can make 48 copies per minute. How much more efficient is the newer machine in copies per minute? in efficiency percentage?

5-3 Scale

A draftsperson uses triangular rulers that contain many different scales. This combination of scales spares the draftsperson the necessity of calculating the size to be drawn when working to a ratio or scale other than actual or *full size*.

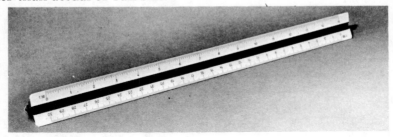

The architect's scale, shown above, includes several useful scales that are in a ratio to 1 ft. The scales represent the ratios of the drawing size to the actual object. Some of the more useful scales are shown at the right.

Scale
1 in. : 1 ft
$\frac{1}{2}$ in. : 1 ft
$\frac{1}{4}$ in. : 1 ft

The $\frac{1}{2}$ in. : 1 ft and 1 in. : 1 ft scales are shown below. Notice how two different line segments can be drawn representing an *actual length* of 2 ft 6 in. using the two different scales.

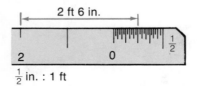

$\frac{1}{2}$ in. : 1 ft

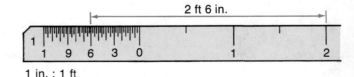

1 in. : 1 ft

The most useful scales of a metric scale are shown at the right. The ratios of drawing size to actual size are represented. The diagrams below show how line segment AB, 35 mm long, can represent different actual lengths, depending on the scale used.

Scale	Ratio
1 : 100	(1 mm : 100 mm)
1 : 50	(1 mm : 50 mm)
1 : 20	(1 mm : 20 mm)

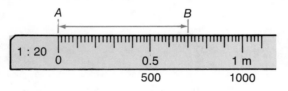

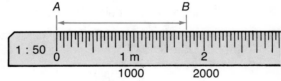

The smallest division represents 20 mm. The actual length of $\overline{AB}$ is 700 mm.

The smallest division represents 50 mm. The actual length of $\overline{AB}$ is 1750 mm.

Example 1

One dimension of a storage shed is 2 ft 3 in. Draw a line to represent this length using the $\frac{1}{2}$ in. : 1 ft scale.

Solution

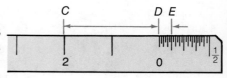

The distance *CD* at the right represents 2 ft and the distance *DE*, in the small subdivisions at the end of the scale, represents 3 in. Thus, distance *CE* represents the 2 ft 3 in. dimension on the storage shed.

Example 2

One dimension of a boat trailer is 1.55 m. Draw a line to represent this length using the 1 : 50 scale (1 mm : 50 mm).

Solution

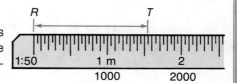

Recall that 1.55 m = 1550 mm. The distance *RT* at the right represents 1550 mm or 1.55 m on the boat trailer since the smallest division on the 1 : 50 scale represents 50 mm.

Example 3

The metal panel below was drawn to a ratio of 1 mm : 20 mm. Use an architect's scale to determine the actual dimensions *A* and *B* of the panel in millimeters.

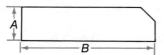

Solution

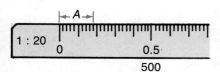

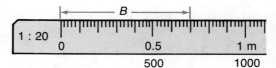

The actual length represented by *A* is 180 mm and the actual length represented by *B* is 700 mm since the smallest division on the 1 : 20 scale represents 20 mm.

EXERCISES

A Use triangular rules if you have them. Otherwise, reproduce the scales at the back of the book to do the following questions.

1. The actual dimensions of a two-room house addition are given below. Draw a line to represent each length using the 1 in. : 1 ft scale.

   a. 18 in.   **b.** 5 ft   **c.** 3 ft 6 in.   **d.** 2 ft 9 in.

2. The actual dimensions of a bridge are given below. Draw a line to represent each length using the $\frac{1}{2}$ in. : 1 ft scale.

   a. 12 ft   **b.** 9 ft   **c.** 6 ft 5 in.   **d.** 2 ft 11 in.

Use a metric scale for each.

3. A few of the actual dimensions of a shed are given below. Draw a line to represent each length using a 1 : 20 scale (1 mm : 20 mm).

   a. 2 m   **b.** 1.5 m   **c.** 0.8 m   **d.** 1.25 m

4. The actual dimensions of a parking lot are given below. Draw a line to represent each length using the 1 : 100 scale (1 mm : 100 mm).

   a. 4 m   **b.** 9 m   **c.** 6 m   **d.** 10 m

Use the $\frac{1}{4}$ in. : 1 ft scale to find the actual length represented by each line segment.

5. ————————————————

6. ——————————————————————————

Use the 1 : 50 scale (1 mm : 50 mm) to find the actual length represented by each line segment.

7. ——————————————————————

8. ————————————————

Draw a line to represent each actual length using the given scale.

9. Scale 1 in. : 1 ft

   a. 1 ft 6 in.   **b.** 5 ft   **c.** 4 ft 3 in.

10. Scale 1 : 50 (1 mm : 50 mm)

   a. 7 m   **b.** 3.5 m   **c.** 0.4 m

B **11.** Make a scale drawing of a rectangular plate having a length of 8.2 m and a width of 4.0 m. Use the 1 : 100 scale of a metric rule.

12. The floor plan below has been drawn to a 1 : 100 scale. Find each of the following dimensions in meters.

    a. the length and width of the outside of the house
    b. the lengths and widths of the inside of rooms *A*, *B*, *C*, and *D*

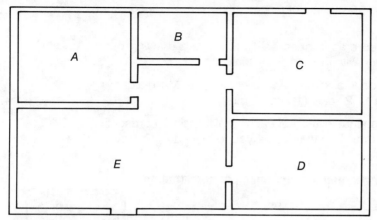

13. The table leg below has been drawn to a scale of 1 in. : 1 ft. Find the dimensions *P*, *Q*, *R*, *S*, and *T* of the actual table leg in feet and inches.

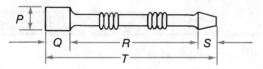

14. The drawing for a small parking area has been drawn with a 1 : 100 scale. Find the actual dimensions *R* and *S*.

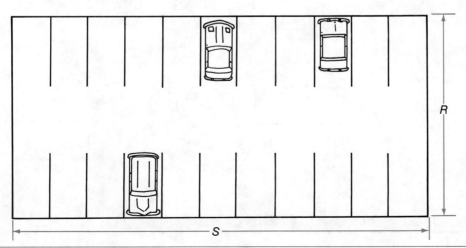

15. Make a sketch of your bedroom floor. Then measure its length and width and place these dimensions on your sketch.

    a. Make a drawing of your bedroom floor using the 1 : 50 scale.

    b. Make a drawing of your bedroom floor using the 1 : 100 scale.

    c. Which scale is more appropriate for a sheet of $8\frac{1}{2}$ in. by 11 in. drawing paper?

16. Follow the same procedure as in question 15 to make a drawing of the kitchen floor at your home with the $\frac{1}{4}$ in. : 1 ft scale.

C **17.** Below is a drawing of a door lockplate in actual, or *full size*. Make a drawing of the lockplate using a 1 : 2 scale (1 mm : 2 mm).

Adapt the 1 : 20 scale so that instead of each small division representing 20 mm, it represents 2 mm. This will enable you to have a 1 : 2 scale.

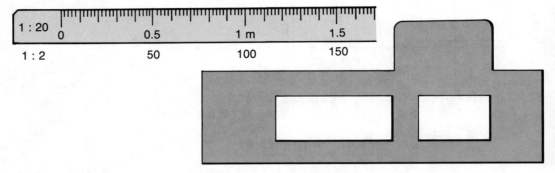

18. Below is a drawing of a light-switch plate in actual size. Make a drawing of the length and width of the switch plate using a 1 : 2 scale.

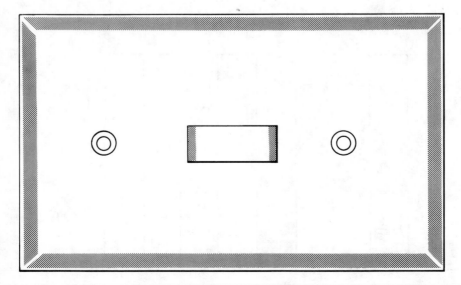

Self-Analysis Test

Use a calculator where possible.

1. What is the ratio of 4 in. to 3.5 ft in simplest colon form?

Find the ratio of rise to run for each roof below in simplest fraction form.

2.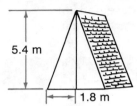

5.4 m

1.8 m

3.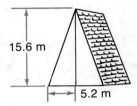

15.6 m

5.2 m

4. Eight hundred kilowatts of electrical power are input to a motor. If the output of the motor is 720 kW, what is the efficiency ratio of the motor in simplest colon form?

5. An airplane engine has an input of power in the form of gasoline amounting to 324 kW. The power output of the airplane engine is 120 kW. What is the percent efficiency of the airplane engine?

6. A submersible sump pump has 1.5 hp of power in the form of electricity input to it. If the pump has an efficiency of 67%, what is the amount of power the pump is expected to output in horsepower?

Use a triangular rule if you have one. Otherwise copy the scales at the back of the book to do the following questions.

7. One dimension of an air duct is 4 ft 3 in. Represent this length using the 1 in. : 1 ft scale of an architect's scale.

Use a metric scale to draw a line to represent each length and scale given.

8. 1.7 m (scale 1 : 100)

9. 4.7 m (scale 1 : 100)

10. 1.25 m (scale 1 : 50)

11. 3.45 m (scale 1 : 50)

12. 0.32 m (scale 1 : 5)

13. 0.15 m (scale 1 : 5)

Use an architect's scale to find the actual length represented by the line segment using the following scales: **a.** $\frac{1}{2}$ in. : 1 ft; **b.** $\frac{1}{4}$ in. : 1 ft; **c.** 1 in. : 1 ft.

14. ───────────────────

Use a metric scale to find the actual length represented by the line segment using the following scales: **a.** 1 : 100; **b.** 1 : 50; **c.** 1 : 20.

15. ──────────────────────────

Proportion

5-4 Using Rate and Proportion

A comparison of two *unlike* quantities is called a **rate**. Some examples of rates are 85 mi/h, $60/day, and 120 m/s. Many problems deal with rates. Such problems can be solved by first writing a statement of equality between two rates. This kind of statement is called a **proportion**. Consider the example below.

Example

On the map at the right, the straight-line distance between Chicago and St. Louis is 1.4 in. What would the actual distance in miles be?

Note: For this text, the numbers on a map scale are considered to be exact.

Solution

Set up a proportion using the scale of 0.5 in. to 100 mi as shown below. Let x be the unknown number of miles. Notice that the map distances (in inches) are in the numerators and the actual distances (in miles) are in the denominators. Then solve for x using a calculator.

$$\frac{\text{map distance}}{\text{actual distance}} \quad \rightarrow \quad \frac{0.5 \text{ (in.)}}{100 \text{ (mi)}} = \frac{1.4 \text{ (in.)}}{x \text{ (in.)}}$$
$$0.5x = 1.4 \times 100$$
$$x = 280$$

Scale: 0.5″ : 100 mi

`1 · 4 × 1 0 0 ÷ · 5 =` 280

The straight-line distance between Chicago and St. Louis is 280 mi. The result is accurate to two significant digits.

EXERCISES

Use a calculator where possible.

A **1.** An automobile can travel 150 km on 12 L of gasoline. How far can the automobile travel on 160 L?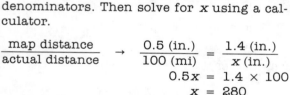

2. The cost of 25 ft of half-inch-thick neoprene rubber sheeting is $1176. What is the cost per foot (or the *unit rate*)?

$$\frac{25\ (\text{ft})}{1176\ (\$)} = \frac{1\ (\text{ft})}{x\ (\$)}$$

3. Under standard conditions, 10.0 ft of copper wire provides 0.0160 Ω of resistance. What is the resistance of 240.0 ft of copper wire?

$$\frac{10.0\ (\text{ft})}{0.0160\ (\Omega)} = \frac{240.0\ (\text{ft})}{x\ (\Omega)}$$

B **4.** The scale for the map at the right is 1.5 in. : 300 mi. If you could travel in a straight line, what would the actual distances between the cities below in miles be? (Use a rule showing tenths of an inch.)

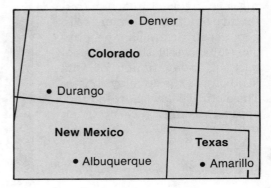

   a. Denver to Amarillo
   b. Durango to Albuquerque
   c. Amarillo to Durango

5. One day, the exchange rate for buying Canadian dollars was $0.74. This means that for each $1.00 (Cdn.) bought, $0.74 (U.S.) was paid. How many Canadian dollars, to the nearest cent, could have been bought for $50 (U.S.)?

6. If 80.0 kg of cement are used to make 400.0 kg of concrete, how much cement is needed to make 1600 kg of concrete?

7. A machine produces 2070 widgets in 8.50 h. How many widgets will the machine produce in 40.0 h?

8. Find the weight of a 180.0 cm by 120.0 cm rectangular sheet of aluminum if the weight of a 100.0 cm² piece of the aluminum is 71.0 g.

9. An aerial photographer takes pictures that can be used to find distances on the ground. The photographer uses the following proportion to determine unknown distances.

$$\frac{\text{height of airplane}}{\text{focal radius of camera}} = \frac{\text{distance on the ground}}{\text{distance on the photograph}}$$

   a. Find the distance on the ground if the focal radius of the camera is 9.5 in., the height of the airplane is 2000.0 ft, and the distance on the photograph is 3.5 in.
   b. What is the distance on the photograph if the focal radius of the camera is 9.5 in., the height of the airplane is 2000.0 ft, and the distance on the ground is 1250 ft?

C **10.** The profits of a small boutique are divided between its two partners in the ratio 3 : 5. How much should each partner receive from weekly profits of $2500?

5-5 Rates Used in the Manufacturing Process

Many different processes, such as etching, welding, and painting, are involved in the manufacture of a product in order to produce its desired shape, physical characteristics, and properties. Today, much of this kind of work is done by robots. The picture below shows a robot spray-painting a pick-up truck bed.

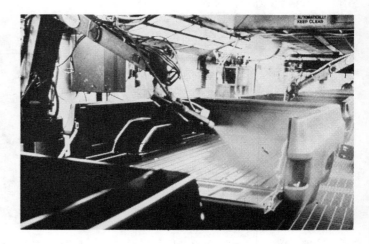

Often problems relating to rate and proportion come up as these manufacturing processes occur. Consider the example below.

Example

Skill(s) _18, 19, 28_

A car manufacturing plant produces 60 *complete* pick-up trucks per hour with the aid of industrial robots. If the plant is in operation for two 8.0 h work shifts each day, how many *complete* pick-up trucks are produced per day?

Solution

Set up a proportion involving the rate of 60 pick-up trucks/h. Let 16 h represent the two eight-hour shifts each day.

$$\frac{\text{pick-up trucks}}{\text{time}} \rightarrow \frac{60 \text{ (pick-up trucks)}}{1 \text{ (h)}} = \frac{x \text{ (pick-up trucks)}}{16 \text{ (h)}}$$

$$x = 60 \times 16$$
$$x = 960 \text{ (pick-up trucks)}$$

`6` `0` `x` `1` `6` `=` 960

Altogether, 960 *complete* pick-up trucks are produced per day, to two significant digits.

EXERCISES

Use a calculator where possible.

A **1.** How many pick-up trucks can be produced in the given times for the car manufacturing plant described on the previous page? (Use the rate, 960 pick-up trucks/day, to set up each proportion.)
   a. one five-day work week
   b. 52 five-day work weeks

B Solve each problem by the proportion method.

2. A gas cutting torch is capable of cutting a certain type of steel at a rate of 18 in./min. How many inches of steel can be cut in one hour?

3. A building with 1500.0 ft^2 of wall area is to be painted. One gallon of paint will cover 400.0 ft^2. How much paint is needed to paint the building?

4. One gallon of white glue weighs 11.0 lb and sells for $5.44. What is the price per pound of this glue to the nearest cent?

5. It takes 40.0 s to etch a printed circuit for a television set. A company wishes to produce 10,000 sets. How much time must be allowed for the etching of the circuits?

6. A current of 20.0 A/ft^2 is needed to nickel-plate a surface. How many amperes of current are needed to plate a surface of 3.5 ft^2?

7. When making bread, a baker places five loaves of bread in each pan. Then sixteen pans at a time are put on a rack to go into the oven.

   a. If three racks of bread are baking at one time, how many loaves of bread are in the oven?
   b. If the bread is made between 5 A.M. and noon and it takes 140 min to completely make a rack of bread, how many loaves are baked in one day with three racks baked at a time?

C **8.** A painter applies a primer coat at the rate of 150 ft^2/h and is paid $14.50/h. How much will it cost to have the painter cover 1200 ft^2?

9. Rosin-core solder for use in printed circuits contains 60% tin and 40% lead. How much tin is contained in a 1.0 lb spool of this solder?

10. A gas cutting torch will burn for 4.0 h on one tank of gas. If the torch is capable of cutting six pieces per minute, how many pieces can be cut with one tank of gas?

5-6 Mechanical Advantage

Machines make work easier to perform. When a machine is used, a very heavy load can be lifted or moved by applying a small effort. For example, a heavy car can be lifted by applying a small force upon a jack. This capacity for a machine to *multiply* a small input force into a large output force is called its **mechanical advantage**. It is described mathematically by the following **force ratio**, comparing output force to input force.

$$\text{Mechanical Advantage} = \frac{\text{output force (weight of car)}}{\text{input force (force applied on the jack)}} \quad \text{or} \quad MA = \frac{F_O}{F_i}$$

In its simplest forms, as studied in this text, mechanical advantage computations ignore friction. Such computations produce what is known as *Ideal Mechanical Advantage* (*IMA*). If friction is considered, the computations produce *Actual Mechanical Advantage* (*AMA*). This text will deal with IMA only, and it will be referred to simply as *MA*.

Example 1

Skill(s) __18__

The portion of an automobile that must be lifted to allow for removal of a wheel weighs 600.0 lb. The jack used requires a hand force of 25 lb. What is the mechanical advantage of the jack?

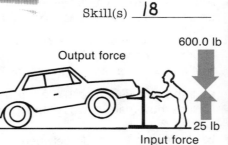

600.0 lb

Output force

25 lb

Input force

Solution

Substitute into the force ratio and solve using a calculator.

$$\text{Mechanical advantage} = \frac{\text{output force (weight of car)}}{\text{input force (force applied on the jack to lift the car)}}$$

$$= \frac{600.0 \text{ (lb)}}{25 \text{ (lb)}}$$

6 0 0 ÷ 2 5 = 24

The mechanical advantage of the jack is 24. There are no units of measure in stating mechanical advantage. This means that the mechanical advantage of the jack will *multiply* the input force of 25 lb by 24 to lift the 600.0 lb car.

Mechanical advantage may also be computed by the distance ratio, which compares the relative distances the two forces move. The ramp (inclined plane) shown at the right makes it easier to move a heavy load 4.0 ft up and onto a platform.

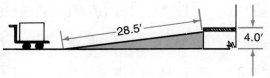

Mechanical Advantage = $\dfrac{\text{distance moved by input force}}{\text{distance moved by output force}}$ or $\dfrac{D_i}{D_o}$

Example 2

What is the mechanical advantage of the ramp shown above?

Solution

$$MA = \dfrac{D_i}{D_o} \text{ or } \dfrac{28.5 \text{ (ft)}}{4.0 \text{ (ft)}}$$

| 2 | 8 | · | 5 | ÷ | 4 | = | 7.125

The mechanical advantage of the ramp is 7.1.

EXERCISES

Use a calculator where possible.

A 1. A chain hoist can lift a 750 lb steel crate when the operator applies a 30 lb force to the chain. Find the mechanical advantage of the hoist.

2. What is the mechanical advantage of a machine with an input distance of 9.0 cm and an output distance of 3.0 cm?

3. To get a 550 lb crate up a ramp and onto a platform, a laborer must push the crate with a force of 110 lb. Find the mechanical advantage of the ramp.

B 4. The mechanical advantage of a wrench is 15. A nut requires a force of 180 lb to release it. How much force must the mechanic apply to the wrench to loosen the nut?

5. A torque wrench requires a 22 lb pull to tighten a headbolt on a tractor engine. If the mechanical advantage of the wrench is 9.5, how many pounds of force are placed on the headbolt?

6. The tip of a hammer handle used to pull a nail travels through an arc of 10.0 in. while pulling out the nail. The nail moves 1.5 in. Find the mechanical advantage of the hammer.

7. If the force applied to the hammer of question 6 was 32 lb, what is the holding power of the nail?

5-7 Simple Machines and Mechanical Advantage

Simple machines use one of three principles (lever, inclined plane, and hydraulic press) to multiply a small input force into a greater output force. There are several simple machines, but this lesson will only deal with the mechanical advantage of the *lever* and the *inclined plane*.

A **lever** is a rigid bar that rotates about a fixed point called a fulcrum.

A fixed **inclined plane**, or a ramp, is used to move heavy objects that cannot be easily moved by lifting up onto a platform.

Computations involving simple machines use a proportion that equates the force ratio and the distance ratio. When three of the four terms in the proportion are known, the fourth term can be solved.

$$\frac{\text{Output force}}{\text{Input force}} = \frac{\text{Distance moved by input force}}{\text{Distance moved by output force}} \quad \text{or} \quad \frac{F_o}{F_i} = \frac{D_i}{D_o}$$

Example

Skill(s) _18, 19, 28_

A man wishes to move a 300 lb boulder using the lever and fulcrum arrangement shown in the diagram at the right. What force must be applied to the lever to move the boulder?

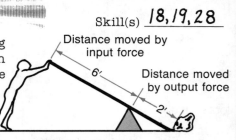

Distance moved by input force

Distance moved by output force

Solution

Substitute into the proportion equating the force ratio and the distance ratio to find the unknown input force, F_i.

$$\frac{\text{Output force}}{\text{Input force}} = \frac{\text{Distance moved by input force}}{\text{Distance moved by output force}}$$

$$\frac{300 \text{ (lb)}}{F_i \text{ (lb)}} = \frac{6 \text{ (ft)}}{2 \text{ (ft)}}$$

`3 0 0 × 2 ÷ 6 =` 100

The man must use 100 lb of force to move the boulder.

EXERCISES

Use a calculator where possible.

A 1. A tree stump embedded firmly in the ground may require 800.0 lb of force to remove it. How much force must be applied to the input end of the lever to move the stump?

2. A 500.0 lb weight to be lifted is placed 1.5 ft from the fulcrum of a 10.0 ft lever. How much input force is needed to lift the 500.0 lb weight?

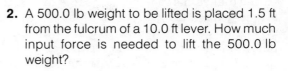

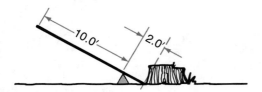

B 3. Refer to the tinsnips at the right.

a. If a force of 12 lb is applied at point A, what force is output at point B?

b. If the distance moved by the input force, D_i, were 9.8 in., what force is output at point B?

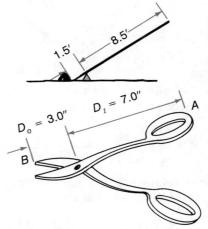

4. For the ramp shown below, what force is necessary to roll a 488 lb drum to the top of the ramp?

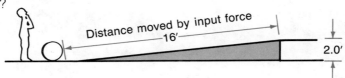

5. A crate weighing 372 lb must be placed on a platform shown below. What is the force needed to push the crate up the ramp (neglecting friction)?

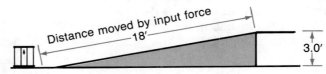

C 6. A 6.0 ft lever is to be used to lift a 153 lb block of wood. Where must the fulcrum be placed relative to the block if the input force is to be 76.5 lb?

7. A 300 lb drill press is to be moved up a ramp on frictionless dollies to a truck bed that is 4 ft high. If the ramp is 12 ft long, how much force is needed to push the drill press up to the truck bed?

8. A man pries up one end of a sidewalk slab that weighs 340.0 lb using a 6.0 ft pry bar. If he exerts 40.0 lb of force and lifts only half the weight of the sidewalk, how far from the sidewalk slab is the fulcrum?

5-8 Gears and Pulleys

In the diagram at the right, gear A drives the larger gear, B. If gear A turns at a speed of 25 r/min, gear B turns at 12 r/min. The table below shows how the smaller driver gear *reduces* the number of revolutions per minute of the driven gear.

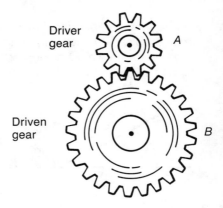

Driver gear

A

Driven gear

B

Speed of gear A (r/min)	25	50	75	100
Speed of gear B (r/min)	12	24	36	48

The speeds at which these two gears turn is *inversely related* to the number of teeth in the gears.

(r/min of **driver** gear A) $\dfrac{25}{12} = \dfrac{25}{12}$ (number of teeth in **driven** gear B)
(r/min of **driven** gear B)   (number of teeth in **driver** gear A)

This inverse relationship is summarized by the following proportion, which equates the gear speed ratio and the number of teeth ratio.

$$\frac{\text{Speed of } \textbf{driver} \text{ gear (r/min)}}{\text{Speed of } \textbf{driven} \text{ gear (r/min)}} = \frac{\text{Number of teeth in } \textbf{driven} \text{ gear}}{\text{Number of teeth in } \textbf{driver} \text{ gear}}$$

Example 1

Skill(s) <u>19, 28</u>

The driver gear, A, in the diagram at the right, has 32 teeth, and the driven gear, B, has 60 teeth. If gear A is turning at 300 r/min, how many revolutions per minute does gear B output? (In 300 r/min, assume all digits, including zeros, to be significant.)

Solution

Let s be the speed of gear B.

$$\frac{\text{Speed of driver gear } A}{\text{Speed of driven gear } B} = \frac{\text{Teeth in driven gear } B}{\text{Teeth in driver gear } A}$$

$$\frac{300 \ (\text{r/min})}{s \ (\text{r/min})} = \frac{60}{32}$$

`3 0 0 × 3 2 ÷ 6 0 =` 160

Gear B outputs 160 r/min.

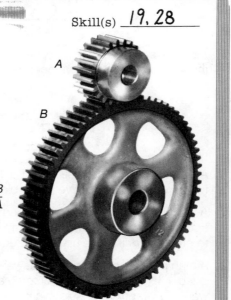

A

B

A pulley system, shown at the right, is similar to a gear system. The smaller sheave, A, drives the larger sheave, B. The speeds at which these two sheaves turn is *inversely related* to the size of the diameters of the sheaves.

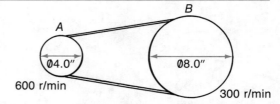

$$\frac{\text{(r/min of \textbf{driver} sheave } A)}{\text{(r/min of \textbf{driven} sheave } B)} \quad \frac{600}{300} = \frac{8.0}{4.0} \quad \frac{\text{(diameter in inches of \textbf{driven} sheave } B)}{\text{(diameter in inches of \textbf{driver} sheave } A)}$$

This relationship is summarized in the following proportion that equates the sheave speed ratio and the diameter ratio.

$$\frac{\text{Speed of \textbf{driver} sheave (r/min)}}{\text{Speed of \textbf{driven} sheave (r/min)}} = \frac{\text{Diameter of \textbf{driven} sheave}}{\text{Diameter of \textbf{driver} sheave}}$$

Example 2

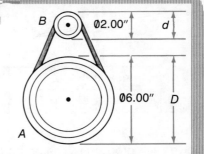

If driver sheave A at the right is rotating at 300 r/min, what is the speed output by sheave B? (In 300 r/min, assume all digits, including zeros, to be significant.)

Solution

Let x be the speed of the driven sheave, B. Set up a pulley speed and diameter proportion and use a calculator to solve the problem.

$$\frac{\text{Speed of driver sheave } A}{\text{Speed of driven sheave } B} = \frac{\text{Diameter of driven sheave } B}{\text{Diameter of driver sheave } A}$$

$$\frac{300 \,(\text{r/min})}{x \,(\text{r/min})} = \frac{6.00 \,(\text{in.})}{2.00 \,(\text{in.})}$$

[3][0][0][×][2][÷][6][=]   100

The output of the driven sheave B is 100 r/min.

EXERCISES

Use a calculator where possible.

A Use the given diagram to complete each table. The numbers in each table represent complete turns.

1.

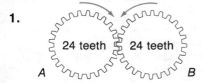

24 teeth   24 teeth

A   B

Driver gear A	1	24	100	450
Driven gear B	?	?	?	?

2.

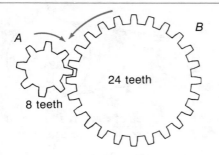

8 teeth
24 teeth

Driver gear A	3	6	9	30
Driven gear B	?	?	?	?

3.

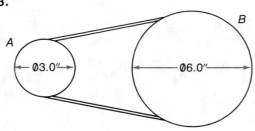

Ø3.0"   Ø6.0"

Driver sheave A	2	?	40	?
Driven sheave B	1	20	?	60

4.

Ø2.0"   Ø8.0"

Driver sheave A	4	?	100	?
Driven sheave B	1	50	?	150

B Solve each problem by the proportion method.

5. In a gear system, driver gear, A, makes 450 r/min and the driven gear, B, makes 150 r/min. How many teeth does gear A have if gear B has 42?

6. For the pulley system below, sheave B is turning at 400 r/min. How fast is sheave A turning?

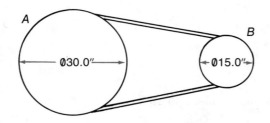

Ø30.0"   Ø15.0"

7. A smaller driver gear below is set on the shaft of a motor running at 80 r/min. How many revolutions per minute are output by the driven gear?

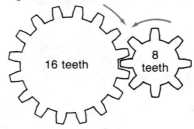

16 teeth   8 teeth

8. The motor used to run a drill press has a pulley system inside that turns a driver sheave at 1950 r/min. The driver sheave, which is 10.0 cm in diameter, drives a sheave that has a diameter of 15.0 cm. The driven sheave is joined to a shaft that turns the drill bit. At what speed does the drill bit revolve?

9. A 26-tooth gear is on the shaft of an electric motor running at 3600 r/min. The gear drives a 60-tooth gear. What is the speed of the driven gear?

10. In a pulley system, the driven sheave has a diameter of 30.0 cm and the driver sheave has a radius of 8.0 cm. If the driven sheave is turning at 500 r/min, how fast is the driver sheave that is connected to a motor turning?

11. In the gear sprocket system at the right, the driver gear has 23 teeth and the driven gear has 60 teeth. What is the speed of the driver gear if the driven gear is turning at 1200 r/min?

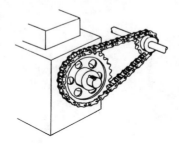

Tricks of the Trade

The Drill Press

Because of the large number of clever and practical accessories available today, a drill press becomes a versatile tool that can be used to shape, rout, buff, grind, mix paint, cut dovetails, as well as do the basic job of drilling.

In everyday drilling operations, speed is a critical factor in producing a good job. On most drill presses, speed is varied by moving the drive belt on step pulleys. Generally, a press with a pulley setup will give speeds of 600 r/min to 5000 r/min. This range is good enough to handle most jobs.

The table below gives a range of recommended speeds for different operations and different materials. A good rule-of-thumb to keep in mind is that the larger the diameter of the drill bit and the harder the material, the slower the drill press should be run.

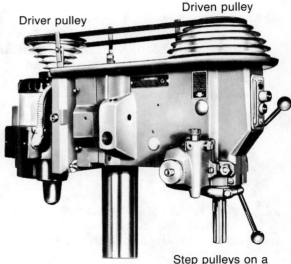

Driver pulley

Driven pulley

Step pulleys on a drill press provide different speeds.

Recommended Drill-Press Operating Speeds		
Material	Operation	Speed (r/min)
Wood	Drilling to 0.6 cm	3800
Wood	Drilling 0.6 cm to 1.3 cm	3100
Wood	Drilling 1.3 cm to 1.9 cm	2300
Wood	Drilling 1.9 cm to 2.5 cm	2000
Wood	Drilling over 2.5 cm	600
Metal	Wire-brushing	3600
Glass, ceramic	Drilling	600

5-9 A Car's Transmission Ratio

Power is transmitted from the engine to the wheels of an automobile by a series of gears. The diagram below shows a cut-away view of an automobile with manual transmission.

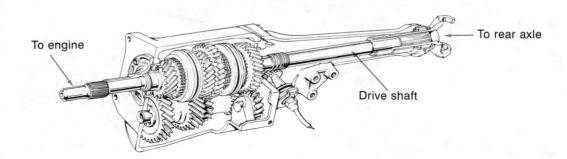

An automobile tachometer indicates the speed of the engine in revolutions per minute. The speed of the drive shaft is changed by the gears in the transmission. A car's **transmission ratio** is a comparison of the engine speed to the drive-shaft speed.

Transmission ratio: $\dfrac{\text{engine speed (r/min)}}{\text{drive-shaft speed (r/min)}}$

Example

Skill(s) _19, 28_

The tachometer in a manual transmission car shows the engine is running at 3000 r/min as the car travels in third gear. If the transmission ratio for this situation is 1.800 to 1.000, what is the speed of the drive shaft to four significant digits? (Assume all digits in 3000 r/min, including zeros, to be significant.)

Solution

Let s be the speed of the drive shaft. Set up a proportion using the transmission ratio to solve the problem.

$$\frac{\text{engine speed (r/min)}}{\text{drive-shaft speed (r/min)}} \rightarrow \frac{1.800}{1.000} = \frac{3000}{s}$$

`3` `0` `0` `0` `÷` `1` `·` `8` `=` 1666.6667

The speed of the drive shaft is 1667 r/min.

EXERCISES

Use a calculator where possible.

A Write the transmission ratio to two significant digits.

	Tachometer Reading of Engine Speed (r/min)	Drive-Shaft Speed (r/min)
1.	3000	1000
2.	4000	2000
3.	3500	3150
4.	3000	2300

B **5.** The tachometer of a manual transmission car shows the engine speed is 2400 r/min. If at the same time the drive-shaft speed is 800 r/min, what is the transmission ratio to two significant digits?

6. When a car is traveling at a speed of 30 km/h in first gear, the tachometer shows the engine speed is 3000 r/min. If the transmission ratio for this situation is 3 : 1, what is the speed of the drive shaft?

7. While driving in second gear, the tachometer of a car shows an engine speed of 2800 r/min. If the speed of the drive shaft is 1750 r/min, what is the transmission ratio to two significant digits?

8. A manual transmission car is traveling in fourth gear at a speed of 90 km/h. The tachometer shows an engine speed of 3000 r/min. If the transmission ratio for this situation is 1 : 1, what is the speed of the drive shaft?

Check Your Skills

Find equivalent fractions.

1. $\dfrac{11}{7} = \dfrac{?}{49}$ **2.** $\dfrac{7}{2} = \dfrac{28}{?}$ **3.** $\dfrac{9}{5} = \dfrac{?}{20}$ **4.** $\dfrac{3}{4} = \dfrac{?}{100}$ (Skill 10)

There are twelve resistors, nine capacitors, and four transistors on a circuit board. Write the ratios of these items as fractions in simplest form.

5. resistors to capacitors   **6.** transistors to resistors   (Skill 18)

Solve for x in each proportion.

7. $\dfrac{10}{x} = \dfrac{5}{2}$ **8.** $\dfrac{x}{41} = \dfrac{3}{246}$ **9.** $\dfrac{6}{7} = \dfrac{18}{x}$ (Skill 19)

Write each percent as a decimal. Write each decimal as a percent.

10. 75.2%   **11.** $18\frac{3}{4}\%$ **12.** 0.175   **13.** 4.5   (Skill 22)

Solve for x in each equation.

14. $2x = 5.6$ **15.** $5x = 2.25$ **16.** $11x = 33.66$ (Skill 28)

Self-Analysis Test

Use a calculator to solve the problems.

1. An automated auto-painter can paint 54 cars in an hour. How long will it take to paint 1000 cars?

2. A cutting torch can cut metal at the rate of 0.5 in./s. How long will it take the torch to cut 28 ft of metal?

3. A glue-gun operator can fasten 250 boxes in an hour. How long will it take three persons to glue 2500 boxes?

4. A house with a floor area of 1800 ft² costs $86,400 to build. Use a proportion to find the cost to build a house with 1500 ft² of floor area.

5. A ramp has a mechanical advantage of 4.50. A 420.0 lb box is placed on a 98.0 lb cart to be pushed up the ramp. How hard must a laborer push on the load to get it up the ramp?

6. What is the mechanical advantage of a lever 18 in. long if the fulcrum is to be 4 in. from the weighted end?

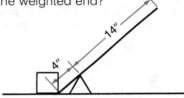

7. In the gear sprocket system below, what is the speed of the smaller driver gear if the speed of the driven gear is 400 r/min?

8. In the pulley system below, what is the speed of sheave *B* if the driver sheave, *A*, is turning at 80 r/min?

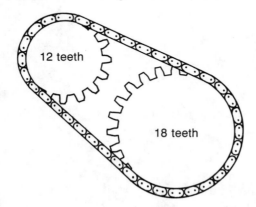

12 teeth

18 teeth

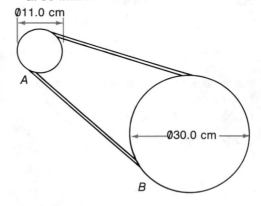

Ø11.0 cm

A

Ø30.0 cm

B

9. The tachometer of a car shows a reading of the engine speed at 3200 r/min. At the same time, the drive shaft turns at 1780 r/min. What is the transmission ratio in simplest form to two significant digits?

A wholesale grocery company is interested in determining which of its four trucks provide the best gasoline mileage. For convenience sake, we will identify them as truck A, truck B, truck C, and truck D.

The following BASIC program may be used to quickly calculate the gasoline mileage for each truck so comparisons can be made. When the number of gallons of gasoline used, G, and the two odometer readings R1 and R2 are entered into the program, it calculates the miles per gallon, M.

Notice the output if this gasoline mileage program is RUN with the data below input by the user.   G   = 12.3   R1 = 7846.6   R2 = 8109.8

```
130  PRINT "ENTER THE NUMBER OF GALLONS USED"
140  INPUT G
150  PRINT "ENTER THE ODOMETER READINGS"
160  INPUT R1,R2
170  M = ABS (R1 - R2) / G
180  PRINT "NUMBER OF GALLONS USED ";G
190  PRINT "ODOMETER READINGS ";R1" ";R2
200  PRINT "MILES PER GALLON ";M
210  PRINT "DO YOU WANT TO DO ANOTHER PROBLEM?"
220  PRINT "TYPE YES OR NO"
230  INPUT A$
240  IF A$ = "YES" THEN 130
```

```
]RUN
ENTER THE NUMBER OF GALLONS USED
?12.3
ENTER THE ODOMETER READINGS
?7846.6,8109.8
NUMBER OF GALLONS USED 12.3
ODOMETER READINGS 7846.6 8109.8
MILES PER GALLON 21.3983739
DO YOU WANT TO DO ANOTHER PROBLEM?
TYPE YES OR NO
?NO
```

Note that in line 170 the formula $M = \frac{R1 - R2}{G}$ is used to calculate the average number of miles per gallon. The ABS before the right side of the equation indicates that we are using the **absolute value** of the expression R1 − R2. This means that the computer uses the positive value of the two numbers R1 − R2 and R2 − R1 in the formula. This allows the user to enter the odometer readings in any order in line 160.

If the program were RUN, how many miles per gallon, to the nearest whole number, would each of the wholesale grocery company's four trucks use? The data for each truck are given in the table below.

		R1	R2	G
1.	Truck A	14623.7	15044.1	18.6
2.	Truck B	38423.2	38796.3	21.2
3.	Truck C	27463.6	27747.5	8.5
4.	Truck D	82614.6	89072.8	263.6

The U.S. automotive aftermarket is a multi-billion–dollar industry. Companies of all sizes manufacture and distribute automotive replacement parts, accessories, tools, and equipment to the service trade to keep the millions of cars in this country moving.

Today's automotive aftermarket industry provides a wide array of exciting and challenging job opportunities for young women and men. Jobs range from sales and marketing to administration and management.

Americans insist on rapid service and repair of their vehicles. The automotive aftermarket makes it possible.

Job Description

What do automotive parts merchandisers do?

1. When working in the area of inside sales, an automotive parts merchandiser works behind a wholesaler's counter or at a desk, in a customer service department. These merchandisers have the know-how to talk with customers, interpret their needs, use catalogs or computer terminals to select correct items, and process orders.

2. Some merchandisers work in outside sales, meaning they go to the customer and evaluate the customer's needs and requirements. Frequently, they are called on to demonstrate new products, materials, and services.

3. An automotive parts field service representative applies a deep knowledge of vehicle and product technology to solve service problems. They often instruct at clinics to update and upgrade technicians.

4. An automotive parts merchandiser may write brochures that will help sell products or may instruct users in the care of the products.

5. Other jobs in automotive parts merchandising might involve controlling inventory. These merchandisers work with management and salespeople to ensure that the goods arrive at correct locations in proper volume.

Qualifications

It is beneficial to have training provided by a community college's automotive parts merchandising program. Students in such a program get ''hands-on'' experience with parts merchandising in a computer lab, an automotive engine lab, an electrical diagnosis lab, an auto body lab, a wheel alignment and brakes lab, and an applied science lab. In addition, many companies will provide on-the-job training.

Automation has become an integral part of the auto industry today. The mass production of the automobile has been characterized by hundreds of boring, repetitive tasks. The auto worker who performs only a minute of the same work on hundreds of jobs a day has been plagued by boredom. Because of the tedious, and often injurious, nature of the work involved in mass-producing cars, the auto industry has focused on robotic applications.

In one of the nation's most sophisticated automobile assembly plants, 162 robots and nearly 2000 intelligent devices can produce more than 1000 full-sized automobiles in two eight-hour shifts. The use of robots in the auto industry today has led to a reduction in production costs and unskilled manpower along with an increase in quality and skilled manpower.

As global competition increases, business and industry must seek new ways to develop and maintain a competitive edge. One solution lies in the use of automated production. The introduction of automation to the workplace can lead to more export opportunities and to reductions in the amount of imported goods.

Use the information in the tables below to answer the following questions.

	Robot A	Robot B	Robot C	Alice	Betty	Carl
Maximum Number of Pieces Produced Per Hour	90	95	105	70	80	75
Actual Number of Pieces Produced Per Hour	60	62	82	68	72	70

1. For robot A, what is the ratio of actual number of pieces produced per hour to the maximum number of pieces produced per hour?

2. What is the ratio of maximum number of pieces produced to actual number of pieces produced for Betty?

3. Find the ratio of the actual number of pieces produced by the robots to the actual number of pieces produced by the workers.

4. Find the ratio of the maximum number of pieces produced by the workers to the maximum number produced by the robots.

Chapter 5 Test

Use a calculator where possible.

1. Write the ratio of 750 cm to 2.50 m in simplest colon form. (5-1)

2. The current, I, in amperes is the ratio of the power, P, in watts to the voltage, V, in volts. If $P = 460$ W, $V = 110$ V, find I. (5-1)

3. A profit of $6752 is divided by shareholders in the ratio 3 : 4. Calculate the amounts received to the nearest cent. (5-1)

4. The power input to an electric motor is 6400 W. The power output by the motor is 4800 W. What is the percent efficiency of the motor? (5-2)

5. An oil pump has a power output of 3.0 hp. If the pump is 78% efficient, find the power input. (5-2)

6. A power turbine is 85% efficient. The power delivered by the turbine is 0.92 kW. Find the power input. (5-2)

Use a triangular rule if you have one. Otherwise, reproduce the scales at the back of the book to do the following questions.

Use an architect's scale to find the actual length represented by each line segment using the following scales: **a.** $\frac{1}{4}$ in. : 1 ft; **b.** $\frac{1}{2}$ in. : 1 ft; **c.** 1 in. : 1 ft. (5-3)

7. ─────────────────────────────

8. ──────────────

9. ──────────────────────────────────────

Use a metric scale to find the actual length represented by each line segment using the following scales: **a.** 1 : 100; **b.** 1 : 50; **c.** 1 : 20. (5-3)

10. ───────────────────────────

11. ─────────────────────

12. ───────────────────────────────

Draw a line segment to represent each length using the given scale. (5-3)

13. 1 in. : 1 ft

　　a. 5.5 ft 　　　　**b.** 6 ft 　　　　　**c.** 3.25 ft

14. 1 mm : 20 mm

　　a. 2 m 　　　　　**b.** 1.7 m 　　　　　**c.** 0.8 m

15. If it takes 8775 lb of gravel to make 5.0 yd^3 of concrete, how much gravel is needed for 12.5 yd^3 of concrete? (5-4)

16. If 20.0 A of land yields 840 bu (bushels) of soybeans, how many acres are needed to raise 3500 bu of beans? (5-4)

17. A drive shaft makes 21,000 revolutions in 3.5 min. Calculate the number of revolutions in 60 min. (5-4)

18. A grinding wheel cuts through a 5 cm steel plate at a rate of 12 cm/s. How long does it take to complete the job? (5-5)

19. An automated painter can paint 135 units per minute. How many units can be painted in an 8 h shift? (5-5)

20. Twenty-one staples are needed to assemble each container. How many staples are needed for 25,000 containers? (5-5)

21. A well-drilling machine has a mechanical advantage of 30. If the engine of the machine can apply a maximum force of 650 lb, what is the greatest load that the machine can lift? (5-6)

22. In a block and tackle system of several pulleys, the rope pulled through by the operator measured 7.3 m and the weight being lifted moved 0.61 m. Find the mechanical advantage of the system. (5-6)

23. A 530 lb lathe is to be raised to a truck bed by using a ramp 39 ft long. If the truck bed is 9.8 ft above the ground, how much force would be needed to move the lathe up the ramp? (5-7)

24. If driver gear, *A*, in the diagram below turns at a rate of 600 r/min, how many revolutions per minute will the driven gear make? (5-8)

25. In the pulley system below, sheave *A* is rotating at 450 r/min. What is the speed of sheave *B*? (5-8)

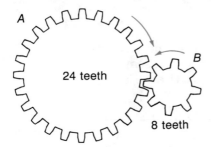

A   24 teeth   B   8 teeth

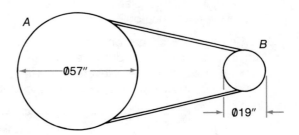

A   Ø57"   B   Ø19"

26. The tachometer of a car shows the engine speed is 3600 r/min. If the drive-shaft speed is 2400 r/min, what is the transmission ratio to one decimal place? (5-9)

Use a calculator where possible.

1. Write an expression for the perimeter of the triangular panel shown at the right.

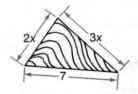

Write and solve an equation for each situation.

2. The selling price less the discount of $8.55 is $46.45. What is the selling price?

3. Eighteen defective items in a lot is 4% of the total lot. How many items are in the total lot?

4. The number of hours worked times $16.40 plus $87.75 in material costs equals $210.75. How many hours were worked?

Write a formula for each rule.

5. The air speed is the difference between the ground speed and the headwind speed. Let V_{air} be the air speed; V_{ground} be the ground speed; and $V_{headwind}$ be the headwind speed.

6. The percent of change is the quotient of the amount of change and the original amount. Let p be the percent of change; a be the amount of change; and o be the original amount.

Find the solution to the problem to the appropriate number of significant digits.

7. How many volts of electricity are needed to produce a current of 5.0 A across a resistance of 2.4 Ω?

8. When two objects are on a seesaw, they will balance if the products of their weights and their distance from the pivot point are the same. This is summarized in the law of levers formula, $w_1 d_1 = w_2 d_2$. Use the formula to find the unknown weights, w, or distances, d.

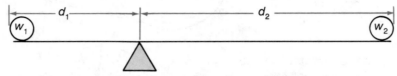

   a. Find w_2 if w_1 is 5.0 lb, d_1 is 10.0 in., and d_2 is 2.0 in.
   b. Find d_2 if w_1 is 60.0 g, w_2 is 18.0 g, and d_1 is 12.0 cm.
   c. Find d_1 if w_1 is 32.0 lb, w_2 is 20.0 lb, and d_2 is 15.0 in.

9. The perimeter of a rectangular driveway is 48.8 m. If the width is 5.4 m, find the length.

10. Find the diameter of a tree trunk that has a circumference of 23.55 in.

11. What is the brake power of an engine if the length of the braking arm is 8.5 ft, the force is 125 lb, and the engine speed is 3500 r/min?

12. Find the speed of the cutting edge of a grinding wheel 5.5 in. in diameter that is rotating at 1500 r/min.

13. A rectangular template has an area of 45.9 cm². If the width is 4.5 cm, find the length.

14. A trapezoidal wooden panel has an area of 36 ft² and bases that are 8 ft and 16 ft long. Find its height.

15. The region below is to be covered with insulation. If each bag of insulation covers 45 ft² of area, how many bags are needed?

16. Find the area of the shaded region in the metal plate shown below.

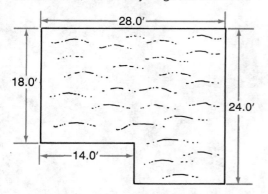

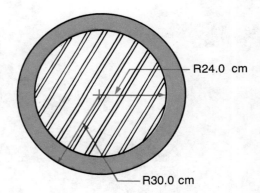

17. The deepest points in the Pacific and Atlantic oceans are 35,400 ft and 30,200 ft respectively. Write the ratio of the depths in simplest colon form.

18. A grinder is 35% efficient. If the power input to the grinder in the form of electricity is 0.72 hp, what is the power output?

19. Use an architect's scale to draw a line to represent each dimension using the $\frac{1}{4}$ scale.

    a. 24 in.   **b.** 12 ft   **c.** 8 ft 6 in.

20. A farmer can seed 220.0 A in a 10.0 h day. How long will it take to seed 800.0 A?

21. If a gallon of paint will cover 375 ft² of area, how many gallons are needed to paint 2500 ft² of area?

22. In a pulley system, a 12 in. driven sheave running at 600 r/min is connected by a belt to an 8.0 in. driver sheave. Calculate the speed of the 8.0 in. sheave.

23. A driver gear with 15 teeth is turning at 250.0 r/min. What is the speed of the driven gear with 42 teeth?

24. A transmission ratio is 1.6 to 1. If the engine speed is 3600 r/min, find the drive-shaft speed.

Chapter 6
Plane Geometry

After completing this chapter, you should be able to:

1. Understand the use of a protractor, compass, straightedge, T-square, set squares, and carpenter's square in drawing geometric figures.

2. Apply geometric constructions to the solution of problems.

3. Use congruent and similar triangles to solve measurement problems.

Angles, Lines, Circles, and Polygons

6-1 Measuring and Drawing Angles

In most technical drafting work, you will find a need for drawing angles. Drafters can use a drafting machine or protractor to measure or draw angles of specific sizes. The arms of the drafting machine at the right can be adjusted to draw and measure angles. A protractor, shown below, can also be used for the same purpose.

∠ PTR is a **right angle** since it measures 90°. →

∠ QTS is an **obtuse angle** since it →
measures between 90° and 180°.

∠ PTQ is an **acute angle** since it →
measures less than 90°.

∠ PTS is a **straight angle** →
since it measures 180°.

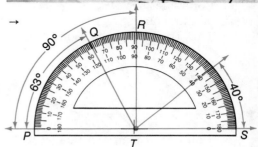

On technical drawings, the size of an angle is indicated by its arc, as shown above. The **vertex** of an angle is the common endpoint of the two rays that form the sides of an angle. A **ray** is a line with one endpoint. Point T above is the vertex of several angles. Point T is also the endpoint of the rays that form the sides of the angles.

Example 1

Measure the size of the angle in the drawing at the right.

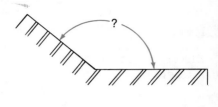

Solution

1. *Estimate* the measure as between 90° and 180°.
2. Place the center of the protractor at the vertex of the angle.
3. Place the base line of the protractor so that zero lies on one side of the angle to be measured.
4. Read the number that lies on the other side of the angle. Since the angle was estimated as between 90° and 180°, read the larger number scale. The angle measures 140°.

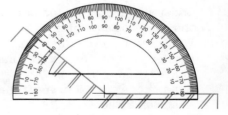

Example 2

With a compass and straightedge, construct a copy of ∠ ABC in the drawing at the right.

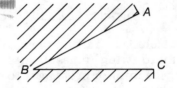

Solution

Follow these steps to copy the angle.
1. Construct ray *PT*.
2. On the original angle, place the needle of the compass at vertex *B*. Construct arc *DE*.
3. Without changing the radius of the compass, place its needle at vertex *P* of the new angle. Construct an arc intersecting ray *PT* at *R*.

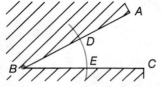   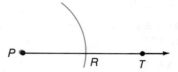

4. Using the original angle, adjust the compass radius to measure the distance *DE*.
5. Place the needle of the compass on the new angle at *R*, and construct a small arc intersecting the previously drawn arc at *Q*.
6. Construct a ray from *P* through *Q*. This will complete the copied angle, ∠ *QPR*. ∠ *QPR* ≅ ∠ *ABC*. (The ≅ symbol means "is congruent to".)

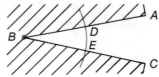   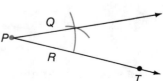

Example 3

Bisect ∠ *A* at the right with a compass and straightedge.

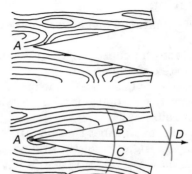

Solution

1. Construct arc *BC* by placing the needle of the compass at center *A*.
2. With *B* and *C* as centers and with a radius greater than half of arc *BC*, construct a pair of arcs intersecting at *D*.
3. Construct a ray from *A* through *D*. ∠ *BAC* has been bisected. ∠ *BAD* ≅ ∠ *DAC*.

EXERCISES

A *Estimate* the given angle as acute, obtuse, or right. Then measure the indicated angle with a protractor. Extend each side of the angle with a sheet of paper to make it easier to measure.

1. 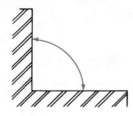 **2.** 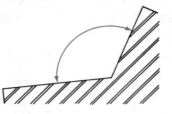 **3.**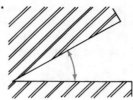

Draw an angle for each given measure with a protractor.

4. 45° **5.** 75° **6.** 90° **7.** 125° **8.** 112°

B Draw a figure that resembles the one given. Then use a compass and straight-edge to copy ∠*DEF* to make ∠*QPR*. (Locate the position of point *R* outside the figure.)

9. 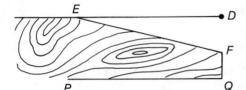 **10.**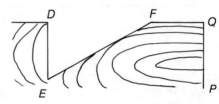

11. 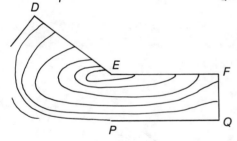 **12.**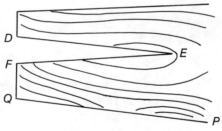

Draw a figure that resembles the one given. Then use a compass and straight-edge to bisect the indicated angle.

13. 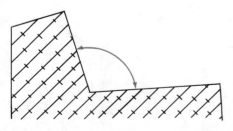 **14.**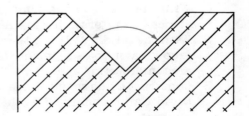

C **15.** Draw a 90° angle with a protractor. Then use *only* a compass and straight-edge to construct a 45° angle, a 22.5° angle, and a 135° angle.

6-2 Perpendicular Lines

Two lines that intersect at right angles are called **perpendicular lines**. A T-square, a pair of set squares, or a carpenter's square, shown below, can be used to make and test perpendicular lines, when a drafting machine is not available.

T-square and a pair of set squares:

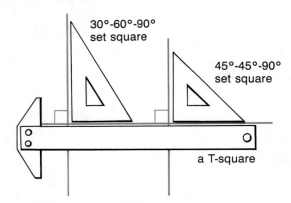

30°-60°-90° set square

45°-45°-90° set square

a T-square

Carpenter's square:

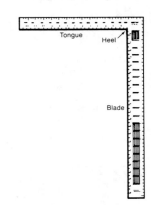

Tongue   Heel

Blade

When the above equipment is not available, you can use a compass and a straightedge to construct the required figures.

Example 1

In the drawing at the right, a line segment perpendicular to ray AB at point C is required. Construct the perpendicular with a compass and straightedge.

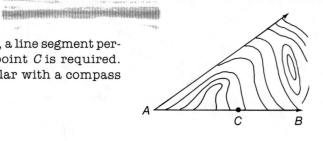

A   C   B

Solution

1. With the needle of the compass at C, construct arcs intersecting ray AB at D and E.
2. With the needle at D and then E and a radius greater than segment CD, construct arcs intersecting at F.
3. Construct a line segment passing through C and F and touching the other side of the angle at G.
   $\overrightarrow{CG} \perp \overrightarrow{AB}$ (The symbol means "is perpendicular to.")

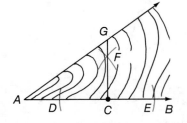

G

F

A   D   C   E B

162

Example 2

The drawing at the right requires a line segment perpendicular to segment QR with an endpoint at P. Construct the perpendicular with a compass and straightedge.

Solution

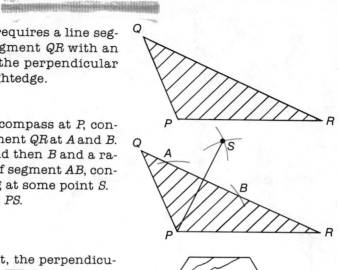

1. With the needle of the compass at P, construct arcs cutting segment QR at A and B.
2. With the needle at A and then B and a radius greater than half of segment AB, construct arcs intersecting at some point S.
3. Construct line segment PS.
   $\overline{PS} \perp \overline{QR}$

Example 3

In the drawing at the right, the perpendicular bisector of line segment WR is needed. Use a compass and straightedge to do this construction.

Solution

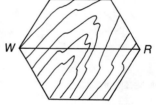

1. With the needle of the compass at W and a radius greater than half of segment WR, construct arcs above and below segment WR.
2. With the needle at R and the same radius, construct arcs intersecting the other arcs at X and Y.
3. Construct a segment from one side of the hexagon to the other that passes through X and Y. Label the contact points with the hexagon U and V.
   $\overline{UV} \perp \overline{WR}$ and $\overline{UV}$ bisects $\overline{WR}$.

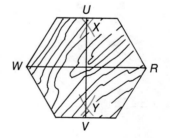

EXERCISES

A Is the indicated angle a right angle? Use an appropriate instrument to measure the given angle.

1.

2.

3.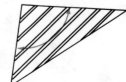

B Draw a figure that resembles the one given. Then construct a line segment perpendicular to segment *MN* at point *P*. Use a compass and a straightedge.

4.

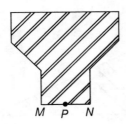

5.

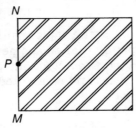

Follow the same directions as in questions 4 and 5. Use a pair of set squares and a T-square if you have them. Otherwise, reproduce the pair of set squares in the back of the book and construct a T-square.

6.

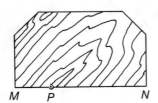

7.

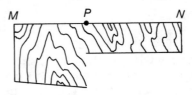

For questions 8 to 14, draw a figure that resembles the one given. Construct a line segment perpendicular to segment *CD* that passes through point *P*. Use a compass and a straightedge.

8.

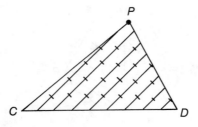

9.

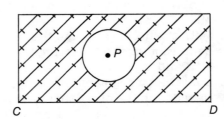

Follow the same directions as in questions 8 and 9 using only a pair of set squares and a T-square.

10.

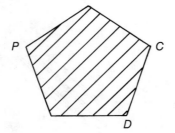

11.

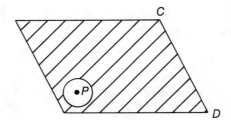

Construct a line segment that is the perpendicular bisector of segment *JK*. Use a compass and a straightedge.

12.

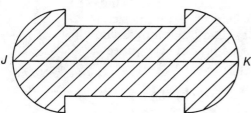

13.

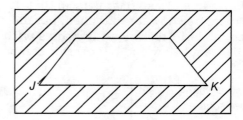

14. Draw the perpendicular bisectors of the sides of each.

a.   **b.**   **c.**

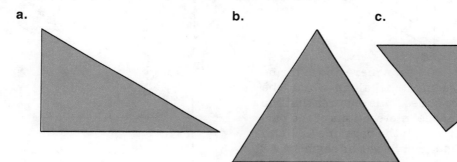

15. A **median** of a triangle is a line drawn from a vertex to the midpoint of the opposite side. Trace each drawing above. Draw the three medians of each triangle.

C 16. Draw a line segment 5 in. long. Use a compass and straightedge to divide the line segment into four equal parts.

17. Draw a rectangle having a length of 9 cm and a width of 4.5 cm with a compass and ruler.

18. Use a compass and ruler to draw a square having sides 3 in. long.

Tricks of the Trade

Miter Cuts

To join two pieces of stock at a 90° angle, a carpenter can use a miter square to cut the stock at 45° angles. The pieces fit as shown below.

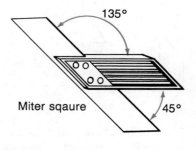

135°

Miter sqaure   45°

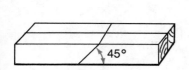

45°

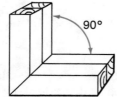

90°

6-3 Parallel Lines

Lines in a plane (flat surface) that do not intersect are called **parallel**. No matter how far extended, parallel lines never meet. Such lines can be made in technical drawings with a variety of instruments. The diagram below shows how this can be done with a pair of set squares.

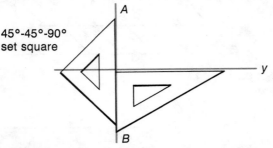

45°-45°-90° set square

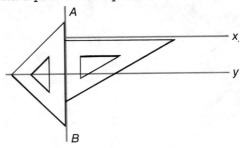

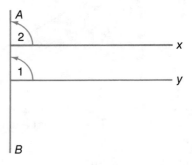

The 45° – 45° – 90° set square above forms the base line segment *AB* along which the other set square slides. This ensures that the angles formed at the intersections of the parallel lines *x* and *y* with the base line (∠1 and ∠2) are the same size. Base line segment *AB* can also be called a **transversal**. When two or more parallel lines are cut by a transversal, several pairs of corresponding angles are created that are the same size, or **congruent**. In the diagram, ∠1 and ∠2 are corresponding congruent angles. In this case, both are right angles.

The diagram below shows how parallel lines can be made with a T-square and a set square using the property of equidistance. Points on the base line are first marked using the rule along the T-square. Equal angles are formed at the intersections of the parallel lines with the base line along the T-square, or the transversal.

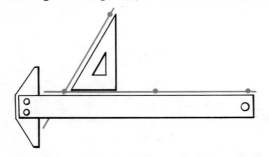

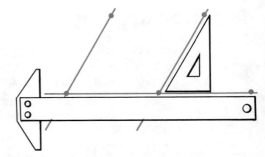

When a drafting machine or set squares are not available, parallel lines can be constructed with a compass and straightedge.

Example 1

The drawing at the right requires two equally spaced line segments 12 mm apart that are parallel to segment *VT*. Use a compass and a ruler.

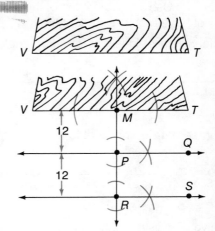

Solution

1. Choose any point *M* on segment *VT* and construct a perpendicular to segment *VT* at *M*.
2. Use a ruler to locate points *P* and *R* so that segments *MP* and *PR* measure 12 mm.
3. Construct perpendiculars to line *MR* at *P* and *R*. These lines are parallel to line *VT*.

Note that the parallel lines *PQ* and *RS* can also be drawn by copying ∠ *PMT* at points *P* and *R*, as explained in lesson 6-1.

(Dimensions are in millimeters.)

Example 2

The drawing at the right shows an enlargement of two screw threads. Complete the drawing to include the left side of the next thread using a pair of set squares.

Solution

1. Extend the sides between the two threads to locate the vertex of the angle. Label the angle, ∠ *BAC*, as shown at the right.
2. Position the pair of set squares so that one side lies along side *AC* of the angle, as shown below.
3. Hold the bottom set square in place and move the top one to point *P*. Draw a line parallel to side *AC* through *P*.

1.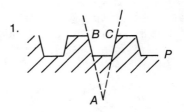

2.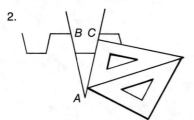

3.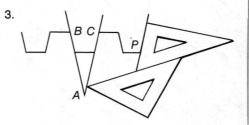

EXERCISES

A Trace the given line. Then use a pair of set squares and a ruler to draw the following. Use a reproduction of the set squares in the back of the book if you do not have a pair.

1. four parallel lines that form 90° angles with transversal *MN*

2. three parallel lines that form 45° angles with transversal *ST*

3. five equidistant parallel lines that form 60° angles with transversal *PQ*

4. six equidistant parallel lines that form 30° angles with transversal *CD*

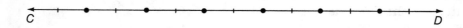

B Trace each given drawing. Then use a compass and straightedge to construct two lines, one through *J* and one through *F*, that are parallel to segment *AB*.

5.

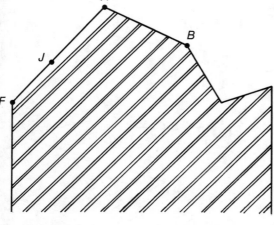

6.

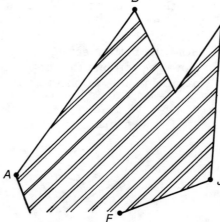

7.

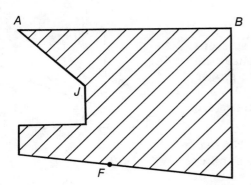

8.

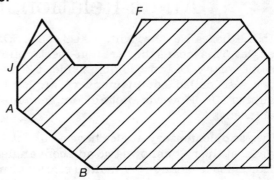

C 9. Draw a figure that resembles the one below.

    a. In the space above line segment *AB*, and within the partial rectangle, draw four lines 15 mm apart that are perpendicular to segment *AB*.

    b. In the space below line segment *AB*, draw two lines 10 mm apart that are parallel to segment *AB*.

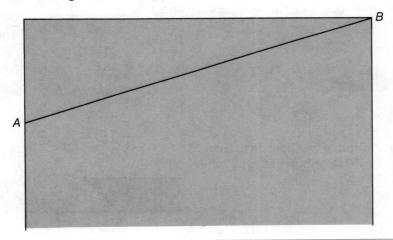

Tricks of the Trade

A carpenter has a variety of tools for making or measuring angles. The **bevel square** adjusts to any angle and is held tight by a wing nut. If the exact degree of an angle to be cut is known, the bevel square can be placed on a protractor and preset to the desired degree.

The **centering square** helps locate the center of a circle quickly and then doubles as a protractor to measure angles.

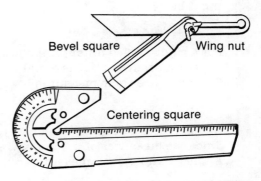

6-4 Angle Relationships

In the diagram at the right, ∠ *XOZ* and ∠ *ZOY* are **adjacent angles**. This means they are non-overlapping and have the same vertex, *O*, and a common side, $\overrightarrow{OZ}$. The two angles are also called **complementary angles** because their sum is 90°. However, not all complementary angles are adjacent.

Two **supplementary angles** have a sum of 180°. The diagram at the right shows two adjacent angles whose exterior sides $\overrightarrow{OM}$ and $\overrightarrow{ON}$ make a straight line. The sum of the two supplementary angles *MOP* and *PON* is 180°. Supplementary angles also need not be adjacent.

The two intersecting lines in the diagram at the right form four angles. The angles *opposite* each other (∠ 1 and ∠ 3; ∠ 2 and ∠ 4) are called **vertical angles** and are congruent.

It is often useful to know these angle relationships on the job.

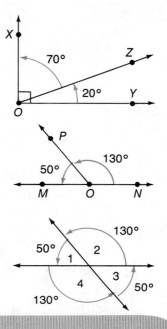

Example 1

Suppose we want to splice two boards to form a tight joint. If ∠ *ABC* is 45°, what must the measure of ∠ *DEX* be?

Solution

Points *X*, *E*, and *C* lie in a straight line and ∠ *ABC* and ∠ *DEX* are supplementary. The sum of their measures is 180°. Thus, ∠ *DEX* must be 135°.

Example 2

For a right-angled miter joint shown, assuming the measure of ∠ 1 is equal to the size of ∠ 2, what are their measures?

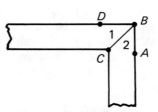

Solution

Angles 1 and 2 are complementary. The sum of their measures is 90°. Since the measures of ∠ 1 and ∠ 2 are equal, each must be half of 90°. Angles 1 and 2 each measure 45°.

EXERCISES

For questions 1 to 3, sketch each solution, labeling the angles and their measure.

A **1.** Angles *ABT* and *TBC* are complementary. Find the measure of ∠*TBC* for the given measure of ∠*ABT*.

    a. ∠*ABT* = 35°   **b.** ∠*ABT* = 84°   **c.** ∠*ABT* = 66.3°

2. Angles *CEG* and *GED* are supplementary angles. Find the measure of ∠*GED* for the given measure of ∠*CEG*.

    a. ∠*CEG* = 46°   **b.** ∠*CEG* = 108°   **c.** ∠*CEG* = 146.5°

3. Angle 4 is a vertical angle. Find the measure of the other three angles if ∠ 4 has the given size.

    a. ∠4 = 28°   **b.** ∠4 = 110°   **c.** ∠4 = 62.5°

B **4.** If ∠*ABC* measures 75°, how large is ∠*DEF*? (Assume that there is one board under the carpenter's square.)

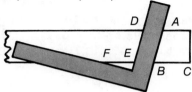

5. For the dovetail joint shown below, how large is ∠2 if ∠1 = 82°?

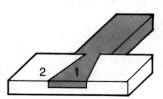

6. A carpenter's square can be used to size the lumber for a rafter. If the plate angle below is 40°, what is the size of the ridge angle?

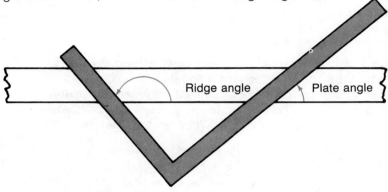

6-5 Circles

Circles and arcs of circles are commonly used in technical drawings.

Example 1

The top view drawing of a cooling tank at the right requires a line that touches the outer circle. Use a compass and straightedge to construct the required line, which is tangent to the circle. (A **tangent** line touches a circle at only one point and is perpendicular to the radius drawn to that point.)

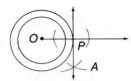

Solution

1. Locate P, the required point of tangency.
2. Draw ray OP. (O is the center of the circle.)
3. Construct a line PA perpendicular to ray OP at P. The line PA is tangent to the circle at P.

Example 2

How can the center of the circular piece of wood be located?

Solution (Using a Compass and Straightedge)

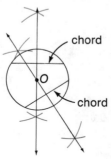

1. Draw any two non-congruent chords on the circle. (A **chord** is any line segment that has its endpoints on the circle.)
2. Construct the perpendicular bisector of each chord. Since the perpendicular bisector of a chord always passes through the center of the circle, the center is the intersection point O.

Solution (Using a Carpenter's Square)

1. Hold a carpenter's square against the piece of wood and carefully mark points 1 and 2 on the wood where it touches.
2. Rotate the piece so that only one mark touches the square. Carefully mark the point 3 as shown. Rotate again, and mark the point 4. Connect points 1 and 3 and points 2 and 4. The center is where the two lines intersect.

EXERCISES

A Use a compass and straightedge to construct a circle with center O and radius OX. Construct a line tangent to the circle at X.

1. At what point does the tangent line touch the circle?

2. What angle is formed by the tangent line and the radius OX?

B 3. Trace a circular object. Locate the center of the traced circle using a carpenter's square.

For questions 4 and 5, make a drawing that resembles the given figure, only larger. Locate the center of each circle using a compass and straightedge.

4.

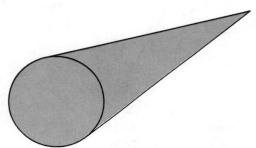

5.

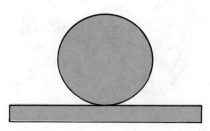

6. a. Construct the perpendicular bisector of segment AB so that it intersects the top of the circle at point C and the bottom of the circle at point D.

b. Construct line w tangent to the circle at point A, line x tangent to the circle at point C, line y tangent to the circle at point B, and line z tangent to the circle at point D.

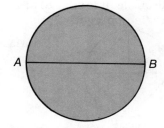

c. Extend line w to intersect line x at point P, x to intersect line y at point Q, line y to intersect line z at point R, and line z to intersect line w at point S.

d. What type of polygon is formed by points P, Q, R, and S?

C 7. The square corner of a piece of wood drawn at the right needs to be rounded. Follow the steps below to draw an arc tangent to side a and side b.

a. Draw arcs with a convenient radius and center at the corner of the wood cutting at D and E respectively.

b. With centers D and E and the same radius, draw intersecting arcs. Label the intersection point, O.

c. With center O and the same radius, draw the required arc that will round the corner.

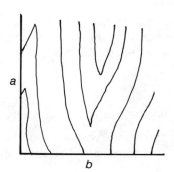

6-6 Polygons and Circles

The circumference of a watch is divided into 12 congruent *arcs* (between the hour numbers). These arcs correspond to 12 congruent **central angles** formed by the hands. The circle shown below to the right is divided into congruent parts by 12 central angles. The central angles also divide the circumference into 12 equal arcs. The arcs and the central angles have the same measure, 360° ÷ 12, or 30°.

 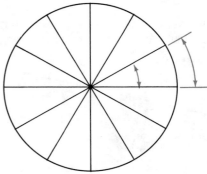

A **polygon** is a closed figure with three or more sides that are line segments. By making congruent central angles and/or equally spaced points on the circle, it is easy to draw a **regular polygon** (an equal-sided and equal-angled polygon).

Square:  Equilateral Triangle: Octagon:

Example

Where can a car designer position the five lugs in the metal rim shown at the right, needed to hold the wheel to the car?

Skill(s) ___4___

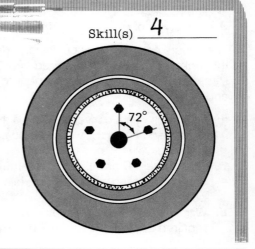

Solution

1. Locate the center of the wheel (lesson 6-5).

2. Use the measure of each of the five central angles to position the lugs: 360° ÷ 5, or 72°.

The five lugs can be placed at 72° intervals.

EXERCISES

A Use a protractor to find the measure of each central angle in the given figure.

1.

2.

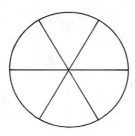

3.

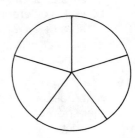

Determine the measure of each arc in the given figure.

4.

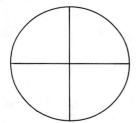

5.

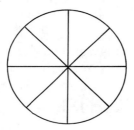

6.

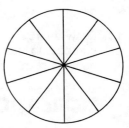

Trace each figure below.

a. Find the measure of each angle in the given polygon.
b. Find the sum of the angles in each polygon.

7.

8.

9.

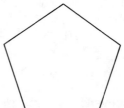

10.

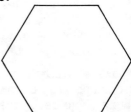

B 11. A graphic artist wants to draw a logo that includes a square inscribed in a circle. Show how this can be done by drawing a circle and any diameter. Then draw another diameter that is perpendicular to the first diameter at the center of the circle.

12. A jewelry designer wants to position eight pearls around a center rhinestone in a circular pin. What measure of arc is needed between each pearl?

13. A circular wheel on a tractor has six holes that are equally spaced on a circle. What measure of arc is between each hole?

14. Show how a designer could inscribe an equilateral triangle in a circle.

15. Draw a circle and any diameter *AB*.

    a. Pick a different point *C* on the circle and draw △*ABC*. What is the measure of ∠*C*?

    b. Pick a different point *C* on the circle and draw △*ABC*. What is the measure of ∠*C*?

    c. Repeat with two different locations for point *C* on the circle. What conclusion can you make?

16. The star polygon at the right is used by an artist in designing a logo for a new product. Explain how to draw the star polygon, using a protractor. (Points *A*, *B*, *C* ,*D*, and *E* are equally spaced.)

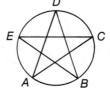

17. Construct a star polygon in a circle using a compass and a straightedge. (Hint: Open the compass to a radius equal to the radius of the circle.)

C **18.** Draw a circle with center *O* and radius *OA*. Starting at point *A*, use the compass to mark off successive arcs with chords equal to the radius. Connect the points in order with chords. What figure results?

Technology Update

A Facsimile System

In simple words, a facsimile (fax) is a communications system that sends copies of any document over ordinary telephone lines from one fax unit to another in a matter of seconds. Fax copies are high-resolution duplicates of an original document, and can include images, text, graphics, and any printed or handwritten material. With such a system, a document can be transmitted from Tokyo to Chicago in less than 60 s.

A fax system uses laser light. A very fine beam scans a document row by row, and the beam is reflected into a photo device that converts successive light values into electrical impulses. The impulses are encoded and sent across a phone line. At the receiving end of the line, the impulses are decoded and applied to a facsimile that reproduces the document.

Research Project: What advantage does a fax machine have over a courier service? disadvantage?

Original document

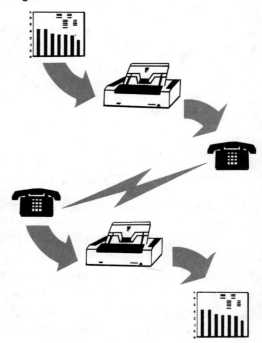

Fax copy

Self-Analysis Test

For questions 1 to 6, make a drawing that resembles the one given.

1. Bisect ∠ABC.

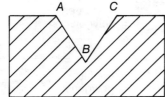

2. Refer to the screw-thread enlargement below. Construct ∠GHI congruent to ∠DEF. (You are to locate the position of point I.)

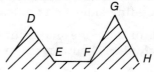

3. Construct a line perpendicular to ray BC at point C. Use a compass and a straightedge.

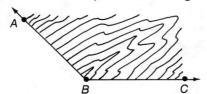

4. Construct the perpendicular bisector of segment AB.

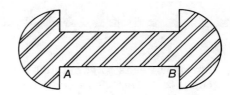

5. Refer to the drawing of the tapered rod below. Construct parallels to the center line at C and D.

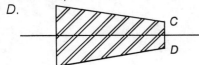

6. A draftsperson must construct three equidistant parallels that form 60° angles with a segment AB below. Construct the parallels at A, B, and the midpoint of segment AB.

7. Refer to the drawing of the wooden frame below. Find the measures of angles 1 and 2.

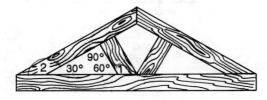

8. Trace arc AB below and construct a circle that contains the arc.

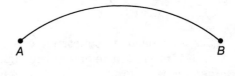

9. Trace around the edge of a circular can to make a circle.

   a. Locate the center of the circle using a compass and a straightedge.
   b. Inscribe a square in the circle by constructing a diameter and its perpendicular bisector.

Similar Polygons

6-7 Congruent Polygons and Similar Polygons

Polygons of the same shape and size are **congruent**. Their corresponding sides and angles are congruent.

Polygons of the same shape but not necessarily the same size are **similar**. Their corresponding angles are congruent and their corresponding sides are proportional.

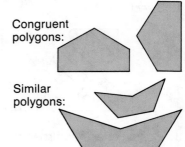

Congruent polygons:

Similar polygons:

Example 1

A drafter needs to construct a triangle congruent to $\triangle ABC$. Use a compass and straightedge to do this.

Solution

1. Draw a base line and mark point D on it.
2. With radius AC and the compass needle at D, locate point F so that segment DF is congruent to segment AC.
3. With radius AB and the compass needle at D, mark the length AB with an arc.
4. With radius BC and the compass needle at F, draw an intersecting arc at point E.
   $\triangle DEF \cong \triangle ABC$

Example 2

A designer needs to construct a parallelogram that is similar to the one at the right and has sides twice as long.

Solution

1. Draw a base line and mark point J on it. Locate point I so that segment JI is twice the length of segment FE.
2. Copy $\angle F$ at J and $\angle E$ at I (lesson 6-1).
3. Locate points G and H so that segments JG and IH are twice the length of segment DE.
4. Draw segment GH. Parallelogram $GHIJ$ is similar to parallelogram $CDEF$.

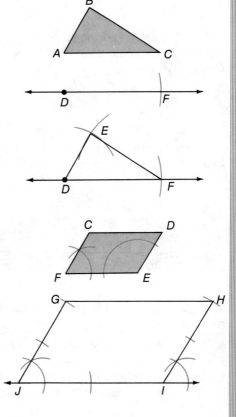

EXERCISES

A For questions 1 to 6, draw a polygon that resembles the one given. Construct a polygon *congruent* to the polygon you drew using a compass and straightedge.

1.

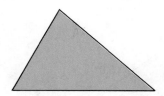

2.

3.

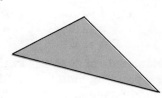

Construct a *similar* polygon to the polygon you drew that has sides twice as long using a compass and straightedge.

4.

5.

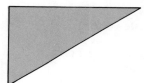

6.

B **7.** A graphic artist used this procedure for constructing a polygon similar to polygon *ABCDE* with dimensions twice as large. Trace polygon *ABCDE* and duplicate the procedure.

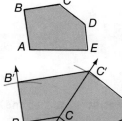
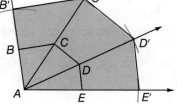

   i. Draw diagonals *AC* and *AD* and extend the rays from *A*.
   ii. Mark off segment *BB′* congruent to segment *AB*, segment *CC′* congruent to segment *AC*, segment *DD′* congruent to segment *AD*, and segment *EE′* congruent to segment *AE*.
   iii. *AB′C′D′E′* is the desired polygon.

Draw a polygon that resembles the one given. Construct a similar polygon having:

   a. sides twice as long;
   b. sides half as long.

8.

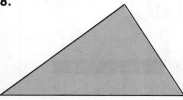

9.

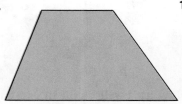

10.

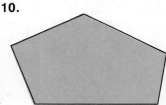

179

6-8 Dimensions of Similar Polygons

The vinyl panels at the right are **similar** triangles. Their corresponding angles are congruent.

$$\angle D \cong \angle A$$
$$\angle E \cong \angle B$$
$$\angle F \cong \angle C$$

Each side of $\triangle ABC$ is twice as long as its corresponding side in $\triangle DEF$. Thus, the corresponding sides are proportional, or have equal ratios.

$$\frac{DE}{AB} = \frac{FD}{CA} = \frac{EF}{BC}$$

$$\frac{3}{6} = \frac{4}{8} = \frac{5}{10} \quad \leftarrow \text{All ratios equal } \frac{1}{2}.$$

If $\triangle MNO$ is similar to $\triangle ABC$, the length of side NO can be found by the proportion method.

$$\frac{AB}{MN} = \frac{BC}{NO} \text{ or } \frac{6}{9} = \frac{10}{x} \text{ or } x = \frac{9 \times 10}{6}$$

Side NO is 15 in. long.

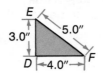

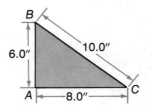

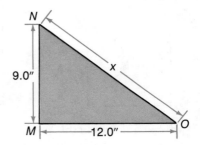

Example

Skill(s) ___19___

A graphic artist wants to make a "blow up" of an 8.0 in. by 10.0 in. photo. If the width of the "blow up" is to be 44.0 in., what will be the height?

Solution

Since the two figures are similar, the corresponding sides have equal ratios. Set up a proportion and solve for the missing term using a calculator.

		width:	height:
original	$\rightarrow$	$\dfrac{8}{44}$	$= \dfrac{10}{h}$
enlargement	$\rightarrow$		

$$h = \frac{44 \times 10}{8}$$

$$h = 55$$

$$\boxed{4}\;\boxed{4}\;\boxed{\times}\;\boxed{1}\;\boxed{0}\;\boxed{\div}\;\boxed{8}\;\boxed{=} \quad 55$$

The height of the "blow up" will be 55 in., to two significant digits.

EXERCISES

A Find dimension *x* for each pair of similar polygons.

1.
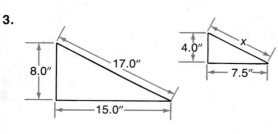

2.
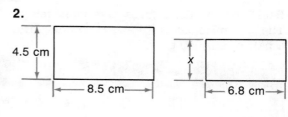

3.

4.
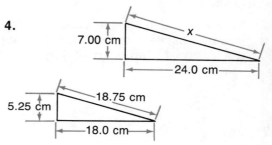

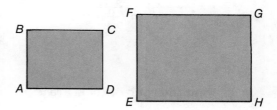

B For each case below, find the missing dimensions.

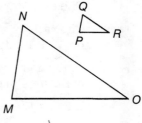

	AB	DA	EF	HE
5.	3.0 in.	5.0 in.	3.72 in.	?
6.	4.20 in.	?	5.46 in.	8.06 in.
7.	8.0 in.	10.0 in.	?	15.0 in.

For each situation, find the unknown lengths.

	PQ	RP	QR	MN	OM	NO
8.	4.0 in.	6.0 in.	7.0 in.	8.0 in.	?	?
9.	18.0 in.	15.0 in.	12.0 in.	4.50 ft	?	?
10.	6.0 in.	7.5 in.	5.5 in.	9.0 in.	?	?
11.	?	?	2.50 cm	6.00 cm	9.00 cm	10.0 cm
12.	?	?	12 cm	8.0 cm	7.0 cm	4.0 cm
13.	?	?	15 m	7.0 cm	5.0 cm	3.0 cm

14. An enlargement of the sketch of a house revealed the dimensions shown at the right. What are the following dimensions of the house in the original sketch?

a. dimension *x*
b. dimension *y*
c. dimension *z*

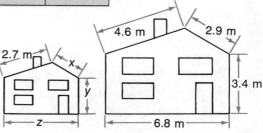

6-9 Areas of Similar Polygons

Recall that similar figures have congruent angles and proportional sides. The examples below illustrate another interesting relationship between a pair of similar polygons.

Example

Skill(s) **3, 10, 18, 30**

The two wooden panels below are similar figures. What relationship do you find between the ratio of their areas to the ratio of the squares of the lengths for any pair of corresponding sides?

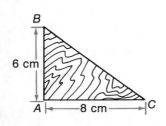

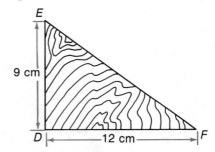

Solution

1. Find the ratio of their areas.

   For $\triangle ABC$: Area $= 0.5 \times 8 \times 6$
   $= 24$

   For $\triangle DEF$: Area $= 0.5 \times 12 \times 9$
   $= 54$

   The area is 24 cm².

   The area is 54 cm².

   The ratio of the areas of the two triangular wood panels is 24 : 54, or 4 : 9 in simplest form.

2. Find the ratio of the squares of the lengths for any pair of corresponding sides.

   $AB^2 : DE^2 \rightarrow$ $6^2 : 9^2$
   $36 : 81$
   $4 : 9$ (simplest form)

   $CA^2 : FD^2 \rightarrow$ $8^2 : 12^2$
   $64 : 144$
   $4 : 9$ (simplest form)

   The ratio of the squares of the lengths for a pair of corresponding sides is 4 : 9.

   > In general, the ratio of the areas of the similar triangular figures is *equal* to the ratio of the squares of the lengths for any pair of corresponding sides. This relationship is true for *any pair* of similar polygons.

EXERCISES

A Express each ratio in simplest form.

   1. 36 : 54   **2.** 64 : 80   **3.** 45 : 51   **4.** 60 : 135

B For each pair of similar figures below, answer the following questions.

   a. What is the ratio of their areas in simplest form?
   b. What is the ratio of the squares of the lengths for a pair of corresponding sides in simplest form?
   c. Are the two ratios the same?

5.
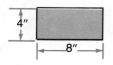
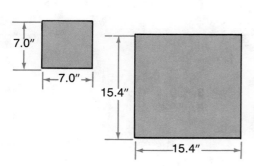

6.
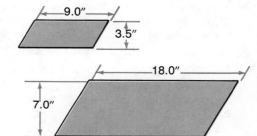
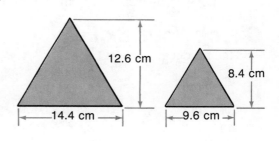

7.

8.

9. What is the ratio, in simplest form, of the areas of two square table tops when:

   a. the ratio of the squares of the lengths for a pair of corresponding sides is 1 : 2?
   b. the ratio of the squares of the lengths for a pair of corresponding sides is 2 : 3?
   c. the sides measure 4.0 in. and 16.0 in.?
   d. the sides measure 1.4 m and 0.8 m?

10. In simplest form, what is the ratio of the squares of the lengths of the sides of a pair of similar window frames when:

   a. the ratio of the areas is 60 : 75?
   b. the ratio of the areas is 20 : 26?
   c. the sides measure 3.6 ft and 4.8 ft?
   d. the sides measure 4.5 ft and 6.0 ft?

11. In simplest form, what is the ratio of the areas of two triangular wall hangings when:

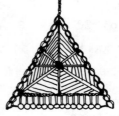

    a. the ratio of the squares of the lengths for a pair of corresponding sides is 10 : 14?

    b. the ratio of the squares of the lengths for a pair of corresponding sides is 1.2 : 1.0?

    c. two corresponding sides measure 28 in. and 84 in.?

    d. two corresponding sides measure 18 cm and 54 cm?

12. A rectangular wooden panel has an area of 32 m². One of its sides is 4.0 m long. Find the area of a similar wooden panel having a corresponding side of the given length.

    a. 2 m   **b.** 3.6 m   **c.** 0.5 m   **d.** 6.0 m

13. A triangular garden plot has an area of 144 ft². One of its sides is 9.00 ft long. Find the area of another garden plot similar to the first having a side corresponding to each length given.

    a. 3.0 ft   **b.** 4.5 ft   **c.** 21.0 ft   **d.** 14.5 ft

14. On a new housing development map, the length of a piece of property is 6.5 cm. In actuality, the length of the property is 60.0 m. The area of the property on the map is 12.5 cm². What is the area of the piece of property?

Check Your Skills

Find the product.

  1. 25 × 32   **2.** (44)(117)   **3.** 50 × 375   (Skill 3)

  4. 281(507)   **5.** (611)(352)   **6.** 5920(236)

Divide. Express any remainders that occur as fractions.

  7. 135 ÷ 7   **8.** $\frac{360}{8}$ **9.** 2185 ÷ 35   (Skill 4)

10. $\frac{1800}{600}$ **11.** 3162 ÷ 18   **12.** 2170 ÷ 1050

Express in the simplest form.

13. $\frac{36}{92}$ **14.** $\frac{123}{336}$ **15.** $\frac{1452}{2904}$ (Skill 10)

16. 14 : 7   **17.** $4^2 : 12^2$ **18.** 35 : 100 : 80   (Skill 18)

Find the value of x that would make the equation true.

19. $\frac{3}{5} = \frac{9}{x}$ **20.** $\frac{x}{7} = \frac{21}{49}$ **21.** $\frac{3}{x} = \frac{36}{60}$ (Skill 19)

22. $\frac{50}{105} = \frac{x}{21}$ **23.** $\frac{105}{315} = \frac{2}{x}$ **24.** $\frac{x}{1024} = \frac{4}{16}$

Simplify.

25. 9^2 **26.** 7^3 **27.** 18^2 (Skill 30)

28. $(0.4)^2$ **29.** $(1.5)^2$ **30.** $(6.21)^2$

31. $\left(\frac{4}{5}\right)^3$ **32.** $\left(\frac{11}{16}\right)^2$ **33.** $(0.9)^3$

Self-Analysis Test

Draw a polygon that resembles the one given. Then construct a polygon similar to it with sides twice as long.

1.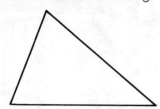

2.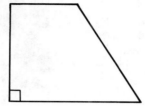

3.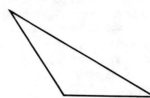

Use a calculator where possible.

4. A surveyor uses the similar triangles below to calculate inaccessible distances across a river.

   a. Find the distance from point *B* to point *C*.
   b. Find the distance from point *A* to point *B*.

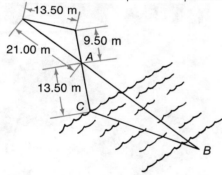

5. A flagpole has a shadow 16.0 ft long at the same time a yardstick has a shadow 1.0 ft long. Find the height of the flagpole.

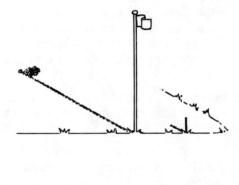

6. A photographic slide is 34.0 mm wide and 22.0 mm high. The projection of the slide on a screen yields an image that is 520.0 mm high. Find the width of the projected image.

7. Rectangles *P* and *Q* are similar. Find the ratio of their areas.

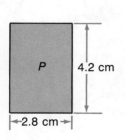

8. Triangles *ABC* and *DEF* are similar. Find the lengths of sides *DF* and *EF*.

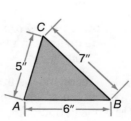

The volume of commercial photography used in industry has grown rapidly in recent years. At the same time, simplified cameras and film cartridges have made amateur photography a very popular hobby. Processing the large amounts of professional and amateur photos requires the services of trained photo lab technicians. Opportunities in this field are expected to increase faster than the average of most occupations.

Job Description

What do photo lab technicians do?

1. In small labs, a photo lab technician develops film by placing it in a "bath" of chemical developers and stabilizers. The technician then dries the film and transfers the images to photo print paper through a series of lenses and light densities. Next, the photo lab technician places the photo print paper into another chemical "bath", dries the paper, and cuts it to the specified photo size.

2. In large photographic labs, a photo lab technician operates machines that develop film automatically.

3. A photo airbrush artist restores damaged and faded photographs and color drawings to simulate photographs.

4. A photograph retoucher alters negatives to accentuate the desired features or removes undesirable features of a photograph.

5. A colorist applies oil colors to portrait photographs to create a natural appearance.

Qualifications

Graduation from high school is usually required. Courses in mathematics and chemistry are helpful. Trade courses in photography or hobby experience in film developing are also useful.

Photo lab technician often get on-the-job training. In many cases they begin as assistants to experienced workers. Some community colleges offer programs leading to two-year diplomas in photographic technology.

From five operating radio stations in the U.S. in 1921, there are now more than 10,000 stations on the air. Of these, about 40% are commercial FM stations, almost half are commercial AM stations and about 13% are educational radio stations.

Radio broadcasting provides a wide range of services, including news, music, and foreign languages, to the listening public. Virtually 100% of the population has access to a radio both at home and in a vehicle.

In 1941, there were two operating television stations. Now, there are more than 1300 stations on the air. Of these, about three fourths are commercial television stations, while the remaining fourth are educational television stations.

The Public Broadcasting System (PBS) provides educational and cultural programming to more specialized audiences. PBS is funded by the federal government, by grants from business and industry, and by public donations.

Other educational stations are funded by universities and colleges. Such stations provide courses for home study, transmit courses to other campuses, and provide lectures of interest to limited audiences.

Use the graph below to answer the questions.

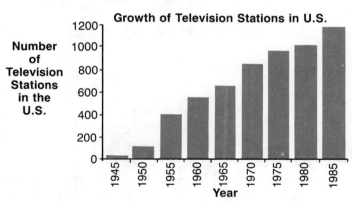

Growth of Television Stations in U.S.

1. Estimate the number of television stations in the U.S. in 1950, 1960, 1970, and 1980.

2. During which five-year period was the increase in the number of stations the largest?

3. During which five-year period was the percentage increase in the number of stations the largest?

4. Which five-year period showed the smallest percentage increase in the number of stations?

Chapter 6 Test

For questions 1 to 4, make a drawing that resembles the given one.

1. Copy ∠*XYZ* with a compass and straight-edge. (6-1)

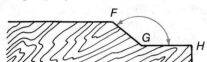

2. Bisect ∠*FGH* using a compass and straight-edge. (6-1)

3. Construct a line segment perpendicular to line segment *AB* at point *X* with a compass and straightedge. (6-2)

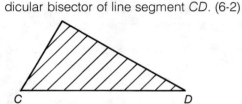

4. Construct a line segment that is the perpendicular bisector of line segment *CD*. (6-2)

5. Trace line *AB*. Then draw four lines 30 mm apart and intersecting segment *AB* to form 45° angles. (6-3)

6. Draw a parallelogram having sides that are 10 cm and 5 cm long and angles that measure 120° and 60°. (6-3)

7. Refer to the diagram at the right. (6-4)

   a. If ∠*PVR* measures 135°, what is the size of ∠*PVU*?

   b. If ∠*QVR* measures 90°, what is the size of ∠*PVQ*?

   c. What is the size of ∠*TVS*?

   d. What is the size of ∠*UVT*?

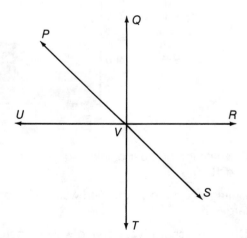

8. With a compass and straightedge, construct a circle with center *O*. Draw radius *OB*. Construct a tangent to the circle at point *B*. (6-5)

9. Trace a circular object and locate the center of the circle drawn. (6-5)

10. If there are six equally spaced spokes on a wheel, how large is the central angle between two adjacent spokes? (6-6)

11. A graphic artist needs to draw a circle and to inscribe a regular decagon (ten-sided polygon) in it. Use a protractor and a straightedge to do this. (6-6)

For questions 12 and 13, make a drawing that resembles the one given.

12. Make a copy of the triangle you drew using a compass and straightedge. (6-7)

13. Draw a trapezoid *similar* to, but smaller than, the one you drew with a compass and straightedge. (6-7)

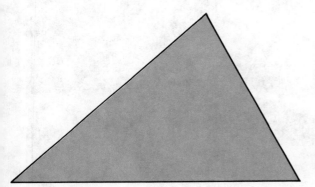
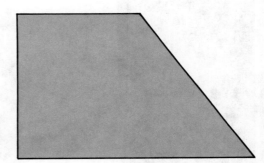

14. The two triangular panels below are similar. Find the length of segment *JK* and segment *JL*. (6-8)

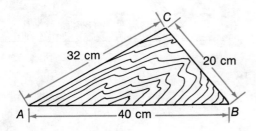
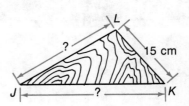

15. The two triangular pieces of carpet are similar. What is the area of the larger triangle? (6-9)

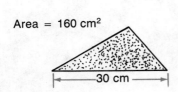
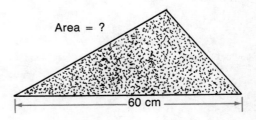

Chapter 7
Solid Geometry

After completing this chapter, you should be able to:

1. Find the surface area of objects having the shape of prisms, pyramids, cylinders, cones, and spheres.

2. Calculate the volume of three-dimensional objects.

3. Work with the common units of weight and capacity measurement.

4. Apply volumes of cylinders to problems dealing with engine displacement and compression ratios.

5. Apply volumes of cylinders to problems about hydraulic and pneumatic systems.

Surface Area

7-1 Prisms and Pyramids

When three-dimensional objects are manufactured, it is important to know how much material is needed. The diagrams below show some common three-dimensional objects that can be classified as *prisms*. A **prism** has two congruent and parallel polygonal bases and three or more rectangular faces.

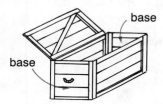

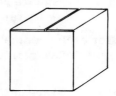

The wooden crate has the shape of a of a **rectangular prism**. It has 6 faces in all, 4 faces and 2 bases.

The cardboard box has the shape of a **cube**. It has 6 faces of the same size.

The tent has the shape of a **triangular prism**. It has 5 faces in all, 3 faces and 2 bases.

To find the amount of material needed to manufacture the objects above, calculate the surface area. The surface area of a right prism is equal to its lateral area (areas of the sides) plus the areas of its bases.

Surface Area of a Prism = Lateral Area + Areas of Bases

Example 1

Skill(s) __27, 28__

How much plastic is needed to make the tissue box shown at the right? Allow 16.0 in.² for the opening.

2.25″
4.50″
10.0″

Solution

Lateral Area = 2(4.50 × 2.25) + 2(10.0 × 2.25)

$\boxed{2} \times \boxed{4} \cdot \boxed{5} \times \boxed{2} \cdot \boxed{2} \boxed{5} = + \boxed{2} \times \boxed{1} \boxed{0} \times \boxed{2} \cdot \boxed{2} \boxed{5} =$

65.25 *   The lateral area is 65.3 in.², to three significant digits.

Area of Bases = 2(4.50 × 10.0)

$\boxed{2} \times \boxed{4} \cdot \boxed{5} \times \boxed{1} \boxed{0} = 90$

The area of the bases is 90.0 in.², to three significant digits.

SA = Lateral Area + Area of the Bases − Allowance for opening

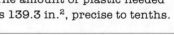 139.3

The amount of plastic needed is 139.3 in.², precise to tenths.

* For calculators with built-in order of operations.

Each paper ornament below is a pyramid. A **pyramid** has one base that is a polygon and three or more lateral faces that are triangles.

   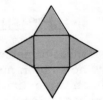

The above ornament is in the shape of a **triangular pyramid**. It has 4 faces in all, 3 lateral faces and 1 base.

The above ornament is in the shape of a **square pyramid**. It has 5 faces in all, 4 lateral faces and 1 base.

To find the amount of paper needed for each ornament, calculate the surface area. The surface area of a pyramid is equal to its lateral area plus the area of its base.

Surface Area of a Pyramid = Lateral Area + Area of Base

Example 2

How much canvas is needed for the floor and sides of the pyramid-shaped tent at the right?

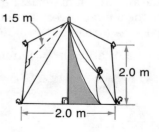

Solution

There are 5 faces in all, 4 lateral faces and 1 base.

SA = Lateral Area + Area of Base
= [4(0.5 × 2.0 × 1.5)] + [2.0 × 2.0]

| 4 | × | . | 5 | × | 2 | × | 1 | . | 5 | = | + | 2 | × | 2 | = | 10*

* For calculators with built-in order of operations.

The amount of canvas needed is 10.0 m², precise to the tenths place.

EXERCISES

In this text, assume all prisms to be right and all pyramids to be regular.

A Find the lateral area of each object.

1.

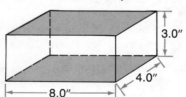

2.

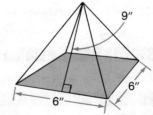

B **3.** A hollow plastic cube for displaying photographs has edges that are 7 in. long. How much plastic is used in the construction of the six sides of the cube?

4. A carton in the shape of a cube is constructed using cardboard. If the edges of the carton are 45.00 in. long, how much cardboard is needed for all six sides?

5. How many square meters of plywood are needed to make a six-sided packing crate that has dimensions of 3.2 m, 2.4 m, and 1.5 m?

6. A metal storage shed is a rectangular prism with dimensions of 12.0 ft, 10.0 ft, and a height of 8.0 ft. How much metal sheeting is needed for the walls and roof of the shed if 42.0 ft^2 are allowed for a door and windows?

7. A skylight is shaped like a square pyramid. The base of the pyramid has sides that are 6.5 ft long. The slant heights of the skylight's triangular sides are 3.6 ft. Calculate the area of the glass needed for the skylight.

8. How many square feet of roofing material are needed for a garage roof that is a square pyramid with 24.0 ft sides and that has a slant height of 12.5 ft?

9. The walls of the barn shown below are to be covered with plywood sheeting, including the triangular wall. How many square feet of sheeting are required? Allow 64 ft^2 for the doors.

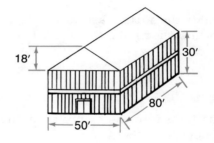

10. How much canvas is needed to construct the tent shown below, given that the base is square? Assume the bottom of the tent is canvas.

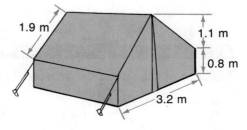

C **11.** The wooden road barrier shown at the right is used to separate traffic during construction work. The barrier is painted with reflective paint. If one gallon of paint covers about 275 ft^2 and the paint comes in 1.0 gal cans, estimate the number of cans of paint needed for 50 road barriers. (Assume that the bottoms are not painted.)

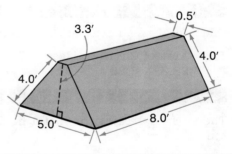

7-2 Cylinders, Cones, and Spheres

By definition, a **cylinder** is a three-dimensional figure with two congruent and parallel circular bases joined by a curved rectangular surface.

The circumference is $2\pi r$.

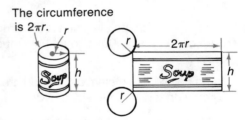

The soup can at the right has the shape of a cylinder. Its lateral surface is a rectangle with a length equal to the circumference of its circular base and a width equal to the height of the cylinder. When the label of the soup can is removed, the rectangle is seen.

Surface Area of a Cylinder = Area of Circular Bases + Area of Rectangle
$$= 2\pi r^2 + (2\pi r)h, \text{ where } r \text{ is the radius and } h \text{ is the height}$$

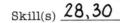

Example 1

Skill(s) _28,30_

What is the surface area of metal used to make the can shown at the right if the radius is 1.00 in. and the height is 4.50 in.?

Solution

Substitute into the formula for the surface area of a cylinder using 3.14 as π.

Area of Circular Bases = $2\pi r^2$
$$= 2 \times 3.14 \times (1.00)^2$$

| 2 | × | 3 | · | 1 | 4 | × | 1 | × | 1 | = | 6.28

The area of the two circular bases is 6.28 in.², to three significant digits.

Lateral Area = $(2\pi r)h$
$$= (2 \times 3.14 \times 1.00)4.50$$

| 2 | × | 3 | · | 1 | 4 | × | 1 | × | 4 | · | 5 | = | 28.26

The lateral area is 28.3 in.², to three significant digits.

SA = Area of Circular Bases + Lateral Area
$$= 28.3 + 6.28$$

| 2 | 8 | · | 3 | + | 6 | · | 2 | 8 | = | 34.58

The surface area of the metal in the can is 34.6 in.², precise to the tenths place.

A **cone** has one circular base and a vertex directly above (or below) the center of the base. The amount of plastic needed to make the hollow cone at the right is the sum of the area of its base and its lateral surface area.

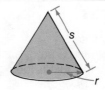

$$\text{Surface Area of a Cone} = \text{Area of Base} + \text{Lateral Area}$$
$$= \pi r^2 + \pi rs, \text{ where } r \text{ is the radius and}$$
$$s \text{ is the slant height}$$

A **sphere** has the shape of a ball. All points on the surface of a sphere are an equal distance from its center. The amount of leather needed to cover the surface of the baseball at the right is four times the area of its major circle.

Area of major circle $= \pi r^2$
Surface Area of a Sphere $= 4\pi r^2$, where r is the radius

Example 2

What is the surface area of a cone-shaped exhaust collector with a 5.00 ft radius and a 8.00 ft slant height? The exhaust collector has no base.

Solution

Find the lateral area of the cone.

| 3 | · | 1 | 4 | × | 5 | × | 8 | = | 125.6

The surface area of the exhaust collector is 126 ft², to three significant digits.

Example 3

Find the surface area of a spherical fuel tank if it has a radius of 10.2 m.

Solution

Surface Area $= 4\pi r^2$

| 4 | × | 3 | · | 1 | 4 | × | 1 | 0 | · | 2 | × | 1 | 0 | · | 2 | = | 1306.7424

The surface area of the fuel tank is 1310 m², to three significant digits.

EXERCISES

Use a calculator.

A Find the surface area of each cylinder with the given dimensions. (Assume all cylinders in this text to be right.)

1. $r = 5.0$ in., $h = 10.0$ in.

2. $r = 8.5$ in., $h = 12.4$ in.

3. $r = 1.5$ cm, $h = 3.2$ cm

4. $d = 3.4$ m, $h = 1.5$ m

Find the surface area of each sphere with the given dimension.

5. $r = 4.0$ in.

6. $r = 6.25$ ft

7. $r = 9.4$ cm

8. $d = 8.150$ m

Find the surface area of each cone with the given dimensions. (Assume all cones in this text to be right.)

9. $r = 2.0$ in., $s = 5.0$ in.

10. $r = 10.50$ in., $s = 12.42$ in.

11. $r = 80.5$ cm, $s = 150.3$ cm

12. $d = 6.184$ m, $s = 8.240$ m

B 13. How many square feet of metal are needed to make the water tank shown below?

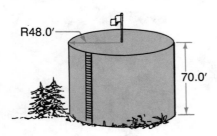

14. A trough is to be built of metal in the shape of the half cylinder shown below. Calculate the amount of metal needed.

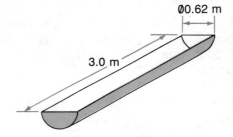

15. How many square millimeters of sheet metal are needed to make the cone-shaped funnel like the one shown below? Allow 400.0 mm² for the handle. Subtract 3100 mm² for the tip of the cone, which is not used.

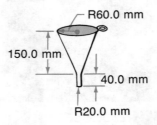

16. The plumb bob shown below has the form of two attached cones. Find the surface area of the plumb bob.

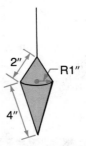

C 17. A spherical gas tank has a diameter of 20.0 ft. How many gallons of paint would it take to paint the tank if one gallon covers 200.0 ft²?

196

Self-Analysis Test

Use a calculator.

1. A tool box is a rectangular prism with dimensions 16 in. by 8.5 in. by 9.0 in. Find the surface area.

2. A stone landscaping wall is 42.0 ft long, 3.5 ft high, and 9.0 in. thick. Calculate the surface area excluding the base.

3. A cube-shaped footstool has edges that are 20 in. long. How much vinyl material is needed to cover it, excluding the base?

4. The top of a greenhouse has the shape of a square pyramid. Each edge along the square is 2.5 m and the slant height is 1.8 m. How much clear plastic is needed to form the top of the greenhouse? The base of the square pyramid is not included.

How many squares of shingles are needed for all sides of each roof shown below? (One square of shingles covers 100 ft² of roof.)

5.   **6.**

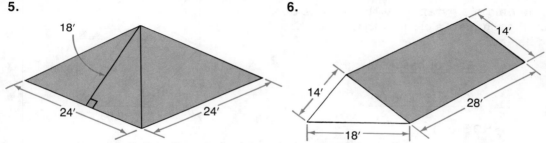

How much metal is required to make each object below? The top of each is not included.

7.   **8.**

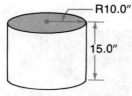
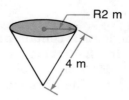

9. Find the surface area of a spherical bead with a radius of 0.35 in.

10. What is the surface area of a spherical ornament with a radius of 42 mm?

11. An open-topped, cone-shaped hopper used to load sand has a radius of 5.6 ft and a slant height of 9.8 ft. Find the surface area of the hopper.

12. The propane storage tank shown at the right is a cylinder with hemispheres at each end. Find the surface area of the tank.

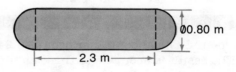

197

Volume, Weight, and Capacity

7-3 Volumes of Prisms and Pyramids

To store or ship a three-dimensional object, it is often important to know the amount of space it occupies, or its **volume**. Volume is measured in *cubic units*, such as cubic inches (in.³), cubic feet (ft³), cubic meters (m³), and cubic centimeters (cm³).

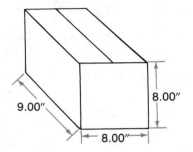

To find the volume, V, of the box in the shape of a prism at the right, we use the following formula.

$$V_{prism} = \text{Area of base} \times \text{height}$$
$$= (9.00 \times 8.00)8.00$$
$$= 576 \ (\text{in.}^3)$$

The volume of a pyramid, at the right, is *one third* the volume of a prism with the same base and height.

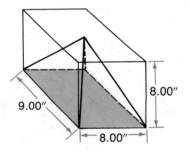

$$V_{pyramid} = \frac{\text{Area of base} \times \text{height}}{3}$$
$$= \frac{(9.00 \times 8.00)8.00}{3}$$
$$= 192(\text{in.}^3)$$

Example

Skill(s) __7, 28__

One of the industrial containers, shown at the right, measures 2.80 ft long, 2.50 ft wide, and 3.50 ft high. How much space does one container occupy in the warehouse?

Solution

Use the formula for the volume of a prism and a calculator to solve the problem.

$$V_{prism} = \text{Area of base} \times \text{height}$$
$$= 2.80 \times 2.50 \times 3.50$$

| 2 | · | 8 | × | 2 | · | 5 | × | 3 | · | 5 | = | 24.5 |

One industrial container occupies 24.5 ft³ of space in the warehouse. The result is accurate to three significant digits.

EXERCISES

Use a calculator.

A Find the amount of space occupied by each rectangular prism with the given length, width, and height.

1. l = 8.0 in., w = 6.0 in., h = 5.0 in.　　**2.** l = 1.5 ft, w = 0.9 ft, h = 2.0 ft

3. l = 0.95 m, w = 0.45 m, h = 0.30 m　　**4.** l = 0.55 m, w = 0.40 m, h = 0.65 m

What is the amount of space occupied by each rectangular pyramid with the given dimensions?

5. l = 14.40 in., w = 14.40 in., h = 22.50 in. **6.** l = 0.25 m, w = 0.35 m, h = 0.45 m

B **7.** How much space is occupied by the refrigerator carton shown at the right if it measures 32 in. by 29 in. by 62 in.?

8. The freezer of the refrigerator measures 24 in. by 18 in. by 9.0 in. What is the volume of the freezer?

9. What is the total volume of the two vegetable bins inside the refrigerator if each bin measures 12 in. by 13 in. by 4.5 in.?

10. A decorative candle is made in a pyramid shape with a square base. The height of the candle is 12.5 cm, and the sides of the base are 5.6 cm long. How much wax would be needed to make 25 candles in this shape?

11. How many cubic yards of dirt must be excavated to build a basement 35 ft by 20 ft by 8 ft? (1 yd^3 = 27 ft^3)

12. In building a swimming pool, a contractor excavates an area measuring 59 ft by 118 ft by 13 ft. If the excavated soil is used as landfill, how many cubic yards does the excavation yield? (1 yd^3 = 27 ft^3)

13. One of the Egyptian pyramids has a square base that is approximately 240 yd on a side. If the height is about 160 yd, estimate the volume of the pyramid.

C **14.** A trophy in the shape of a square-based pyramid is carved from a block of oak that is 4.0 in. long, 4.0 in. wide, and 6.5 in. high. If the finished trophy has the same dimensions as the block of oak, how much wood is cut away?

7-4 Volumes of Cylinders, Cones, and Spheres

To find the amount of space occupied by an object in the shape of a cylinder, cone, or sphere, use one of the following volume formulas.

The volume, V, of the cylindrical can below is the area of its base times its height.

$$V_{cylinder} = \text{Area of base} \times \text{height}$$
$$= (\pi r^2)h$$
$$= (3.14 \times (1.5)^2)4.0$$
$$= 28 \ (\text{in.}^3, \text{ to two significant digits})$$

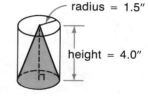

radius = 1.5"

height = 4.0"

The volume of a cone is *one third* the volume of a cylinder with the same base and height.

$$V_{cone} = \frac{\text{Area of base} \times \text{height}}{3}$$
$$= \frac{(\pi r^2)h}{3}$$
$$= \frac{(3.14 \times (1.5)^2)4.0}{3}$$
$$= 9.4 \ (\text{in.}^3, \text{ to two significant digits})$$

radius = 1.5"

height = 4.0"

Example

Skill(s) __7, 28__

How much space is occupied by the cement roller shown at the right, if its radius is 2.80 ft and its height is 4.60 ft?

Solution

Use the formula for the volume of a cylinder and a calculator to solve the problem.

$$V_{cylinder} = (\pi r^2)h$$
$$= (3.14 \times (2.80)^2) \times 4.60$$

| 3 | · | 1 | 4 | × | 2 | · | 8 | × | 2 | · | 8 | × | 4 | · | 6 | = | 113.24096 |

The space occupied by the cement roller is 113 ft³, to three significant digits.

The volume of a sphere is found with the following formula.

$$V_{sphere} = \frac{4\pi r^3}{3}$$

$$= \frac{4 \times 3.14 \times (1.5)^3}{3}$$

$$= 14 \text{ (in.}^3\text{, to two significant digits)}$$

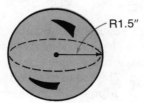

R1.5"

EXERCISES

Use a calculator where possible.

A Find the volume of each cylinder with the given dimensions.

1. $r = 10.0$ in., $h = 8.0$ in.

2. $r = 4.0$ in., $h = 18.0$ in.

3. $r = 2.50$ cm, $h = 5.00$ cm

4. $d = 0.180$ m, $h = 0.120$ m

Find the volume of each cone with the given dimensions.

5. $r = 10.0$ in., $h = 6.0$ in.

6. $r = 1.80$ ft, $h = 1.20$ ft

7. $r = 1.5$ cm, $h = 5.0$ cm

8. $d = 50.0$ cm, $h = 20.0$ cm

Find the volume of each sphere having the radius given.

9. $r = 12.0$ in.   **10.** $r = 6.0$ in.   **11.** $r = 0.3$ m   **12.** $r = 184.0$ cm

B 13. A cone-shaped sand pile has a radius of 11.0 ft and a height of 18.0 ft. What is the volume of sand in the pile?

14. What amount of space is taken up by a spherical water tank with a radius of 24.0 ft?

15. A cone-shaped hopper at a mixing plant has a radius of 4.25 ft and a height of 14.0 ft. Find the volume of the hopper.

16. An engineer takes cylindrical core samples from a roadbed. If the cores have a radius of 0.75 in. and a height of 17.5 in., find the volume.

17. How much steel is used to make 15,000 ball bearings each with diameters of 0.650 in.?

C 18. The machine part below is milled from a 1.0 in. diameter rod. How much metal is wasted?

19. A sphere with a radius of 4 cm is milled from an 8 cm cube, as shown below. How much material is wasted?

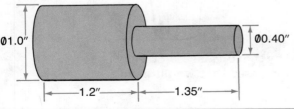

Ø1.0"   Ø0.40"

|←——1.2"——→|←——1.35"——→|

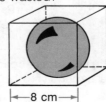

|←—8 cm—→|

7-5 Weight, Mass, and Capacity

The **weight** of an object is the amount of force that gravity exerts on the object. This force varies in the universe. For example, larger planets have larger gravitational forces than smaller ones. The basic unit of weight is the pound. The **mass** of an object is the amount of material of which an object is composed. The basic unit of mass is the gram.

The vise at the right weighs 25 lb. This weight can be converted to kilograms using the proportion method. (The Equivalent Weight Units table in the back of the book states that 1 lb is about 0.454 kg.)

$$\frac{1 \ (\text{lb})}{0.454 \ (\text{kg})} = \frac{25 \ (\text{lb})}{x \ (\text{kg})}$$

$$1x = 0.454 \times 25$$
$$x = 11 \ (\text{kg, to two significant digits})$$

Weight = 25 lb or 11 kg

The **capacity** of an object is a measure of the amount of material, usually liquid, it can hold. Liquids and materials that pour are measured in gallons and liters.

For example, the can at the right has a capacity for 0.80 L of liquid. The capacity can be converted to gallons. (The Equivalent Weight Units table in the back of the book states that 1 gal is about 3.785 L.)

$$\frac{1 \ (\text{gal})}{3.785 \ (\text{L})} = \frac{x \ (\text{gal})}{0.80 \ (\text{L})}$$

$$x = \frac{0.80}{3.785}$$
$$x = 0.21 \ (\text{gal, to two significant digits})$$

Capacity = 0.80 L or 0.21 gal

In the metric system, it is convenient to change from units of volume to units of capacity and mass. The table below shows the equivalences among three kinds of measure for water at 4°C.

Equivalent Measures		
Volume	Capacity	Mass
1 cm³ =	1 mL =	1 g
1 dm³ =	1 L =	1 kg
1 m³ =	1 kL =	1 t (tonnes)

For example, the volume of space occupied by the full glass is 804 cm³. The glass has a capacity to hold 804 mL of water that would have a mass of 804 g.

Example

Skill(s) _19_

A dishwasher has an internal volume of 0.162 m³. How many kiloliters of water will it hold if it is completely filled? how many gallons?

Solution

a. Find the capacity in kiloliters.

$$\frac{1 \text{ m}^3}{1 \text{ kL}} = \frac{0.162 \text{ m}^3}{0.162 \text{ kL}}$$

The dishwasher's capacity is 0.162 kL.

b. Find the capacity in liters.

$$\frac{1 \text{ kL}}{1000 \text{ L}} = \frac{0.162 \text{ kL}}{162 \text{ L}}$$

The dishwasher has a capacity for 162 L.

c. Find the capacity in gallons. The Equivalent Measures Units table in the back of the book states that 1 gal is about 3.785 L.

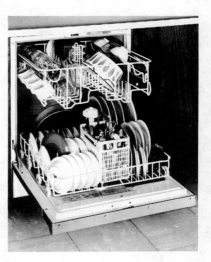

$$\frac{1 \text{ (gal)}}{3.785 \text{ (L)}} = \frac{x \text{ (gal)}}{162 \text{ (L)}}$$

The dishwasher has a capacity for 42.8 gal, to three significant digits.

EXERCISES

A Use the tables in the back of the book to complete each conversion.

1. How many pounds are there in a truck load weighing 2.75 t?

2. How many milligrams is 454 g of salt?

3. How many pounds are in 2.7 kg of hamburger meat?

4. How many quarts is 72 oz of orange juice?

5. How many liters is 25 mL of cologne?

6. How many liters is 22.5 gal of gasoline?

7. How many quarts is 2500 mL of a liquid?

8. How many cubic meters does 18 t (metric tonnes) of water occupy?

9. How many liters of water have a mass of 3050 g?

B 10. If 0.5 kg of nails cost $0.58, what is the cost per pound of the nails to the nearest cent?

11. If gasoline is listed at $1.229/gal, what is the cost per liter to the nearest cent?

12. Each solid rocket booster on a space shuttle has a mass of 589,570 kg. What is the mass in metric tonnes?

13. A cylindrical can has a volume of 525 cm³. How many liters of water (at 4°C) will it hold?

14. A commercial vehicle scale has a maximum capacity of 25,000 lb. What is its capacity in metric tonnes?

15. A tank has a capacity of 1000 L. How many kilograms of water (at 4°C) can it hold?

C 16. A spherical storage tank has a radius of 6.00 m.

    a. How many liters of water (at 4°C) will the tank hold? how many gallons?
    b. How much would the water weigh in metric tonnes if the tank were filled to capacity? How much would it weigh in pounds?

Technology Update

Fiber Optics

As a result of recent technological advances, fiber optics is used in many industries today. Fiber optics uses beams of laser light to transmit signals through very thin glass fibers. At the terminal end of the fibers, a receiver converts the laser light into electrical impulses, which are then used to reconstruct the sound or image. A 1 in. diameter cable of optical fibers can transmit as much information as a 5 in. diameter copper cable. An optical transmission system is even faster and provides less distortion than a system that uses copper cable.

Optical fibers are used in medicine as very small cameras, which, when inserted into an artery, can be manipulated into many of the body's organs. The "camera" then transmits an image of the organ to a video screen. A cardiologist, for example, can insert a fiber into the heart and watch the heart beat on a video screen.

The automobile industry uses fiber optic cables to replace some of the copper wiring. This results in lighter, less expensive cables. Industrial robots in the auto industry use "eyes" of fiber optic materials to monitor their control systems.

Research Project:
Find three other uses of fiber optics.

Self-Analysis Test

Use a calculator where possible.

1. A cement sidewalk is 35.0 ft long, 6.5 ft wide, and 0.4 ft thick. What is the volume of cement in the sidewalk in cubic yards? ($27 \text{ ft}^3 = 1 \text{ yd}^3$)

2. A room measures 42 ft by 31 ft by 11 ft. What is the volume of the room in cubic feet? How many people can comfortably be accommodated in the room if 200 ft³ of space is needed per person?

3. A pyramid-shaped paperweight has a square base, each side measuring 5.5 cm. The height of the paperweight is 6.4 cm. How much space is occupied by the paperweight in cubic centimeters?

4. A birdhouse has the dimensions shown below. What is the amount of space occupied by the birdhouse in cubic centimeters?

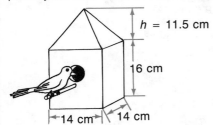

5. A circular tunnel, 400.0 ft long, is dug through a hill as shown below. If the radius of the tunnel is 12.4 ft, how many cubic yards of material must be removed? ($1 \text{ yd}^3 = 27 \text{ ft}^3$)

6. A cylindrical can for tennis balls has a radius of 3.6 cm and a height of 20.0 cm.

   a. Find the volume of the can.
   b. A tennis ball has a radius of 3.2 cm. Find the total volume of three tennis balls.
   c. If three tennis balls are placed in the can, how much air space is left in the can?

7. A wooden cylinder has a radius of 3.5 in. and a height of 7.0 in. The cylinder is turned on a lathe to form a cone with the same radius and height. How much of the wood is wasted?

8. If a water heater has a capacity of 40.0 gal, how many liters will it hold?

9. A hardware store sells bolts for $2.25/lb. What is the cost per kilogram for the bolts to the nearest cent?

10. A water pump has a capacity to pump 42.0 L/min. What is the pumping rate in gallons per minute?

11. A spherical water tank has an inside radius of 4.75 m. How many kiloliters of water (at 4°C) will it hold? What is the capacity of the tank in gallons?

Cylinder Applications

7-6 Car Engine Displacement

Calculating a car's **engine displacement** involves finding the total volume of cylinders of space. The number of cylinders involved depends on the car. Most cars today have four, six, or eight cylinders. A piston moves within each cylinder. The displacement of one cylinder is the amount of space through which a piston travels in one *stroke*. The diagram of a car cylinder in two positions below shows the relationship of the *bore* (the inside diameter of the cylinder) and the stroke to the displacement.

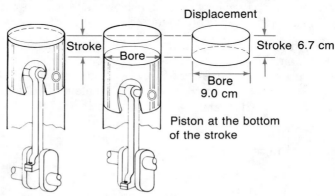

Piston at the bottom
of the stroke

The formula for car engine displacement is derived from the formula for the volume of a cylinder, as shown below.

Volume of a cylinder $= \pi \times r^2 \times h$

Car engine displacement $= (\pi \times (\text{radius of bore})^2 \times \text{stroke}) \times \text{number of cylinders}$

Example

Skill(s) __7.28__

Suppose the diagram above is for a four-cylinder car. What is its engine displacement in cubic centimeters and in liters?

Solution

Car engine displacement $= (\pi \times (\text{radius of bore})^2 \times \text{stroke}) \times \text{number of}$

$= (3.14 \times (4.5)^2 \times 6.7) \times 4$ cylinders

$\boxed{3 \cdot 1 \ 4 \times 4 \cdot 5 \times 4 \cdot 5 \times 6 \cdot 7 \times 4 =}$ 1704.078

The displacement of the four-cylinder engine is 1700 cm³, to two significant digits. Since 1000 cm³ = 1 L, the engine displacement can also be stated as 1.7 L.

EXERCISES

A Write each engine displacement in liters.

1. 1500 cm³   **2.** 2800 cm³   **3.** 5700 cm³

Write each engine displacement in cubic centimeters.

4. 4.7 L   **5.** 2.0 L   **6.** 3.8 L

Identify the stroke and the bore size in the given engine displacement formula. All dimensions are in centimeters.

7. Engine displacement = $(3.14 \times (4.18)^2 \times 8.62) \times 6$

8. Engine displacement = $(3.14 \times (5.8)^2 \times 4.3) \times 8$

9. For question 7, how many cylinders does the car engine have?

Use a calculator.

B **10.** What is the displacement in liters of a four-cylinder engine with an 8.25 cm bore and a 6.95 cm stroke?

11. Find the displacement in liters of a V-6 engine (six cylinders) with a 9.53 cm bore and an 8.57 cm stroke.

12. A V-8 engine (eight cylinders) has a 9.5 cm bore and a 7.25 cm stroke. What is the engine displacement in cubic centimeters?

13. What is the displacement in liters of a V-8 engine with an 8.3 cm bore and an 8.0 cm stroke?

14. What is the displacement in liters of an eight-cylinder engine having an 8.9 cm bore and an 8.25 cm stroke?

A V-8 engine

15. What is the engine displacement in liters of a four-cylinder car with a 12.5 cm bore and a 9.5 cm stroke?

16. Find the displacement in cubic centimeters of a six-cylinder engine with an 8.75 cm bore and a 7.7 cm stroke.

17. A four-cylinder car engine has a 9.5 cm bore and an 8.25 cm stroke. What is the engine displacement in liters?

C **18.** An optional V-6 engine has a displacement of 4850 cm³. If the cylinder bore is 9.25 cm, what is the size of the stroke?

19. A six-cylinder pick-up truck engine has a displacement of 4.3 L. If the cylinder bore is 9.5 cm, what is the size of the stroke?

7-7 Compression Ratio

Mrs. Peters and Mr. Salvador each have a 4200 cm³ V-8 engine in their cars. Mrs. Peters buys regular gas for her car but Mr. Salvador must buy premium, higher octane gas for his car. The difference is that Mr. Salvador's car has a higher compression ratio.

A **compression ratio** is the ratio of the *maximum space* in an engine cylinder to the *minimum space* in a cylinder. The diagram below shows one cylinder of a car engine having a maximum space of 1050.0 cm³ at the bottom of the piston's stroke and a minimum space of 100.0 cm³ at the top of the piston's stroke.

Compression Ratio:

$$\frac{\text{maximum space (cm}^3)}{\text{minimum space(cm}^3)} \quad \frac{1050.0}{100.0} = \frac{10.50}{1} \quad \leftarrow \quad \text{The second term in a compression ratio is always 1.}$$

The *difference* between the maximum and minimum space in the cylinder is the engine displacement. The displacement in each cylinder of the car engine above is 1050.0 cm³ − 100.0 cm³, or 950.0 cm³.

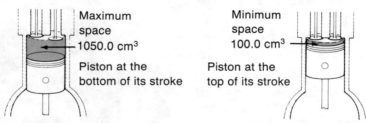

Maximum space
1050.0 cm³
Piston at the bottom of its stroke

Minimum space
100.0 cm³
Piston at the top of its stroke

Example

Skill(s) _18,19_

A cylinder in a car engine has a minimum space of 38 cm³ when the piston is at the top of its stroke and a maximum space of 304 cm³ when the piston is at the bottom of its stroke. What is the compression ratio of the engine cylinder?

Solution

$$\text{Compression Ratio} = \frac{\text{maximum space (cm}^3)}{\text{minimum space (cm}^3)}$$

$$= \frac{304}{38}$$

$$= \frac{8.0}{1}$$

| 3 | 0 | 4 | ÷ | 3 | 8 | = | 8

The compression ratio is 8.0 : 1.

EXERCISES

Use a calculator where possible.

A 1. What is the compression ratio of a car engine if there are 705 cm³ of space when each cylinder piston is at the bottom of its stroke and only 115 cm³ when each piston is at the top of its stroke?

2. A car engine cylinder contains 655 cm³ of space when each piston is at the bottom of its stroke and 82 cm³ when each piston is at the top of its stroke. What is the compression ratio of the car engine?

3. What is the compression ratio for a car engine cylinder that has 520 cm³ of space when each piston is at the bottom of its stroke and 82 cm³ when each piston is at the top of its stroke?

B 4. The compression ratio of an engine is 10.0 : 1. When each piston is at the top of its stroke, there are 98 cm³ of space in the cylinder. How much space is there when each piston is at the bottom of its stroke?

5. An engine has a compression ratio of 8.5 : 1. There are 66 cm³ of space in the cylinder when each piston is at the top of its stroke. How much space is there when each piston is at the bottom of its stroke?

6. A car engine has a displacement of 480.0 cm³ in each cylinder. There are 40.0 cm³ of space in each cylinder when each piston is at the top of its stroke. How much space is there in each cylinder when the piston is at the bottom of its stroke?

C 7. A six-cylinder van has a compression ratio of 9.0 : 1. In each of the six cylinders, there is 590 cm³ of space when the piston is at the bottom of its stroke. What is the displacement for each cylinder in the six-cylinder van?

8. A used car has an engine displacement of 2025.0 cm³ in six cylinders. There are 37.5 cm³ of space in each cylinder when the piston is at the top of its stroke. What is the engine's compression ratio?

7-8 Hydraulic Power

Hydraulic power, used in work-saving devices like lifts and pumps, works because of Pascal's Law. This law states that pressure applied on a confined fluid is transmitted undiminished in all directions and acts with equal force on equal areas and at right angles to them. The diagram at the right shows the principle of hydraulic transmission. A force of 50 lb is applied to piston A. This 50 lb input force is transmitted through piston A into the hy-

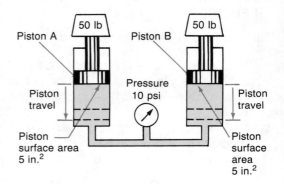

draulic fluid and onto piston B. The amount of fluid pressure, in pounds per square inch (psi), inside the hydraulic cylinder is found by the following formula.

$$\text{Pressure (psi)} = \frac{\text{Input force (lb)}}{\text{Surface area of piston } A \text{ (in.}^2)}$$

$$= \frac{50 \text{ (lb)}}{5 \text{ (in.}^2)}$$

$$= 10 \text{ (psi)} \quad \leftarrow \text{The pressure inside the hydraulic cylinder is 10 psi.}$$

The force output by the hydraulic cylinder is found by the following formula.

$$\text{Output force (lb)} = \text{Pressure (psi)} \times \text{Surface area of piston } B \text{ (in.}^2)$$

$$= 10 \text{ (psi)} \times 5 \text{ (in.}^2)$$

$$= 50 \text{ lb} \quad \leftarrow \text{The output force of the hydraulic cylinder is 50 lb.}$$

In the above case, the output force is the same as the input force because the surface areas of the two pistons are the same. The example below shows how a hydraulic cylinder can be used more advantageously.

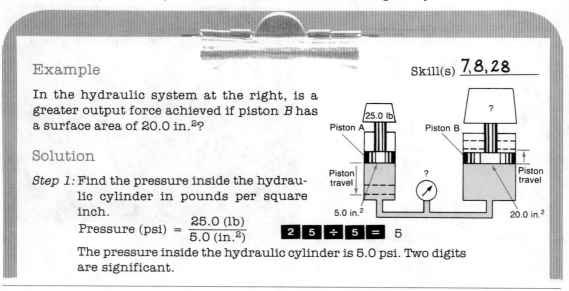

Example

In the hydraulic system at the right, is a greater output force achieved if piston B has a surface area of 20.0 in.^{2}?

Skill(s) __7, 8, 28__

Solution

Step 1: Find the pressure inside the hydraulic cylinder in pounds per square
inch.

$$\text{Pressure (psi)} = \frac{25.0 \text{ (lb)}}{5.0 \text{ (in.}^2)}$$

[2] [5] ÷ [5] = 5

The pressure inside the hydraulic cylinder is 5.0 psi. Two digits
are significant.

> **Step 2:** Find the output force in pounds.
>
> Output force (lb) = Pressure (psi) × Surface area of piston B (in.^{2})
> = 5.0 × 20.0   [5] [×] [2] [0] [=]   100
>
> The output force is 100 lb, to two significant digits. It is 4 times greater than the input force because the surface area of the output piston is 4 times greater.

EXERCISES

Use a calculator where possible.

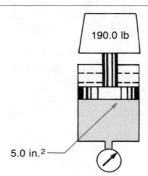

190.0 lb

5.0 in.^{2}

A 1. The smaller piston in a hydraulic cylinder has a surface area of 5.0 in.^{2}. The piston receives a force of 190.0 lb. What is the pressure on the liquid in the cylinder in pounds per square inch?

2. In a hydraulic cylinder, an input force of 20.0 lb acts on a piston with a surface area of 2.0 in.^{2}. What is the pressure on the liquid inside the hydraulic cylinder in pounds per square inch?

3. The liquid in a hydraulic cylinder has a pressure of 15 psi. The force on the input piston is 45 lb. What is the surface area of the input piston?

4. The pressure in a hydraulic cylinder is 20 psi. The output piston has a surface area of 4 in.^{2}. What is the output force in pounds?

5. The small piston in a hydraulic cylinder has a surface area of 6.0 in.^{2}. This piston receives a force of 30.0 lb. The large piston puts out a force of 120.0 lb. Find the surface area of the large piston.

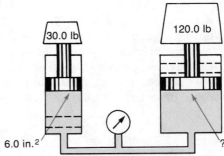

30.0 lb   120.0 lb

6.0 in.^{2} ?

B 6. The input piston for a cylinder of a hydraulic jack has a surface area of 8.0 in.^{2}. The surface area of the output piston is 25.0 in.^{2}. If the input force is 40.0 lb, what is the output force in pounds?

7. For the cylinder in a hydraulic coupler, an input force of 50.0 lb produces an output force of 125.0 lb. The pressure in the cylinder is 25.0 psi. Find the surface area of both the input piston and the output piston.

8. The input piston in the cylinder of a hydraulic pump has a surface area of 6.25 in.^{2} and receives a force of 32.50 lb. The large piston in the cylinder puts out a force of 148.20 lb. Find the surface area of the large piston.

C 9. In a hydraulic lift truck cylinder, an input force of 95.55 lb produces an output force of 876.06 lb. The pressure on the liquid in the cylinder is 27.30 psi. Find the surface area of both the input piston and the output piston.

7-9 Pneumatic Power

The jackhammer shown at the right uses the principle of **pneumatic power** in which power is transmitted through a compressed column of air. The hose in the photograph is connected to an air compressor that transmits compressed air to power the jackhammer.

Boyle's Law clearly defines the relationship between a volume of gas and its pressure. The law states that the absolute pressure of a confined body of gas varies *inversely* as its volume, provided its temperature remains constant.

The diagram below the photograph shows air being compressed in a pneumatic cylinder. The volume of compressed air is reduced to $\frac{1}{3}$. As the air is compressed, it heats up. Boyle's Law says its absolute pressure, after it has been allowed to cool to its original temperature, would be 3 times as great as it was before compression.

You can see in the diagram, however, that the gauge pressure after compression is not exactly three times greater as it was before compression. This is because gauge pressure does not include atmospheric pressure, which is normally 14.7 psi at sea level. Boyle's Law applies only to absolute pressure, which includes atmospheric pressure. Thus in order to calculate gauge pressure and to use Boyle's

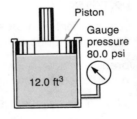

Before column of air is compressed

Piston

Gauge pressure 80.0 psi

12.0 ft³

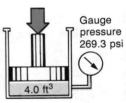

After column of air is compressed

Gauge pressure 269.3 psi

4.0 ft³

Law, atmospheric pressure must be considered. First, a gauge pressure reading should be converted to an absolute pressure by adding 14.7 psi. Then the calculation is made using Boyle's law. After the calculation, the absolute pressure is converted back to a gauge reading by subtracting 14.7 psi.

The example below shows how Boyle's Law is used to calculate the pressure inside a pneumatic cylinder.

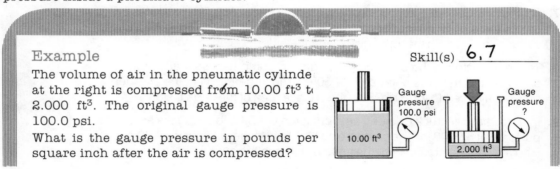

Example

The volume of air in the pneumatic cylinder at the right is compressed from 10.00 ft³ to 2.000 ft³. The original gauge pressure is 100.0 psi.

What is the gauge pressure in pounds per square inch after the air is compressed?

Skill(s) 6, 7

Gauge pressure 100.0 psi

10.00 ft³

Gauge pressure ?

2.000 ft³

Solution

Step 1: Convert the original gauge pressure to its equivalent absolute pressure.

Absolute pressure = Gauge pressure + Atmospheric pressure
= 100.0 (psi) + 14.7 (psi)
= 114.7 (psi, precise to the tenths place)

Step 2: Apply Boyle's Law. Since the original volume is reduced to $\frac{1}{5}$, the new pressure will be 5 times greater than the original pressure.

> ▮ 1 ▮ 1 ▮ 4 ▮ · ▮ 7 ▮ × ▮ 5 ▮ = ▮ 573.5

The new absolute pressure is 573.5 psi, to four significant digits

Step 3: Change the new absolute pressure to gauge pressure.

Gauge pressure = Absolute pressure − Atmospheric pressure
= 573.5 (psi) − 14.7 (psi)
= 558.8 (psi, precise to the tenths place)

The gauge pressure after compression is 558.8 psi.

EXERCISES

Use a calculator where possible.

A 1. Write each absolute pressure as a gauge pressure reading.

    a. 200.0 psi   **b.** 360.0 psi   **c.** 125.0 psi

2. Write each gauge pressure reading as an absolute pressure.

    a. 90.0 psi   **b.** 260.0 psi   **c.** 115.0 psi

B 3. In a pneumatic cylinder, a piston compresses 20.00 ft³ of air to 5.000 ft³. If the pressure gauge reading is 102.3 psi before compression, what is the reading after compression?

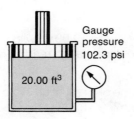
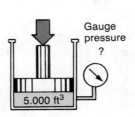

Gauge pressure 102.3 psi   20.00 ft³

Gauge pressure ?   5.000 ft³

4. Air at a 86.3 psi gauge reading is compressed in a pneumatic cylinder so that its new reading is 106.5 psi. If the original volume was 3.0 in.^{3}, what is the new volume?

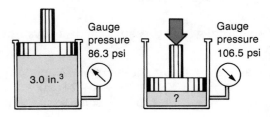

5. What new gauge pressure reading results when 10.00 ft^3 of air at 88.0 psi is compressed to 1.000 ft^3?

6. A volume of air at a gauge reading of 80.0 psi is compressed in a pneumatic cylinder to 3.0 ft^3 with a resulting gauge pressure reading of 269.4 psi. What was the original volume of air?

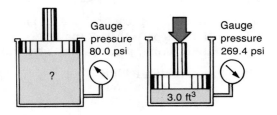

7. A piston compresses 22.50 ft^3 of air at a pressure gauge reading of 90.3 psi to 12.50 ft^3 in a pneumatic cylinder. What is the pressure gauge reading in the cylinder after compression?

8. A pneumatic air hammer operates with an gauge pressure reading of 90.0 psi. If a gauge pressure of 194.7 psi is required after compression, how many cubic feet of air must be compressed to 3.5 ft^3?

9. A volume of air at a gauge reading of 79.9 psi in a pneumatic vise is compressed to 2.50 ft^3 with a resulting gauge reading of 127.2 psi. How much air must be compressed to use the vise?

10. A jackhammer operates with a gauge reading of 95.0 psi acting on 250.0 ft^3 of air. What is the volume of the air in the tank after compression if the new gauge reading is 204.7 psi?

11. If 25.00 ft^3 of air is compressed in a pneumatic vibrator to 10.00 ft^3 with a resulting gauge reading of 167.1 psi, what was the original gauge reading?

12. In a pneumatic press, a volume of air at 92.50 psi is compressed to 6.30 ft^3. The resulting gauge reading is 521.3 psi. What volume of air must be in the cylinder to result in the required new gauge reading?

Tricks of the Trade

Recently, we have seen the development of hydraulic motors. In a hydraulic motor, the force of the hydraulic system is used to turn the motor rather than to move a piston. By controlling the flow of hydraulic fluid to the motor, the speed of the motor can be varied without causing the motor to overheat or otherwise malfunction.

Hydraulic motors are very versatile and are used where mobile power equipment is required. They are not as dangerous in wet or dusty conditions as electric motors may be. Among the many applications of hydraulic motors today are road-paving equipment, cement mixers, conveyors, farm harvesters, mobile cranes, forestry equipment, and ship winch and propulsion equipment.

The next time you drive through an automatic car wash you might notice that most of the brushes are now operated by hydraulic motors. There will be hydraulic lines to the motors rather than electric wires.

Check Your Skills

Evaluate.

1. $2.453 + 6.847$

2. $9.8 + 3.203$

3. $0.10065 + 1.85406$ (Skill 6)

4. $(14.2)(12.3)$

5. $(10.1)(5)$

6. 149.67×42.8 (Skill 7)

7. $\dfrac{18.3}{4}$

8. $129.8 \div 8$

9. $0.3423 \div 0.021$ (Skill 8)

Write each as a *unit rate* in which the second term is 1.

10. 15 turns in 6 s

11. 35 paces in 2 min

12. 200 km in 4 h   (Skill 18)

Solve for the unknown term in the proportion.

13. $\dfrac{x}{12} = \dfrac{4}{3}$

14. $\dfrac{y}{18} = \dfrac{4.5}{6}$

15. $\dfrac{11}{30.25} = \dfrac{q}{93.5}$ (Skill 19)

16. $\dfrac{n}{8.4} = \dfrac{14.25}{19}$

17. $\dfrac{m}{12} = \dfrac{17.5}{30}$

18. $\dfrac{r}{6.4} = \dfrac{6.4}{12.8}$

Solve the following equations.

19. $x + 2.735 = 9.826$

20. $m - 4.323 = 6.827$

21. $w + \dfrac{2}{3} = 1$ (Skill 27)

22. $9.8z = 23.863$

23. $32n = 28.016$

24. $\dfrac{p}{3.61} = 19$ (Skill 28)

Simplify.

25. 6^3

26. 4^4

27. $(0.75)^2$

28. $6.3 \times (1.95)^2$

29. $8.2 \times (8.3)^2$

30. $9.5 \times (2.4)^2$ (Skill 30)

Self-Analysis Test

Use a calculator where possible.

1. What is the engine displacement in cubic centimeters of a V-8 engine if the cylinder bore is 10.0 cm and the stroke is 8.9 cm?

2. Find the displacement in liters of a six-cylinder engine if the bore is 11.0 cm and the stroke is 8.0 cm.

3. A four-cylinder engine has a stroke of 8.2 cm and a bore of 8.9 cm. What is the engine displacement in liters?

4. A car engine cylinder has a maximum space of 655.0 cm^3 when each piston is at the bottom of its stroke and a minimum space of 65.5 cm^3 at the top of its stroke. What is the compression ratio of the car engine?

5. The compression ratio of a car engine is 8.5 : 1. There are 82 cm^3 in the cylinder when each piston is at the top of its stroke. How much space is there when each piston is at the bottom of its stroke?

6. There are 420.0 cm^3 of space in a car engine cylinder when each piston is at the bottom of its stroke. When it is at the top of its stroke, there are 60.0 cm^3 of space in the cylinder. What is the compression ratio of the car engine?

7. An input piston in a hydraulic cylinder receives a force of 32.0 lb. If its surface area is 4.0 in.^{2}, what is the pressure in the cylinder?

8. A hydraulic lift cylinder has an output piston with a surface area of 9.0 in.^{2}. If the pressure in the cylinder is 95.0 psi, what is the output force?

9. In a hydraulic pump cylinder, a force of 50.0 lb acts on a piston with a surface area of 3.5 in.^{2}. What is the pressure on the liquid inside the cylinder?

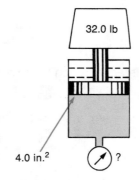

32.0 lb

4.0 in.^{2} ?

10. Air in a pneumatic system at a gauge reading of 81.8 psi is compressed so that its new reading is 178.3 psi. If the original volume was 5.6 ft^3, what is the new volume after compression?

11. In a pneumatic hammer, a 20.00 ft^3 volume of air is compressed to 5.00 ft^3. If the original gauge reading is 87.0 psi, what is the reading after compression?

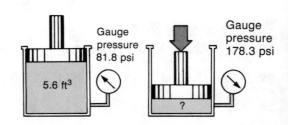

Gauge pressure 81.8 psi

5.6 ft^3

Gauge pressure 178.3 psi

?

Computer graphics was once thought of only as a toy, but today, it is one of the most important applications of computers. Some computers use graphics to run all of their programs.

Engineers use computers to design, in a fraction of the time, things that were previously created on the drafting board . This is called computer-aided design, or CAD. When the computer is used to manufacture the article designed on it, it is called computer-aided manufacturing or CAM. CAD/CAM is becoming one of the fastest growing uses for computers as they become more affordable and easier to use.

Architects use computers in much the same way as engineers except that they design buildings and other large structures. The computer is very useful for this as it enables architects to view the design in three dimensions and from any angle before making a model. The computer image of a bridge design at the right is an example of this.

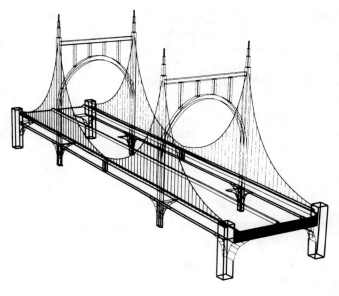

Scientists are using graphics in research to create three-dimensional models of molecules, a process that was very tedious and confusing using the conventional method of joining balls and straws to simulate a molecule.

Image processing is another application of computer graphics that is becoming widespread. Some examples are:

- creation of intricate computer art;
- enhancing X-ray images that are unclear;
- manipulation of satellite photographs.

With image processing, the difference between a computer picture and an actual photograph is becoming more difficult to distinguish.

Graphics are used in many areas of the media, such as in TV commercials and in special effects in movies. Three-dimensional effects that are used in movies and on TV can be made more easily using computers, thereby making more complex images possible.

As the use of computers becomes more widespread, more applications are being discovered. The imagination is the only boundary to the uses of computer graphics.

Tiny electronic chips found in computers, telephones, and many other types of electronic equipment play a vital role in our lives. Some of the many functions these devices facilitate are making phone calls, reserving plane tickets, and billing people for our services. Keeping all types of electronic equipment in good working order is the job of the electronics engineering technician.

Job Description

What do electronics engineering technicians do?

1. Electronics engineering technicians perform preventative maintenance procedures on computers, computer printers, as well as electronically controlled door openers, burglar alarms, thermostats, etc.

2. They troubleshoot equipment failure and repair and/or replace faulty circuit boards, cables, and components. This can involve travel since many technicians may be assigned to several installations.

3. Electronic engineering technicians install cable, new computers, and computer peripherals. Once this procedure is complete, they test the new system to confirm proper operation.

4. They listen to customers' complaints, answer questions, and offer technical advice.

5. Electronic engineering technicians keep detailed records of various aspects of their work, including repair logbooks, expense accounts, and parts inventories.

Qualifications

Electronics engineering technicians must have logical and analytical minds in order to determine where a system malfunction has occurred. They must know how to use specialized tools and test equipment, and must keep up with the technical information and revised maintenance procedures issued by the equipment manufacturers.

Most employers require applicants to have two or three years of post-high school training as an electronics engineering technician. High school courses in mathematics, physics, electronics, and computer programming are useful. New technicians usually receive on-the-job training for the equipment they are maintaining and repairing.

FOCUS ON INDUSTRY

Life in the United States would be quite different without the mining industries. Consider how many things you use each day that depend on the aluminum, steel, or coal industries. Some areas of the country depend on coal-fired electrical power for more than 90% of their electrical needs. Agriculture could not produce enough food to sustain the population without fertilizer produced by the mining industry.

In the coal industry, the output of underground mines has dropped in recent years. However, the output from surface mines has risen dramatically. Surface-mined coal often has a high sulfur content and has received some of the blame for pollution in the form of acid rain. The mining industry has many experimental projects currently under way that one hopes will produce a kind of "super coal." Various chemical and mechanical processes aimed at removing the impurities from coal have been tested.

The following table shows the dollar value of nonfuel mineral production in the ten leading states in a recent year.

Rank	State	Dollar Value	Rank	State	Dollar Value
1	California	$2,094,000,000	6	Michigan	$1,348,000,000
2	Texas	$1,733,000,000	7	Georgia	$946,000,000
3	Florida	$1,559,000,000	8	Pennsylvania	$804,000,000
4	Arizona	$1,550,000,000	9	Missouri	$735,000,000
5	Minnesota	$1,548,000,000	10	New York	$657,000,000

1. The total value of nonfuel minerals produced in the United States was about $23,232,000,000. Make a table showing what percents of that total amount were produced in the ten leading states.

2. In a recent year, the United States produced about 150,000 t of magnesium. What percent is this of the world production of 361,000 t for the same year?

3. The population of California in a recent year was about 26.4 million. Calculate the per capita production of nonfuel minerals for the state of California.

Chapter 7 Test

Use a calculator where possible.

1. What is the surface area of a shoe box without its lid that is 12.0 in. long, 6.0 in. wide, and 5.0 in. high? (7-1)

2. How many square inches of mirror glass are needed to build a decorative square pyramid with a base having 4.0 in. sides and a slant height of 7.5 in. on the triangular sides? (7-1)

3. How many square inches of metal are required to make the cylindrical pail below without the handle? (7-2)

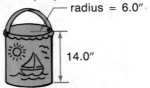
radius = 6.0″

14.0″

4. How many square centimeters of rubber are needed to surface the ball below? The radius of the ball is 3.0 cm. (7-2)

5. How much aluminum is needed for manufacturing the conical roof to a grain storage bin with a radius of 4.2 ft and a slant height of 6.4 ft? (7-2)

6. How much space is occupied by a covered rectangular box with a length of 10.0 in., a width of 9.0 in., and a height of 3.0 in.? (7-3)

7. Find the volume of a glass terrarium in the shape of a square pyramid if the area of the base is 9 in.^{2} and the height of the terrarium is 7 in. (7-3)

8. How much space is occupied by a cylindrical tank having a radius of 8.0 ft and a height of 12.0 ft? (7-4)

9. What is the volume of a cone-shaped dust collector having a radius of 9.0 in. and a height of 20.0 in.? (7-4)

10. An ornament in the shape of a sphere has a radius of 30 mm. What is its volume? (7-4)

11. If film developer costs $1.35/gal, what is the cost per liter? (1 gal = 3.785 L) (7-5)

12. A cylindrical bucket has a capacity for 7.6 L. What is the capacity of the bucket in gallons? (7-5)

13. Find the displacement in cubic centimeters of a V-8 engine with a 9.0 cm bore and an 8.9 cm stroke. (7-6)

14. What is the displacement in liters of a six-cylinder engine whose bore is 9.0 cm and whose stroke is 9.0 cm? (7-6)

15. What is the engine displacement in liters of a four-cylinder car engine with a stroke of 10.8 cm and a bore of 14 cm? (7-6)

16. The compression ratio of a car engine is 9.5 : 1. At the top of each piston's stroke, there are 45 cm^3 of space in the cylinder. How much space does the cylinder contain at the bottom of the stroke? (7-7)

17. There are 3.0 in.^{3} of space in a car engine cylinder when a piston is at the top of its stroke. If the compression ratio is 9.0 : 1, how much space is there when the piston is at the bottom of its stroke? (7-7)

18. A car engine cylinder contains 220 cm^3 of space when the piston is at the bottom of its stroke. There are 25 cm^3 of space when it is at the top of its stroke. What is the compression ratio of the car engine? (7-7)

19. The liquid in a hydraulic cylinder has a pressure of 80.0 psi. The output piston has a surface area of 8.0 in.^{2}. What is the output force? (7-8)

20. The pressure in a hydraulic cylinder is 120.0 psi. The output piston exerts a force of 4000.0 lb. What is the surface area of the output piston? (7-8)

21. In a hydraulic cylinder, piston *B* has a surface area of 3.75 in.^{2} and has a pressure of 90.0 psi applied to it. What is the force that results? (7-8)

22. In a pneumatic cylinder, 30.0 ft^3 of air is compressed to 6.00 ft^3. The gauge reading was 95.1 psi before compression. What is the reading after compression? (7-9)

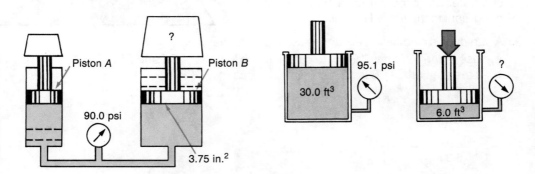

23. If 35.0 ft^3 of air in a pneumatic cylinder with a gauge reading of 85.0 psi is compressed to 5.00 ft^3, what is the reading after compression? (7-9)

Use a calculator where possible.

1. In making a technical drawing, you need a 115° angle. Use a protractor to draw a 115° angle. Use a compass and straightedge to construct the bisector of the angle.

2. The technical drawing also calls for a line segment 8 cm long with perpendiculars at each endpoint. Use a compass and straightedge to do this construction.

3. A drawing requires a line through a point parallel to the edge of a square. Draw a square and select a point outside the square. Construct a line parallel to an edge of the square through the point that you selected.

4. An artist must construct a parallelogram which has a 43° angle. If the angle adjacent to the 43° angle is the supplement of the angle, find the measure of the adjacent angle.

5. Two angles on a navigational map are complementary. If one of the angles has a measure of 62.3°, what is the measure of the other angle?

6. Trace the circle at the right. Use a compass and straightedge to locate the center of the circle.

7. A circular drive wheel must have five equally spaced holes drilled along a circle. Find the number of degrees in the central angles.

8. A commercial artist needs to copy the triangle shown with a compass and a straightedge. Illustrate how this might be done by first tracing the triangle and then using a compass and straightedge to construct a triangle that is congruent to the traced triangle.

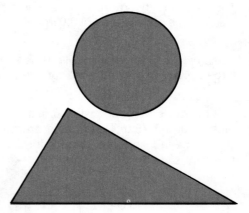

9. A surveyor is calculating distances using similar triangles.
   a. Use triangles shown below to calculate the lengths of *DE* and *EF*.
   b. If the area of △*ABC* is 12.1 m², find the area of △*DEF*.

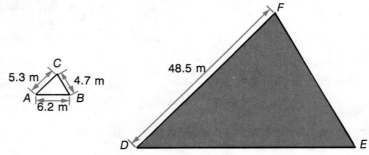

10. A basement recreation room is 23.5 ft by 12.0 ft with a height of 7.0 ft. How much paneling is needed for the four walls of the room? (Ignore the doors and windows.)

11. A cardboard container for packaging shredded Styrofoam is a square pyramid. The edges of the base are 27.0 in. long, and the slant height is 16.5 in. Find the surface area of the cardboard carton.

12. A steel gasoline drum has a diameter of 1.40 m and a height of 2.60 m.

    a. Find the surface area of the drum.
    b. What is the amount of space occupied by the gasoline drum?
    c. If gasoline weighs 0.66 kg/L, find the weight of the gasoline in a full drum.

13. a. A painter must know the surface area of a spherical water tank. If the radius is 5.67 m, find the surface area of the tank.

    b. How much water at 4° C can fill the tank? State your answer in metric tonnes. 1 m³ = 1 t (tonne)

14. The gold bars in a shipment to the mint are 10.0 cm by 6.0 cm by 4.0 cm. If gold weighs 19.3 g/cm³, find the weight of one bar.

15. A sheet of aluminum has a volume of 0.78 m³. If the aluminum weighs 2700 kg/m³, what is the weight of the aluminum sheet in kilograms and in pounds? (1 kg = 2.20 lb)

16. What is the engine displacement in liters of a four-cylinder engine with an 8.0 cm bore and a 7.8 cm stroke?

17. A car engine has a compression ratio of 8.8 : 1. If one of the cylinders has 220 cm³ of space when the piston is at the top of the stroke, how much space is there when the piston is at the bottom of the stroke?

18. Piston A in a hydraulic system has a surface area of 7.5 in.². The surface area of piston B is 22.5 in.². If the input force is 30.0 lb, find the output force.

19. A volume of air in a pneumatic cylinder at a gauge reading of 89.8 psi, is compressed to 2.5 ft³ with a reading of 236.1 psi. What was the original volume of air?

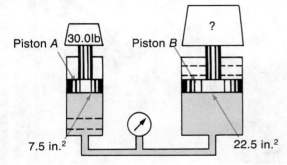

Piston A   30.0 lb   Piston B   ?

7.5 in.²   22.5 in.²

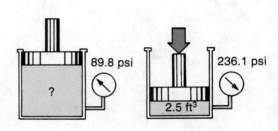

89.8 psi   236.1 psi

?   2.5 ft³

Chapter 8
Publishing a Book

After completing this chapter you should be able to:

1. Understand and use points, picas, and ems in the measurement of type area.
2. Calculate dimensions of reduced or enlarged photographs and illustrations.
3. Estimate the costs involved in the production of a book.
4. Describe the more important kinds of printing methods.
5. Understand the procedures and costs involved in the finishing of a publication.

Preparing a Book for Printing

8-1 Choosing the Type

The production of an attractive, distinctive, and pleasing book involves a wide range of activities. This chapter will attempt to give you an overview of some of these activities, focusing in particular on the production of this book, *Career Mathematics*. The picture on the preceding page shows the graphic designer for this book considering various design options.

The manuscript for this book was entered directly into a microcomputer word-processing system. At an early stage, the book's graphic designer made stylistic decisions about the appearance of the book. The editor then prepared the manuscript on a disk for the typesetter, incorporating the following kinds of choices made by the graphic designer.

- A rough layout of the pages in the book was made that indicated the *format* to be used for the chapter openers, the lesson examples, and the exercise questions.
- The *trim size* (the width and depth of each whole page) was chosen.
- The designer indicated the size of the *type page* (the width and depth of a page including text, illustrations, headings, and footnotes).
- The *typefaces* to be used in the book were determined.

To have a better idea of some of the actual stylistic decisions made by the graphic designer for this book, it is important to understand that type is measured in units called points. A **point** is about 0.0139 in. There are about 72 points to 1 in. Type sizes are designated by points, as shown below.

This is 6-point type.

This is 7-point type.

This is 8-point type.

This is 9-point type.

This is 10-point type.

This is 11-point type.

This is 12-point type.

This is 14-point type.

This is 18-point type.

This is 24-point type.

This is 36-point type.

A **pica** is equal to 12 points, or about $\frac{1}{6}$ in. (0.167 in.) The *pica ruler*, shown below (and reproduced at the back of the book), can be used to measure the width of the text lines on a page. The smallest division on the pica ruler represents 0.5 picas, or 6 points.

A unit of measurement in printing for measuring area is the em. An **em** is the area occupied by the capital letter M in a given type size. The printed capital letter M always takes up a perfectly square space. The diagram at the right shows an em in four different type sizes.

The em can be used to determine how much type will fit in a given area. The example below shows that the em is not a common form of measurement for large areas.

■ Em of 6-point type

■ Em of 8-point type

■ Em of 10-point type

■ Em of 12-point type
The 12-point em is also called the pica em because 12 points are equal to 1 pica.

An indent of one 10-point em →

Natural Resources:
■Precious Metals
Dairy Products
Timber

An indent of two 10-point ems →

Natural Resources:
■■Precious Metals
Dairy Products
Timber

Many different sizes and styles of typefaces are available today. These typefaces are generally compatible with microcomputer word-processing systems used by authors and editors as well as with typesetters' printing requirements. A variety of the typefaces and their point sizes used in *Career Mathematics* are shown below.

This line is set in 10-point American Typewriter roman.
This line is set in 10-point American Typewriter italic.
This line is set in 10-point American Typewriter bold.

This line is set in 10-point Helvetica roman.
This line is set in 10-point Helvetica italic.
This line is set in 10-point Helvetica bold.

This is 18-point American Typewriter.
THIS IS 18-POINT OCR COMPUTER.
This 18-point Kabel bold.
This is 18-point Omega bold.

Example 1

What is the width of the type below in picas? in inches?

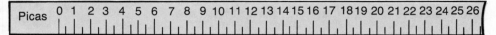

Much copy today is sent to a typesetter on disk.

Solution

Picas 0 1 2 3 4 5 6 7 8 9 10 11 12 13 14 15 16 17 18 19 20 21 22 23 24 25 26

Much copy today is sent to a typesetter on disk.

Inches 0   1   2   3   4

The block of type is 21.5 picas or $3\frac{9}{16}$ in. wide.

Example 2

A block of text beside an illustration is to be 20 picas wide and 30 picas deep, as shown at the right. How many ems of 10-point type are available for this block of text?

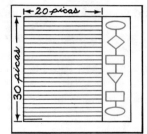

Solution

Step 1: Find the number of points in the 20-pica width and the 30-pica depth. Multiply each by 12 since there are 12 points to one pica.

| 2 | 0 | × | 1 | 2 | = | 240   The width is 240 points.

| 3 | 0 | × | 1 | 2 | = | 360   The depth is 360 points.

Step 2: Find the number of 10-point ems for a 240-point width and a 360-point depth by dividing the width and depth by 10.

| 2 | 4 | 0 | ÷ | 1 | 0 | = | 24

| 3 | 6 | 0 | ÷ | 1 | 0 | = | 36   The block of text is 24 10-point ems wide and 36 10-point ems deep.

Step 3: Find the area of the block of text beside the illustration in 10-point ems by multiplying the width by the depth.

| 2 | 4 | × | 3 | 6 | = | 864   The block of text beside the illustration has 864 ems of 10-point type.

EXERCISES

Use a pica ruler if you have one. Otherwise copy the pica ruler in the back of the book.

A For each line of type below, do the following.
   a. Identify the type as roman, *italic*, or **bold**.
   b. Measure the width of the line of type to the nearest half pica.

1. **Career Mathematics**

2. Graphic designers are familiar with many typefaces.

3. This is an outlined roman face.

4. *Photocomposition is at least 30 years old.*

5. **Metal type is now used mostly for fine-quality work.**

6. A pica em is the square of the body of a 12-point piece of type only.

What is the width of each line of type below to the nearest half pica? Convert your measurement to the nearest tenth of an inch.

7. **A good graphic designer is mathematically precise.**

8. Many specialized tools are used in a design studio.

9. Type can be made from metal or wood.

10. Page numbers are called folios.

Use a calculator where possible. (One pica is 12 points.)

11. Change each measurement to picas.
    a. 48 points   **b.** 96 points   **c.** 108 points   **d.** 72 points

12. Change each measurement to inches. (One pica is about $\frac{1}{6}$ in.)
    a. 18 picas   **b.** 36 picas   **c.** 23 picas   **d.** 48 picas

13. Change each measurement to picas.
    a. 5 in.   **b.** 2 in.   **c.** 8 in.   **d.** 3 in.

B 14. What is the width and depth of each block of type below to the nearest half pica?

   a. Remember the pica system is used horizontally and vertically.

   b. **Type sizes can range in size from six points to seventy-two points.**

Measure the width of each line of type in questions 15 and 16 to the nearest half pica with a pica ruler or a copy of the pica ruler in the back of the book. Then identify the number of ems of the given type size.

15. 10-point type

A good graphic designer has a feeling for balance.

16. 18-point type

Which typeface is this?

17. How many ems of 7-point type are there in a line 217 points wide?

18. How many ems of 8-point type are there in a line 24 picas wide?

19. How many points of width are used by 15 ems of 14-point type?

Measure the width and depth of each block of type below to the nearest half pica. How many ems of the given type size are there in each?

20. 24-point type

It is usually best for preschool children to read from a book that is set in 24-point type.

21. 10-point type

Usually a good choice of type size for a book designed for adult general reading is 10 point.

22. A graphic designer finds that a block of type is 4.5 in. wide and 6.5 in. deep. What is the size of the block of type in picas?

23. A 244-page book has 44 lines to each type page. How many lines of type are in the book?

C **24.** Copy that is 36 picas wide is centered on a page that is 8.5 in. wide. How wide are the margins on either side of the copy in picas? in inches?

25. The trim size width of each whole page of *Career Mathematics* is $7\frac{1}{4}$ in. Measure the depth of each whole page to the nearest eighth of an inch to complete the trim size of *Career Mathematics*.

Trim Size: $7\frac{1}{4}$ in. by ___?___ in.

26. **a.** Give the width of the line of type below to the nearest half pica.
b. What typeface is used?

This typeface is common to this book.

8-2 Sizing the Illustrations

Some of the illustrations and photographs in this book had to be reduced or enlarged so they would fit in the space allowed for them in the layout.

The most accurate way to size illustrations for reduction or enlargement is by using the following formula and a calculator. Measurements of illustrations are generally made in picas.

$$\text{Percent of reduction (or enlargement)} = \frac{\text{reduced (or enlarged) size}}{\text{original size}} \times 100$$

Example

Suppose the photograph at the right is to be used in the rectangular space at the top of this page. What percent enlargement of the photograph would be required?

Skill(s) __7, 8__

Solution

Step 1: Measure the width and depth of the photograph and of the rectangular space above in picas.

The photograph is 8 picas wide and 6 picas deep.
The rectangular space is 12 picas wide and 9 picas deep.

Step 2: Use the percent formula and a calculator to solve the problem.

$$\text{Percent of enlargement} = \frac{\text{enlargement width}}{\text{original width}} \times 100$$

$$= \frac{12 \text{ (picas)}}{8 \text{ (picas)}} \times 100$$

`1` `2` `÷` `8` `×` `1` `0` `0` `=` 150

The width is enlarged to 150%.

Step 3: Check the 150% enlargement of the original depth, 6 picas.

`1` `.` `5` `×` `6` `=` 9   The depth is enlarged to 9 picas.

EXERCISES

Use a calculator where possible.

A Find the percent of reduction (or enlargement) for each to the nearest whole percent.

1. 60 picas to 48 picas   **2.** 50 picas to 58 picas   **3.** 28 picas to 48 picas

4. 24 picas to 10 picas   **5.** 14 picas to 3.5 picas   **6.** 10 picas to 36 picas

For questions 7 to 14, round to the nearest half pica where needed.
Find the reduced (or enlarged) dimension for each.

7. a 54-pica side enlarged to 112%   **8.** a 20-pica side reduced to 25%

9. a 36-pica side reduced to 60%   **10.** a 15-pica side enlarged to 125%

B 11. A photograph that is 12 picas wide and 24 picas deep must be reduced to fit a space that is 6 picas wide. How many picas will the depth measure after the reduction?

12. Suppose each of the two photographs below are to fit in a space the size of the given rectangular space. What percent of reduction or enlargement would be required for each?

a.

b.

13. A photograph measuring 28 picas wide by 36 picas deep must be enlarged to fit a space that is 40 picas wide. What will the enlarged depth of the photograph be?

14. A 36-pica by 48-pica photograph has been reduced to 25% of its original size. What are the dimensions of the reduced size?

C 15. Suppose the photograph in question 12a is required in the rectangular space at the top of the previous page. What percent of reduction or enlargement would be required if some of the depth of the photograph could be removed?

8-3 Costing the Project

A preliminary costing for the publication of a book can be done once its extent (number of pages), type of paper, print quantity, and various other specifications have been decided.

The costing can generally be divided into two main categories. The first is the *non-recurring costs*, which include those expenses that are paid at the first printing but will *not* have to be paid again should the book be reprinted. The second category, *recurring costs*, must be paid again if the book is reprinted.

Below is an example of a preliminary costing sheet for producing 5000 copies of a 304-page book. Note that the preliminary costing is an *estimate* of the total cost of its publication.

Preliminary Costing Sheet

Title: The Diet Book

Extent: 304 pages

Print Quantity: 5000 books

Non-recurring Costs		Recurring Costs	
Editorial work	$6000	Paper	$3500
Graphic design	$2500	Printing	$2460
Typesetting @ $12/page	$3648	Book cover	$1415
Hand-drawn illustrations	$1500	Book binding	$6490
Assembly by graphic designer			
@ $10/page	$3040	**Total recurring costs:**	$13,865
Film of pages	$2432		
Printing plates made from film	$3040		
		Run-on cost per thousand books:	
Total non-recurring costs:	$22,160		$2271

The *run-on cost* is set by the printer and allows the publisher to produce print quantities that are smaller or larger than the quantity originally specified. A run-on cost is part of the recurring costs as it applies only to the cost of paper, printing, the book cover, and binding. This cost is usually given per thousand copies.

The following example illustrates how this preliminary costing sheet can be used to find the total cost per book to the publisher.

Example

A publisher decides that 6000 books can be sold instead of the 5000 print quantity originally requested. What is the total cost per book to the publisher of producing 6000 books based on the costs in the preliminary costing sheet on the previous page? Round your answer to the nearest cent.

Solution

Step 1: Find the non-recurring cost per book.

`2 2 1 6 0 ÷ 6 0 0 0 =` 3.6933333

The non-recurring cost per book is $3.69.

Step 2: Find the new recurring cost per book. To the recurring cost of the original 5000 books, add the run-on cost for the 1000 books above the original print quantity. Then divide the sum by the new print quantity, 6000 books.

`1 3 8 6 5 + 2 2 7 1 = ÷ 6 0 0 0 =` 2.6893333

The recurring cost per book is $2.69.

Step 3: Find the total cost per book.

`3 · 6 9 + 2 · 6 9 =` 6.38

The total cost per book to the publisher is $6.38.

EXERCISES

Refer to the preliminary costing sheet on the previous page. Use a calculator where possible. Round your answers to the nearest cent.

A **1.** What would the typesetting cost be if the per-page rate were increased to $16?

2. What is the non-recurring cost per book if the given number of books is produced?

    a. 5000 books   **b.** 4000 books   **c.** 10,000 books

3. What is the run-on cost per book?

4. What is the run-on cost if the print quantity is increased from the original quantity by the given amount?

    a. 1000 books   **b.** 2000 books   **c.** 3000 books

B **5.** What is the total cost per book to the publisher if the original print quantity were changed to the given number of books?

    a. 7500 books   **b.** 20,000 books

Find the unknown amounts in the preliminary costing sheet below to answer the questions. Round your answers to the nearest cent.

Preliminary Costing Sheet

Title: Keeping Fit

Extent: 256 pages

Print Quantity: 12,000 books

Non-recurring Costs		Recurring Costs	
Editorial work	$7000	Paper	$ 9 600
Graphic design	$3200	Printing	$ 7 200
Typesetting @ $12/page	?	Book cover	$ 5 400
Hand-drawn illustrations	$6000	Book binding	$14,160
Assembly by graphic designer @ $10/page	?	**Total recurring costs:**	?
Film of pages	$2040		
Printing plates made from film	$2560		
		Run-on cost per thousand books:	$2000
Total non-recurring costs:	?		

6. What is the total cost per book for a print quantity of 12,000 books?

7. If the editorial cost decreased from $7000 to $5500, what would be the total cost per book to the publisher?

8. What is the total cost per book to the publisher if the original print quantity were changed to the given number of books?

   a. 15,000 books   **b.** 6000 books

9. How much would it cost to reprint the given number of books?

   a. 12,000 books   **b.** 8000 books

Tricks of the Trade

Desktop Publishing

Publishing today is within easy reach of anyone with the aspiration to express themselves in print. Readily available software on commonplace micro-computers along with laser printers have created a new and exciting phenomenon called desktop publishing.

The ability to produce flyers, newsletters, graphics, magazines, newspapers, and books of varied size and style is surprisingly easy to attain. It is also much more economical to publish at your own desktop than to use the conventional typesetter and printer.

Self-Analysis Test

Use a pica ruler for questions 1 and 2 if you have one. Otherwise copy the pica ruler in the back of the book.

1. Measure the width of the line of type below to the nearest half pica.

This is 14-point type.

2. Measure the width and depth of the block of type below to the nearest half pica. How many ems of 14-point type are there in the block of type?

Reference books are often set in double columns in order to get more material on a page. This reduces the number of pages required and holds down the overall production cost.

3. What is the reduced or enlarged photograph dimension for each? Round to the nearest half pica where needed.

    a. a 48-pica side reduced to 75%   **b.** a 54-pica side enlarged to 120%

4. A photograph measuring 8 in. by 10 in. must be reduced to fit an area measuring 24 picas by 30 picas.

    a. What percent reduction is required?   **b.** What will the final depth be in inches?

Find the unknown amounts in the preliminary costing sheet for *The Blue Book* below to answer the questions. Round your answers to the nearest cent.

Preliminary Costing Sheet				
Title: The Blue Book **Extent:** 64 pages **Print Quantity:** 2000 books				
Non-recurring Costs		**Recurring Costs**		
Editorial work	$1200	Paper		$2500
Graphic design	$ 800	Printing		$1800
Typesetting @ $16/page	?	Binding		$4000
Assembly @ $10/page	?			
Film of pages	$ 512	**Total recurring costs:**		?
Printing plates	$1640			
		Run-on cost per		
Total non-recurring costs:	?	**thousand books:**		$900

5. What is the total cost to the publisher for producing 2000 books? 4000 books?

6. If there is an increase of $320 in the cost of paper from the amount listed above, what would the new total cost per book be for 2000 books?

Printing a Book

8-4 Choosing the Printing Method

There are four main methods used in printing today.

The method of **offset lithography** was chosen for the production of *Career Mathematics*. This process is planographic, meaning the surface of the printing plate is *flat*. Lithography is based upon the fact that grease (ink) and water do not mix. The whole surface has ink and water applied to it. The printing area is chemically treated to accept ink and reject water. Background areas that are not to be printed are chemically treated to accept water and reject ink. Offset lithography is commonly used for books, magazines, brochures, and many other types of publications.

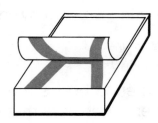

Letterpress is a *relief* method where the image to be printed is *raised* on the printing plate above the background, as in a rubber stamp. This process was the most commonly used form of printing up to the mid-1960s. Letterpress is still used today by some newspapers. This process is also used today for better-quality publications, such as limited-edition books and fine prints.

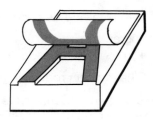

In **gravure** printing, the image to be printed is *etched* into the printing plate. The plate is then coated with ink and wiped off. The paper is then pressed against the printing plate by a rubber-coated roller. This causes the paper to be forced into the recesses of the plate so that it picks up the ink to form the image. The gravure method of printing is best used for long runs as the initial start-up costs for making the plates are high. Some examples of gravure printing are fine-art prints, postage stamps, and photographic books.

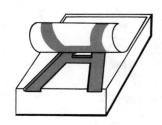

The **silk screen** method is done with a stencil. This highly adaptable method of printing can be used on a variety of surfaces such as wood, plastic, metal, glass, and even electronic circuits.

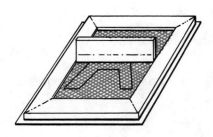

Each of the four printing processes described on the previous page requires paper to be fed into the printing press in either of two ways: sheet fed or web fed.

In **sheet-fed** printing, the paper is cut into sheets of the required size *before* being printed. Usually, several pages of a book are printed on each side of the large sheets of paper that are fed into a sheet-fed press. The picture at the right shows paper being inserted into a small sheet-fed printing press. The press is fed with pre-cut sheets of paper. *Career Mathematics* was printed on a sheet-fed press. A sheet-fed press can run between 4000 and 12,000 sheets per hour.

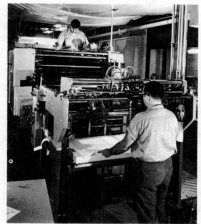

In **web-fed** printing, a large roll of paper (web) is unwound as it passes through the press and is folded at the other end. Printing is done on both sides of the continuous roll of paper. The web is cut into sheets of the required size *after* the printing. A web-fed press is shown at the right. Such a press runs between 15,000 and 50,000 *impressions* per hour. Because web presses run from a continuous roll, the expression *sheets per hour* is replaced by the expression *impressions per hour.*

Example

Skill(s) ___3,4___

A printer runs two sheet-fed presses for three 8 h shifts for five days. The two presses together run a total of 22,000 sheets per hour. How many sheets would both presses print in five days?

Solution

Step 1: Find the number of hours one press is operating in five days.

3 × 8 × 5 = 120

One press is in operation for 120 h in five days.

Step 2: Find the number of sheets two presses can print in 120 h.

1 2 0 × 2 2 0 0 0 = 2640000

Both presses would print 2,640,000 sheets in five days.

EXERCISES

Use a calculator where possible.

A Identify the printing method that is illustrated.

1.

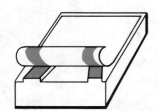

2.

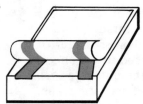

3.

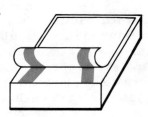

The diagrams below illustrate two kinds of printing presses and also show the usual sequence for four-color printing, each unit being inked with cyan (blue), magenta (red), yellow, and black. Identify the kind of printing press shown.

4.

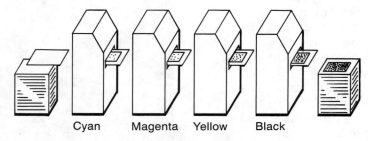

5.

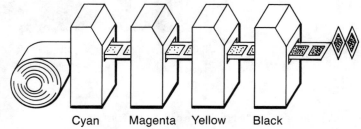

B 6. A printer needs four 6 h shifts to run a book on a press that prints 10,000 sheets per hour. How many sheets will have been printed at the end of the four shifts?

7. A sheet-fed gravure press prints 10,500 sheets per hour and a sheet-fed offset press prints 6000 sheets per hour. If both presses run for 8 h, how many more sheets will be printed by the sheet-fed gravure press?

8. If a sheet-fed press can print 9500 sheets for a brochure in 1 h, how long will it take to print 137,500 sheets?

9. A set of four printing plates for a sheet-fed press costs $240 and the paper costs 36¢ per sheet. What is the cost of the plates and paper for the printing of 35,000 sheets for a pamphlet?

10. A printer charges $100 for two gravure printing plates to be used in producing an advertising brochure. The printer charges 75¢ per sheet for the paper and printing for the first 5000 sheets and 60¢ per sheet for every subsequent thousand sheets. What will the printing cost per copy of 7500 copies of the advertising brochure be?

11. To make an eight-page pamphlet, you need one letterpress printing plate for each page. If each plate costs $80 and the printing and paper costs 18¢ per page, what is the total cost of making 12,000 copies of the brochure?

12. A silk screen press runs 3000 sheets per hour. How many sheets can be printed in three-fourths of an hour?

13. A printer is running two eight-hour shifts per day. How many sheets can be printed by the end of a five-day week on a press that runs 4000 sheets per hour?

14. One hundred fifty thousand copies of a catalog cost 15¢ per copy for the paper and printing.

    a. What is the total cost?
    b. How long will it take a printer to run the job on two web-fed presses that together run 45,000 catalogs per hour?

C 15. A printer has two sheet-fed presses that together run 8000 library fliers per hour and two web-fed presses that together run 40,000 brochures per hour. At the end of three six-hour shifts, how many fliers and brochures will have been printed?

Technology Update

In order to prepare all color artwork, whether it be an illustration or a slide, for four-color printing, it has to be separated into the four basic process colors (yellow, magenta, cyan, and black). This separation process results in four pieces of film, one for each color. In turn, an individual printing plate is made for each color.

Most color separations are now done using electronic scanners. These scanners operate at very high speeds and scan four colors on the original at once, using a high-intensity laser beam. Color filters and a computer built into the scanner convert the signals picked up by the laser beam into film for each of the four colors.

A digital color laser scanner

8-5 Paper

Whenever a book or magazine is printed, several pages are printed on *each* side of a large sheet of paper. This large sheet is then folded and cut to form a **signature**, which makes up one section of the publication. If you examine the binding of *Career Mathematics*, you will be able to see and count its signatures. *Career Mathematics* is made up of 32-page signatures. In other words, 16 pages were printed on one side of a large sheet of paper and 16 pages on the other side. Signatures can have 4, 6, 8, 12, 16, 24, 32, 48, and 64 pages.

Once a printer decides how many pages can be printed on each side of one large sheet, the number of large sheets of paper needed to meet a publisher's print quantity can be calculated. The examples below show how the following two formulas involving quantities of paper are useful.

Number of sheets of large paper formula:

$$\text{Number of sheets of large paper} = \frac{\text{Print quantity} \times \text{Number of pages in publication}}{2 \times \text{Number of pages printed on each side of a large sheet}}$$

Print quantity formula:

$$\text{Print quantity} = \frac{\text{Number of sheets of large paper} \times 2 \times \text{Number of pages printed on each side of a large sheet}}{\text{Number of pages in publication}}$$

Example

Skill(s) *1, 6, 7*

For a 48-page brochure, a printer is printing eight pages on each side of a large sheet of paper. How many sheets are required to print 5000 copies?

Solution

Substitute known values into the number of sheets of large paper formula.

$$\text{Number of sheets of large paper} = \frac{\text{Print quantity} \times \text{Number of pages in publication}}{2 \times \text{Number of pages printed on each side of a large sheet}}$$

$$= \frac{5000 \times 48}{2 \times 8}$$

`5 0 0 0 × 4 8 =` 240000   `2 × 8 =` 16

`2 4 0 0 0 0 ÷ 1 6 =` 15000

For the 48-page brochure, 15,000 sheets of paper are required.

EXERCISES

Use a calculator where possible.

A A ream consists of 500 sheets of paper.

1. How many reams of paper are there in a delivery of 50,000 sheets?

2. How many sheets of paper are in a supply of 12,000 reams?

3. If paper costs $175 per ream, what is the cost of 15 reams?

B 4. Fifteen thousand copies of an 864-page book are to be printed. If 48 pages of the book are printed on each side of a sheet of paper, how many sheets of paper will the printer need?

5. A printer has 22 reams of paper to print 4000 copies of a book that is 96 pages long. Is there enough paper if 16 pages are to be printed on each side of a sheet of paper?

6. A printer is printing 32 pages on each side of a sheet of paper for a 576-page book. If the publisher has ordered 20,000 copies of the book, how many sheets of paper will the printer need?

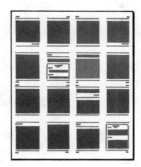

7. A publisher needs 50,000 copies of a 48-page pamphlet.

   a. How many sheets of paper would be required if each of the following number of pages were printed on each side of a single sheet of large paper?

      i. 4 pages   **ii.** 8 pages   **iii.** 12 pages   **iv.** 24 pages

   b. Which number of pages printed on each side of a single sheet of paper uses the least number of sheets?

   c. If the paper required costs $225 per ream, what is the savings between printing four pages and printing 24 pages on each side of a large sheet?

C 8. A printer has 8000 sheets of large paper to print 2000 copies of a book that is 192 pages long. If 32 pages of the book are printed on each side of a large sheet, will the printer have enough paper to print the 2000 copies?

9. A printer has 360 reams of paper and must print 45,000 copies of a 64-page brochure. How many pages would have to be printed on each side of a sheet of paper in order to obtain the 45,000 copies of the brochure?

8-6 Finishing and Binding

Once a publication is printed, there are many subsequent steps involved in creating the final product. These include folding and gathering the paper into signatures, binding the signatures of the book by sewing or gluing, trimming the pages to their final size, and attaching the cover.

When printing a signature, it is important that half of the pages be printed on one side of the sheet and the other half on the other side in a certain arrangement called an **imposition**. This is done to ensure that after the large sheet of paper is folded and cut, the pages in a signature will be in numerical order. For the finishing of *Career Mathematics*, thirteen 32-page signatures were used. The signatures were then sewn and casebound with a laminated case.

Example 1

Suppose the printing of a 32-page pamphlet requires that four large sheets of paper are to be folded into four 8-page signatures. One way to fold an 8-page signature is shown at the right. What would the imposition be for an 8-page signature using this folding pattern?

Skill(s) __3, 4__

Solution

Fold a sheet of paper into one signature as shown above, numbering the pages in order from page 1 to page 8. Open up the folded sheet of paper to reveal the imposition on each side of the sheet of paper.

Front side of imposition:

Back side of imposition:

Example 2

For the production of a 608-page book, 16 pages are being printed in imposition on each side of a large sheet of paper. How many signatures will have to be folded and gathered per book?

Solution

Since 16 pages are being printed on each side of a large sheet of paper, the book has 32-page signatures. Divide the 608 pages by 32 to find the number of signatures in the book.

6 0 8 ÷ 3 2 = 19

There will be 19 signatures per book.

Example 3

A printer calculates bindery costs based on the number of signatures that have to be folded, gathered, and sewn before the book is covered.

Suppose a 256-page book has eight 32-page signatures to be folded, gathered, and sewn. If a printer charges 8¢ per book for folding and gathering each signature and 12¢ per book for sewing, what will the cost be to fold, gather, and sew 2000 copies of the book?

Signatures coming off folding machines at the end of a web-fed press.

Solution

Step 1: Find the cost for folding, gathering, and sewing per book. Multiply the number of signatures, 8, by 8¢ for the folding and gathering cost. Then add in the sewing cost.

$$\boxed{8}\ \boxed{\times}\ \boxed{\cdot}\ \boxed{0}\ \boxed{8}\ \boxed{+}\ \boxed{\cdot}\ \boxed{1}\ \boxed{2}\ \boxed{=}\quad 0.76$$

It costs 76¢ per book for folding, gathering, and sewing.

Step 2: Find the cost of folding, gathering, and sewing 2000 books.

$$\boxed{\cdot}\ \boxed{7}\ \boxed{6}\ \boxed{\times}\ \boxed{2}\ \boxed{0}\ \boxed{0}\ \boxed{0}\ \boxed{=}\quad 1520$$

The cost of folding, gathering, and sewing 2000 books is $1520.

EXERCISES

A Fold a sheet of paper into each kind of 4-page signature as shown below. Make a sketch of the front and back side of the imposition for each.

1.

2.

Sketch the front and back side of the imposition for the given signature.

3. 6-page signature

4. 8-page signature

B **5.** A printer is printing 32 pages of a book on *each* side of a large sheet of paper.
   a. How many pages are in each signature of the book?
   b. How many signatures are there in the book if there are 1216 pages?

6. There are fourteen 32-page signatures in a travel book. How many pages are there?

7. A folding machine can fold 7000 thirty-two-page signatures in an hour. How many signatures will be folded after 8 h?

8. A printer will print eight pages on *each* side of a sheet for a 288-page book. How many signatures are in the book?

9. The bindery cost for the production of a book is 15¢ per book for a print quantity of 800 books. What is the total bindery cost?

10. The finishing of 15,000 copies of a travel brochure requires three folds per brochure. The cost per fold is 1¢.
   a. How much does it cost to fold each brochure?
   b. What is the total cost for folding 15,000 brochures?

C **11.** A printer prepares to fold and gather twenty-four 48-page signatures that make up one book. The cost for folding and gathering is 8¢ per signature.
   a. How many pages are in the book?
   b. What is the total cost for folding and gathering 5000 books?
   c. Would the folding and gathering cost be less if the book were made up of 64-page signatures? By how much?

Check Your Skills

Round each number to the nearest hundredth.   (Skill 1)
   1. 1.0365   **2.** 7.99493   **3.** 12.9244   **4.** 11.99999

Find the product.   (Skill 3)
   5. 23 × 42   **6.** 19 × 27   **7.** (2380)(436)

Divide. Round to the nearest hundredth if necessary.   (Skill 4)
   8. 529 ÷ 16   **9.** $9\overline{)347}$ **10.** $\frac{223}{12}$

   11. $23,150 ÷ 5000   **12.** $95,840 ÷ 2000   **13.** $10,300 ÷ 2500

Find the sum or difference.   (Skill 6)
   14. 8.35 − 2.43   **15.** 9.435 + 6.266   **16.** 9.008 − 1.009

Find the product.   (Skill 7)
   17. 93($3.62)   **18.** 4.653 × 0.157   **19.** 9.62(8.4356)

Divide. Round to the nearest tenth.   (Skill 8)
   20. 405.576 ÷ 43   **21.** 84.6 ÷ 23   **22.** 237,854 ÷ 19,000

Self-Analysis Test

Use a calculator where possible.

1. A sheet-fed press can print 4000 sheets per hour. How long will it take to print 16,000 copies?

2. If paper costs 15¢ per page and one offset printing plate costs $15, how much will it cost to print 3500 copies of a 12-page brochure to the nearest dollar? (Allow one plate for each page.)

3. A silk screen press runs at 4000 sheets per hour. How many sheets can be made in three-fourths of an hour?

4. A printer owns two web-fed presses. One runs 35,000 impressions per hour. The other runs 50,000 per hour. How many impressions can both presses run in 6 h?

5. A ream of paper costs $85. What would it cost to order 82 reams? 150 reams?

6. Two hundred fifty thousand sheets of magazine paper are delivered to a printer.

    a. How many reams of paper are in the delivery?
    b. If the paper costs $200 per ream, what is the total cost of the delivery?

7. For the printing of a catalog, four pages are printed on each side of one large sheet of paper. The catalog is 48 pages long and 12,500 copies are needed. How many sheets of paper are needed? How many reams of paper are needed?

8. How many large sheets of paper (printing on *each* side of a sheet) would be used in the production of a 224-page book made up of the given signature size?

    a. 8-page signature   **b.** 16-page signature

9. A folding machine can fold 8000 sheets in an hour. Each sheet has eight pages of a book printed on *each* of its sides. How many pages can be expected to be folded in 3 h?

10. Gathering machines can gather nine 16-page signatures in five minutes. How many signatures will be gathered in an hour?

11. A book has twenty-four 32-page signatures. The cost of folding and gathering the signatures is 8¢ per signature.

    a. How many pages are in the book?
    b. How much will it cost to fold and gather 65,000 copies of the book?
    c. If 64-page signatures were used, would a saving result in the folding and gathering cost?
    d. What would the saving be?

All forms of printed communication have been graphically designed. This means you will find the work of a graphic designer in every book or magazine you read, all advertising brochures you find in your mailbox, every large billboard advertising display, and all food labels and product packaging.

Graphic designers are employed in all areas of print communication, but they are most often found in the advertising and publishing industries. Large publishing houses and advertising agencies might employ an entire staff of graphic designers.

Job Description

What does a graphic designer do?

1. A graphic designer meets with a client to determine his or her specific needs and to establish an overall concept for the publication to be done.

2. A designer prepares a preliminary layout of the publication to be shown to the client for approval. This could involve writing the necessary text, sketching the illustrations, taking the required photographs, designing the typefaces, and drawing the preliminary layout that shows how the text, illustrations, and photographs fit together.

3. Once the preliminary layout is ready, the graphic designer meets again with the client for approval.

4. The designer then decides on the paper to be used and obtains price quotes from various printing sources.

5. Next, the designer prepares the publication for the printer by having the text typeset, reducing or enlarging photographs, making camera-ready illustrations, and assembling all text and illustrations on art boards.

6. A graphic designer might next check the "press proofs" of a publication from a printer for errors in the color and type.

7. Finally, a designer may have to arrange for the delivery of a printed publication to the client.

Qualifications

A good graphic designer has an artistic flair with a sense of color and design. It is important to be conscientious, adaptable, accepting of criticism, and attentive to detail. Graphic designers also need a good mathematical sense, especially for sizing photographs and costing a project.

Many community colleges offer programs in graphic design. Some secondary schools even offer work experience leading to employment or post-secondary education.

Advertising is a billion dollar industry that promotes products or services, to a large number of people. The United States is greatly involved in advertising and spends more money on it than any other country.

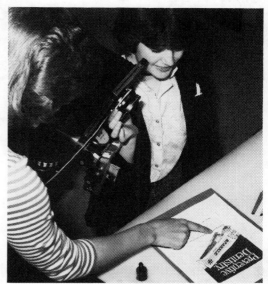

The quantity of sales for a company often depends on the amount of advertising it does. Advertising is the least expensive and most effective method for a company to inform a large number of people about products and services. Many businesses advertise to create a favorable image, and to make their name known. Retail businesses advertise locally to attract customers and increase sales. Manufacturers advertise to build a brand preference on a particular item. Their main goal is to have consumers buy a certain brand of merchandise at the stores where it is sold.

Advertising can be found in almost every aspect of our life. Consequently, it has a great effect on our lifestyle. Our likes and dislikes of food, clothing, products and appliances are influenced by advertising. Advertising also shows us how to obtain more leisure time, by using time-saving appliances. At the same time, it shows us what to do with the extra time we gain. In this way, the advertising industry shapes our tastes, habits, and attitudes.

---

1. In a recent year, advertisers in the United States spent approximately $88 billion on advertising. Find the amount spent on the following expenditures, if
  a. about 21% is spent on television advertising?
  b. about 27% is spent on newspaper advertising?
  c. about 7% is spent on radio advertising?
  d. about 3% is spent on business publications?

2. It costs approximately $150,000 for an advertiser to place a 30-second commercial on a television network. If 60 million viewers watch the commercial, how much does the advertiser pay for each person who sees the advertisement?

3. In a recent year, radio stations earned approximately $6 billion in advertising revenues. If 70% of that was from local advertisers, how much is that in billions of dollars?

4. In a recent year, about 45% of the total advertising expenditures for the year was spent on retail advertising, while 55% was spent on national advertising. If approximately $88 billion is the total expenditures for the year, how much of it was spent on retail advertising and how much on national advertising?

Chapter 8 Test

Use a calculator where possible.
What is the width of the line of type below to the nearest half pica? (8-1)

1. The gutter of a printed page is the space from the printing area to the binding.

2. Measure the width and depth of the block of type at the right to the nearest half pica. How many ems of 14-point type are there in the block of type? (8-1)

> Cover paper is a general term applied to a great variety of papers used for the outside covers of catalogs, brochures, booklets, and similar pieces.

3. A block of type is 6.0 in. wide and 6.5 in. deep. What is the size of the block of type in picas? (8-1)

4. A photograph 18 picas wide and 30 picas deep is to be enlarged to 150%. What is the enlarged width and depth in picas? (8-2)

5. A photograph that is 48 picas wide and 60 picas deep must be reduced to fit a space that is 24 picas wide.

    a. What percent reduction is required?
    b. What is the depth of the reduced photograph in picas? (8-2)

6. A photograph 36 picas wide and 24 picas deep is to be reduced to fit a space 4 in. wide.

    a. What percent reduction is required?
    b. What is the depth of the reduced photograph in inches? (8-2)

7. Refer to the preliminary costing sheet below. Round your answer to the nearest cent. (8-3)

Preliminary Costing Sheet

Title: Travel Brochure
Print Quantity: 10,000 copies

Non-recurring Costs		Recurring Costs	
Writing	$1000	Paper	$1600
Graphic design	$3200	Printing	$ 800
Typesetting @ $12/page	$ 800	Folding	$ 500
Assembly by graphic designer	$ 200		
Film of pages	$ 300	**Total recurring costs:**	?
Printing plates made from film	$ 180		
		Run-on cost per thousand copies:	$ 200
Total non-recurring costs:	?		

a. What is the total cost for publishing 10,000 copies of the brochure?

b. What is the total cost per brochure for 10,000 copies?

c. What would the total cost per brochure be if the original print quantity were changed to the given number of brochures?

    i. 16,000 copies   **ii.** 6000 copies

8. A printer can get 16 pages of a book on one printing plate.

    a. How many plates are needed for a 288-page publication?

    b. If each plate costs $80, what will the total plate cost for the book be? (8-4)

9. A printer has two web-fed presses, each running 50,000 bookmarks per hour. The same printer also has three sheet-fed presses, each running 4000 fliers per hour. How many bookmarks and fliers can be printed after 24 h of continuously running the presses? (8-4)

10. A printer runs two sheet-fed presses in two 6 h shifts for seven days. The two presses run a total of 24,000 sheets per hour. How many sheets will both presses print in seven days? (8-4)

11. a. How many reams of paper are there in a delivery of 65,000 sheets?

    b. If the paper costs $45 per ream, what is the total cost of the delivery? (8-5)

12. How many sheets are required to print 7000 copies of a 256-page book if the printer is printing 16 pages on each side of a sheet? (8-5)

13. A printer has 25,000 sheets of paper and is printing 12 pages of a 144-page book on each side of a large sheet of paper. How many copies of the book will be made? (8-5)

14. A book has eighteen 24-page signatures. How many pages are in the book? (8-6)

15. The twenty-two signatures of a book cost 12¢ per book to sew. If the printer is sewing the bindings of 3700 copies of the book, what is the total cost of sewing the book? (8-6)

16. An 800-page book is to be folded, gathered, and sewn into 32-page signatures. Fifteen hundred copies of the book are required. Folding and gathering costs 18¢ per signature and the sewing costs 21¢ per book.

    a. What is the total folding, gathering, and sewing cost?

    b. What would be the savings in this cost if the book were made up of twelve 64-page signatures and one 32-page signature? (8-6)

Chapter 9
Building a House

After completing this chapter you should be able to:

1. Read house floor plans.

2. Draw house floor plans on grid paper.

3. Determine quantities of materials needed for various house construction projects.

4. Estimate the costs involved in house construction.

Getting Started

9-1 Plans and Estimating Costs

Before you start building anything, you need a plan. It might be a simple sketch or a detailed blueprint. The plan for the ground-floor level of the house pictured on the previous page is shown below. Throughout this chapter, we will refer to this house as the *contemporary house*.

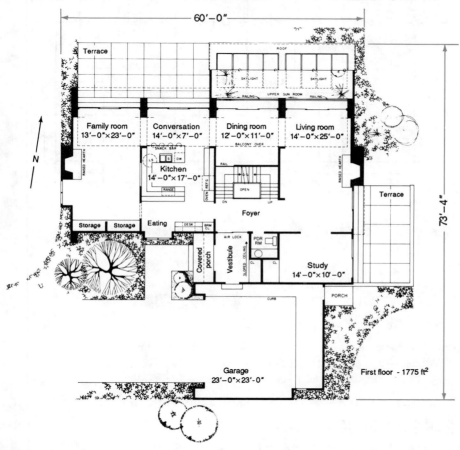

The passive solar design above offers 4200 ft^2 of living space situated on three levels. The primary passive solar energy element in the lower-level sun room admits sunlight for direct-gain heating. The solar warmth collected in the sun room radiates into the rest of the house after it passes the sliding glass doors.

The following example shows how the area of a house can be calculated from a house plan so that an estimate can be made of the cost to build the house.

Example

Below is a floor plan for a small cottage drawn on quarter-inch grid paper to a $\frac{1}{8}$ in. = 1 ft scale. Estimate the cost to build the cottage, to the nearest thousand dollars, if an architect assumes the building cost to be $52/ft^2.

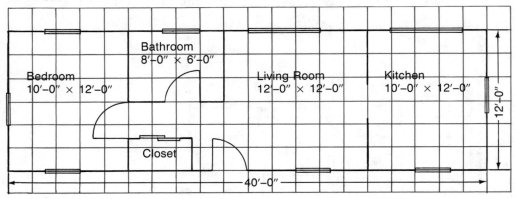

Solution

Step 1: Use the overall cottage dimensions to find its area in square feet.

$$\text{Area (ft}^2) = \text{length} \times \text{width}$$
$$= 40' - 0'' \text{ by } 12' - 0''$$
$$= 40.0 \text{ ft} \times 12.0 \text{ ft} \qquad \boxed{4}\ \boxed{0}\ \boxed{\times}\ \boxed{1}\ \boxed{2}\ \boxed{=}\ 480$$

The area is 480 ft^2. Each of the three digits is significant.

> **Note:** The area can also be found by counting squares on the grid. Since each square represents exactly 4 ft^2 and there are 120 squares, the area is 4×120, or 480 ft^2.

Step 2: Estimate the building cost, to the nearest thousand dollars.

$\boxed{4}\ \boxed{8}\ \boxed{0}\ \boxed{\times}\ \boxed{5}\ \boxed{2}\ \boxed{=}\ 24960$

The estimated cost of building the cottage is $25,000.

EXERCISES

Use a calculator where possible.

A **1.** Refer to the cottage floor plan above.

    a. Count squares to find the area of the kitchen floor.
    b. Count squares to find the area of the bathroom floor.

B 2. a. Make a drawing of the bedroom outlined at the right on quarter-inch grid paper. Use the scale $\frac{1}{4}$ in. = 1 ft. The bedroom is 11.0 ft long and 9.0 ft wide. The length of each window and doorway is 3.0 ft and each is located 2.0 ft from the closest corner.

b. Count squares to find the area of the bedroom floor.

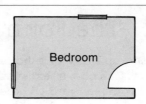

Bedroom

3. a. Make a drawing of the house and patio at the right on quarter-inch grid paper to the scale of $\frac{1}{8}$ in. = 1 ft. The patio is 24.0 ft by 10.0 ft and is 6.0 ft from the nearest corner of the house. The length of the house is 60.0 ft and the width of the house is 26.0 ft.

b. What would the cost of building the house and patio be if an estimate is $48/ft²?

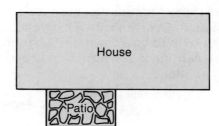

House

Patio

4. Below are estimated rates offered by two different contractors to construct the *contemporary house*. What is the estimate for building the house from each contractor, to the nearest thousand dollars, if there are 4217 ft² for all three levels of the house?

Contractor A	$52/ft²	Contractor B	$58/ft²

5. Below is a sketch of a two-bedroom house.

a. Make a drawing of the house on quarter-inch grid paper to the scale of $\frac{1}{8}$ in. = 1 ft. (The length of each single window and doorway is 3.0 ft. The length of each double window is 6.0 ft.)

b. What is the overall area of the house?

c. What is the cost of building the house, to the nearest thousand dollars, if an estimate is $58/ft²?

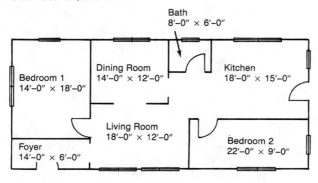

C 6. Design a one-story house on quarter-inch grid paper with 1600 ft² of floor space.

9-2 The Foundation

Before the building of a house can be started, excavation is usually necessary. The cost of excavating depends on the type of material that must be removed and the distance the material must be hauled for disposal. Excavation for a foundation, footing, or basement is usually calculated in cubic yards.

The type of foundation used for a house depends on its structure and location. In areas with a high risk of flooding or earthquakes, houses may be built on a concrete-slab foundation. By building wooden forms, you can more precisely square and level the concrete for the slab foundation.

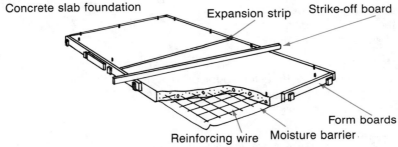

Concrete slab foundation   Expansion strip   Strike-off board

Form boards

Reinforcing wire   Moisture barrier

Otherwise, houses are often constructed with poured concrete foundation footings and basement walls of concrete blocks. Local building codes determine how deep the concrete footings must go; but in general, the footing trench should be dug on solid ground and reach below the frost line. The poured-concrete footing should be twice as wide as the foundation wall.

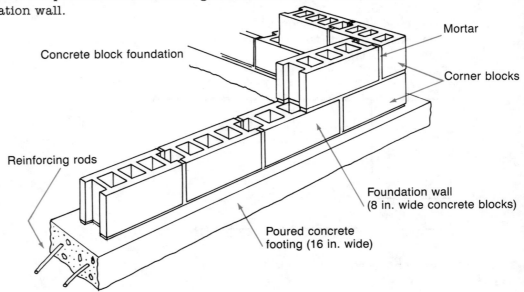

Mortar

Concrete block foundation

Corner blocks

Reinforcing rods

Foundation wall (8 in. wide concrete blocks)

Poured concrete footing (16 in. wide)

Example

Skill(s) __6,7,8__

Estimate the amount of concrete needed for the foundation footing sketched below, to the nearest whole cubic yard.

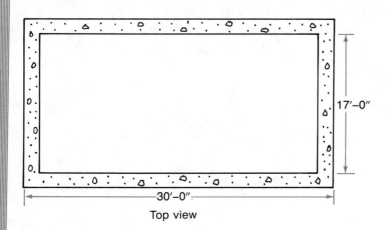

17'–0"

Top view

8"

1'–8"

Side view

30'–0"

Solution

Step 1: Write the dimensions in decimal feet.

Longer sides: 30' – 0" by 1' – 8" by 8" = 30.0 ft × 1.7 ft × 0.7 ft
Shorter sides: 17' – 0" by 1' – 8" by 8" = 17.0 ft × 1.7 ft × 0.7 ft

Step 2: Find the total volume of concrete needed in cubic feet.

Total volume (ft³) = Volume of longer sides + Volume of shorter sides
= 2(30.0 ft × 1.7 ft × 0.7 ft)
+ 2(17.0 ft × 1.7 ft × 0.7 ft)

| 2 | × | 3 | 0 | × | 1 | . | 7 | × | . | 7 | = | + | 2 | × | 1 | 7 | × |

| 1 | . | 7 | × | . | 7 | = | 111.86 * |

*For calculators with built-in
order of operations

The total volume of concrete is about 100 ft³, to one significant digit.

Step 3: Write the volume in cubic yards. Recall that 1 yd³ is equal to 27 ft³.

| 1 | 0 | 0 | ÷ | 2 | 7 | = | 3.7037037 |

About 4 yd³ of concrete are required for the foundation footing.

EXERCISES

Use a calculator where possible.

A Below are drawings of concrete-slab foundations for two small structures. In each case, the concrete slab is to be 3′ − 5″ thick.

a. What is the volume of concrete required, to the nearest whole cubic foot?
b. What is the volume of concrete required, to the nearest whole cubic yard?
c. If concrete costs $62/yd³, what is the estimated cost to the nearest hundred dollars?

1. a garage foundation

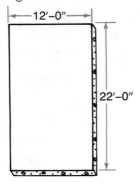

2. a storage shed foundation

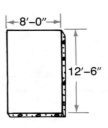

3. An excavation of 750 yd³ is to be done by an earthmover who charges $7/yd³. What will it cost to get the job done?

B **4. a.** The overall dimensions of a house are 64 ft by 35 ft. If the basement for the house requires a 9 ft excavation, about how many cubic feet of earth must be removed? (Round your answer to the nearest whole cubic foot.)
  b. If an earthmover charges $6.50/yd³, what will be the cost of the excavation, to the nearest hundred dollars?

5. a. About how much concrete, in whole cubic yards, is needed for the foundation footing drawn at the right? The footing is to be 18 in. wide and 9 in. thick.
  b. What is the total cost for the concrete, to the nearest hundred dollars, if the rate is $58/yd³?

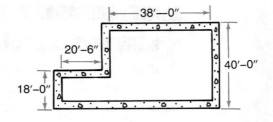

6. How many concrete blocks, 8 in. high by 16 in. long, are needed to build an 8 ft high by 18 ft long wall? Round your answer to the nearest hundred blocks.

7. The *contemporary house* plan requires the following materials for its foundation. Complete the table to find each total cost to the nearest cent.

	Quantity	Material	Unit Cost	Total Cost
	Poured Concrete Footing:			
a.	378 ft	concrete reinforcing rod	$0.18/ft	?
b.	756 ft³	concrete	$58/yd³	?
	Concrete Blocks:			
c.	2096	regular block	$1.39/block	?
d.	178	corner block	$1.55/block	?
	Foundation Drainage:			
e.	242 pieces	4 in. drain tile	24¢/ft	?
f.	15 yd³	pea gravel	$11.50/yd³	?
g.	**Total Cost:**			?

8. The Riveras wish to build a 2.40 m high foundation wall that is 7.60 m long out of 20.0 cm by 40.0 cm concrete blocks. If the blocks cost $1.30 each, how much will the foundation wall cost, to the nearest hundred dollars? Allow 5% for breakage.

9. Ready-mixed concrete for a house footing will be delivered to a construction site for $58/yd³. What will concrete cost for the 65 ft by 24 in. by 12 in. footing, to the nearest hundred dollars?

C 10. The foundation of a house requires 1240 concrete blocks. If concrete blocks cost $1.22 each, plus 50¢/block for installation, how much will it cost to build the foundation, to the nearest hundred dollars? Allow 5% extra in the number of blocks for breakage.

Tricks of the Trade

Board Feet

Lumber is sold either by the *lineal foot* (lin ft) or by the *board foot* (bd ft). Usually, the shorter lengths of wood are sold by the lineal or running foot (regardless of thickness and width), while lumber sold in volume is based on the board foot. A board foot is 1 in. thick, 1 ft wide, and 1 ft long. Board feet are calculated using the formula below from the nominal size of the lumber.

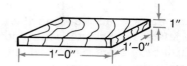

$$\frac{\text{thickness (in.)} \times \text{width (in.)} \times \text{length (ft)}}{12} = \text{board feet}$$

For example, in a 2 in. by 4 in. by 8 ft piece of wood, there are about 5 bd ft.

| 2 | × | 4 | × | 8 | = | ÷ | 1 | 2 | = | 5.3333333 |

9-3 Framing the House

Once the foundation of a house is made, the softwood framework can be started. A contractor orders wood for a house frame in **lineal feet**, or the running length of the wood. The table below shows some of the common dimensions and typical prices of framing lumber. The given dimensions represent the thickness and width of the lumber in inches. For example, 2 by 4 means 2 in. by 4 in.

Spruce, Pine, or Fir				
Size	2 by 4	2 by 6	2 by 8	2 by 10
Price per lineal foot	$0.35	$0.45	$0.68	$0.95
Common framework usage	studs	girders, joists, rafters		joists

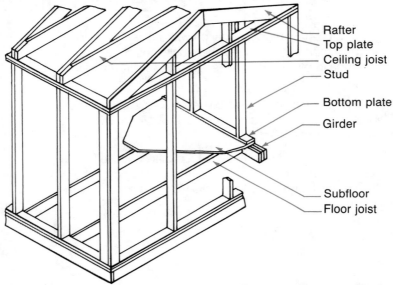

- Rafter
- Top plate
- Ceiling joist
- Stud
- Bottom plate
- Girder
- Subfloor
- Floor joist

There are many places in the construction of a house where pieces of lumber are spaced the same distance apart. This usually happens with floor joists, wall studs, ceiling joists, and roof rafters. In these cases, the first or end piece of wood is flush with the end of the building and the remaining pieces are usually placed 16 in. apart on center. *On center* (o.c.) means that the center of the piece of wood, say a joist, is directly over the pencil mark for the positioning of the joist. The joists at the two ends are flush with the corner supports. Often the corner supports consist of two joists or studs.

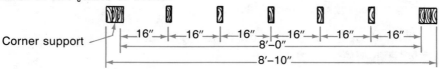

Corner support

Example

The plan for the *contemporary house* shows that ninety-eight 2 by 10s each 16 ft 0 in. long, are needed for floor joists. Estimate the cost of the wood, to the nearest hundred dollars, if spruce is used. (Use the prices given in the table on the previous page.)

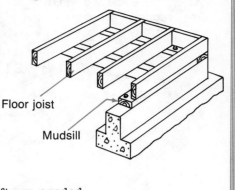

Floor joist

Mudsill

Solution

Step 1: Find the total number of lineal feet needed.

| 9 | 8 | × | 1 | 6 | = | 1568

To three significant digits, 1570 lin ft are needed.

Step 2: Estimate the cost, to the nearest hundred dollars.

| 1 | 5 | 7 | 0 | × | . | 9 | 5 | = | 1491.5

The estimated cost of the floor joists is $1500.

EXERCISES

Use a calculator where possible.

A 1. The exterior walls of the *contemporary house* require 273 studs made of 2 by 4s that are each 8 ft long. How many lineal feet is this?

2. Four hundred and ten studs made of 2 by 4s that are each 8.0 ft long are needed in the *contemporary house* for the interior partitions. Estimate the cost of these studs at 42¢/lin ft, to the nearest hundred dollars.

3. A total of 1900 lin ft of 2 by 4s are required as top and bottom plates in the exterior walls of the *contemporary house*. Would the cost of these plates exceed $300?

B 4. The exterior walls of the *contemporary house* require 3400 ft² of sheathing at 15¢/ft² and 3800 ft² of building felt at 3¢/ft². Estimate the total cost of these materials, to the nearest hundred dollars.

5. Each rafter in the framework of a house requires 22.0 ft of 2 by 6s. The building has 37 rafters on each side of the roof. Estimate the cost of all the required rafters, to the nearest ten dollars. (Use the price table at the beginning of this lesson.)

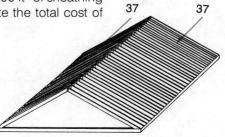

37   37

For questions 6 to 8, consider all board dimensions to be accurate to two significant digits.

6. The wall frame pictured below is made of 2 by 4s placed 16 in. apart o.c. In addition, there are two 2 by 4 studs placed side by side at each end as corner supports. What is the total length of the wall in feet and inches?

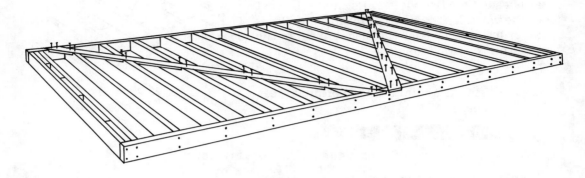

7. How many 2 by 4 studs are needed for the length of the wall frame sketched below? Note that there are two additional studs at each end for corner support.

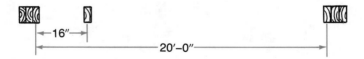

8. In areas where there is no snowfall, the local building codes may allow rafters to be placed 2 ft apart o.c. How many 2 by 6 rafters are needed for a 48 ft 2 in. length on each side of the roof?

9. The framework of the *contemporary house* requires 160 lb of common nails at 92¢/lb and 30 lb of galvanized box nails at $1.05/lb. Estimate the combined cost of both kinds of nails, to the nearest dollar.

C 10. The numbers and lengths of exterior wall studs needed for the *contemporary house* are shown in the table below. Find the missing values in the table. (Use the price table at the beginning of this lesson.)

	Quantity	Length	Lineal Feet	Total Cost
a.	Wall Studs: 273 pieces	2 in. by 4 in. by 8 ft	?	?
b.	83 pieces	2 in. by 4 in. by 10 ft	?	?
c.	Total Cost:			?

Self-Analysis Test

Use a calculator where possible.

1. A 24.0 ft by 40.0 ft cottage is planned to cost $45/ft². What is the estimated cost of construction, to the nearest hundred dollars?

2. Estimate the cost, to the nearest thousand dollars, of building the entire house shown below at $55/ft².

3. A trucker charges $45 to haul one load of earth from an excavation site to a landfill. What will it cost, to the nearest hundred dollars, to haul 3000 yd³ of earth to the landfill if the trucker hauls 10 yd³/load?

4. How many concrete blocks, 8.00 in. by 16.00 in., are needed to build a foundation wall 10 ft high and 40 ft long?

5. What is the cost, to the nearest ten dollars, for the concrete needed for a 55.0 ft by 16.0 ft by 8.0 in. concrete footing at $61/yd³?

6. For the rafters of the *contemporary house*, eighty-five boards 2 in. by 8 in. by 16 ft at 56¢/lin ft are needed. What is the cost for this wood, to the nearest hundred dollars?

Finishing the Job

9-4 Roofing

The roof of a house usually begins with a layer of roofing felt, which is heavy paper saturated with asphalt. The felt is laid down on top of the wooden sheathing that is nailed to the rafters. The roofing felt is laid horizontally across a roof with a 2 in. to 6 in. overlap. Roofing felt comes in rolls 3 ft wide and 144 ft long. So, one roll will cover 3 × 144, or 432 ft².

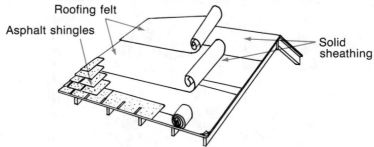

Asphalt shingles are a commonly used roofing material. These come in different qualities, judged mainly by their weight per *square* (the quantity needed to cover 100 ft²). Standard shingle weights run about 215 lb/square to 300 lb/square and are designed to last 15 to 20 years. Usually, three bundles of shingles will cover 100 ft², or one square.

Galvanized steel nails are also needed to attach the shingles. About 2 lb of nails should be allowed for 100 ft² of shingles.

Example

Estimate the number of *whole* squares of shingles needed to cover the gable roof sketched at the right.

Skill(s) 1, 4, 6, 7

Solution

Step 1: Find the total area of the roof, or 2(16.5 ft × 50.0 ft).

> 2 × 1 6 · 5 × 5 0 = 1650

The total roof area is 1650 ft², to three significant digits.

Step 2: Find the number of *whole* squares of shingles needed. Recall that one square covers 100 ft².

> 1 6 5 0 ÷ 1 0 0 = 16.5

About 17 *whole* squares of shingles are needed to cover the gable roof.

EXERCISES

Use a calculator where possible.

A Refer to each roof diagram below.

   a. How many *whole* squares of shingles are needed?
   b. How many *whole* rolls of roofing felt are needed?
   c. If two pounds of nails are needed for each square of shingles, how many *whole* pounds of nails are needed?

1.   **2.**   **3.**

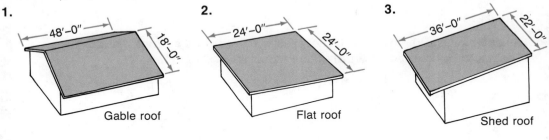

Gable roof   Flat roof   Shed roof

B For questions 4 to 7, round the cost to the nearest ten dollars.

   4. A roof requires 38 squares of shingles at \$26.95/square. Estimate the cost of the shingles.

   5. Estimate the cost of roofing materials for a 1680 ft^2 roof if shingles cost \$24.90/square, roofing felt costs \$0.04/ft^2, and nails are \$0.85/lb.

   6. The materials needed for the roof of the *contemporary house* are shown in the table below. What is the approximate total cost?

Quantity	Material	Unit Cost
5200 ft^2	#15 building felt	\$0.03
$\frac{2}{3}$ squares	asphalt ridge & hip shingles	\$26.95
40 lin ft	asphalt valley roll roofing	\$0.30
38 squares	asphalt self-sealing shingles	\$26.90
80 lb	roofing nails	\$0.90
388 lin ft	metal drip edge	\$0.19
192 pieces	5 in. by 7 in. metal step flashings	\$0.16
5 units	48 in. by 48 in. skylight	\$153.00

C **7.** Refer to the *hip roof* at the right. The height of each triangular and trapezoidal section is 20 ft 6 in.

   a. Estimate the cost of the shingles if they cost \$26.95/square.
   b. If roofing felt is \$12.96/roll and nails are \$0.89/lb, what is the cost of the roofing felt and nails?

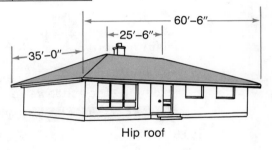

Hip roof

9-5 Wall Coverings

It is important to estimate of the amount of covering needed to finish an exterior or interior wall.

Example

Skill(s) __I,6,7__

The exterior wall below is to be covered with redwood siding. What is the area to be covered, excluding the windows and door?

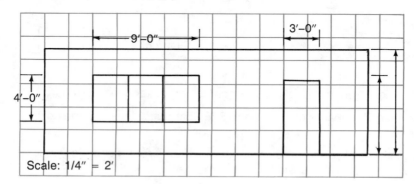

Scale: 1/4" = 2'

Solution

Step 1: Find the area of the total wall, 27' − 0" by 9' − 0", or 27.0 ft × 9.0 ft.

2 7 × 9 = 243

The area of the total wall is 240 ft², to two significant digits.

Step 2: Find the combined area of the windows and door.
Windows: 9' − 0" by 4' − 0", or 9.0 ft × 4.0 ft
Door: 3' − 0" by 6' − 6", or 3.0 ft × 6.5 ft

9 × 4 = + 3 × 6 . 5 = 55.5*

*For calculators with built-in order of operations.

The combined area of the windows and door is 56 ft², to two significant digits.

Step 3: Find the area of wall to be covered by the redwood siding, excluding the windows and door.

2 4 0 − 5 6 = 184

About 180 ft² of wall is to be covered, precise to the tens place.

EXERCISES

A **1.** If redwood siding costs $2.90/ft², what is the cost of covering the exterior wall sketched on the previous page, to the nearest ten dollars?

2. One gallon of exterior paint will cover 375 ft². How many gallons are needed for 2700 ft²?

B Estimate the number of *whole* bricks required for each exterior wall shown below. Allow 700 bricks for 100 ft².

3.

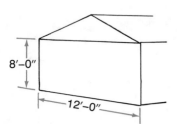

4.

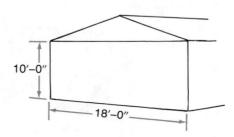

You are to wallpaper each room below. Assume each room has 104 ft² of windows and doors and that one roll of wallpaper covers about 36 ft².

a. How many rolls of *patterned* wallpaper are needed for each room? (Do not subtract for the windows and doors so the pattern will match.)

b. How many rolls of *plain* wallpaper are needed to cover the four walls of each room? (To be on the safe side, subtract half the area of the doors and windows.)

5.

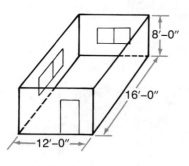

6.

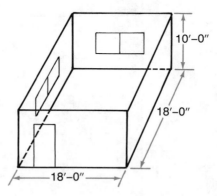

7. The *contemporary house* requires 4800 ft² of wooden siding. What is the total cost of this siding, to the nearest hundred dollars, if the wood costs $1.55/ft² and the cost of installation is $1.45/ft²?

For each wall sketched below, do the following.

a. Make a drawing of the wall on quarter-inch grid paper using the scale $\frac{1}{4}$ in. = 1 ft.

b. Find the amount of wall space in square feet, excluding the windows and doors.

c. What would be the cost of aluminum siding, to the nearest dollar, to cover the exterior of the wall at $1.85/ft^2 installed?

d. A roll of wallpaper covers about 36 ft^2. How many *whole* rolls of plain wallpaper would be required to cover the interior of the wall? (Subtract half of the area of the doors and windows.)

8.

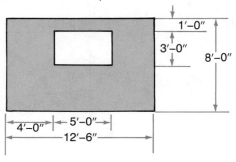

9.

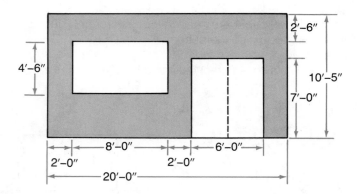

10.

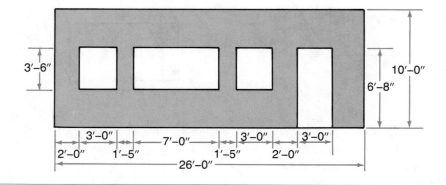

11. What would be the cost of the brick required to cover the exterior of the wall in question 8, to the nearest ten dollars, if each brick costs $0.45 and 700 bricks /100 ft² should be allowed for?

C 12. Using a 1 mm : 200 mm scale, find the number of square meters of wooden siding needed for the given parts of the *contemporary house* shown below. Adapt a 1 : 100 scale on a metric rule, as shown below, so that it can be used as a 1 : 200 scale. (1 mm = 200 mm). Use a triangular rule if you have one; otherwise, copy the 1 : 100 scale at the back of the book.

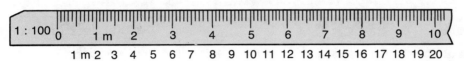

a. triangular panel A   **b.** trapezoidal panel B

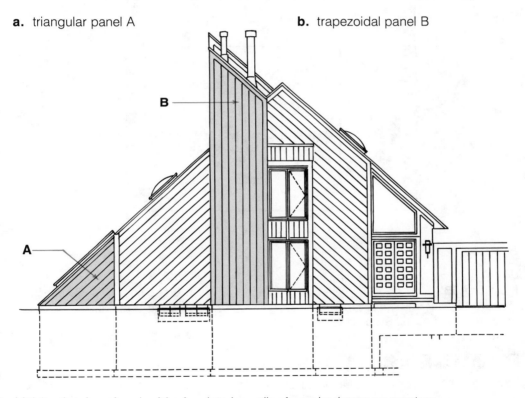

13. Make a drawing of each of the four interior walls of your bedroom on quarter-inch grid paper.

a. How many square feet are taken up by windows and doors?

b. How much wall space is there (excluding windows, doors, and closets)?

c. How many *whole* gallon cans of paint would cover the four walls, excluding the windows, doors, and closets? Assume that one gallon covers 460 ft².

9-6 Floor Coverings

A variety of materials, such as tile, wood, vinyl, or carpet, can be used to finish a floor. The example below illustrates how you can work from a floor plan to estimate the cost of covering a floor.

Example

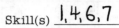

Skill(s) 1, 4, 6, 7

What would be the cost, to the nearest hundred dollars, to carpet the three rooms shown below, including closets and hallway, at $18/yd^2$?

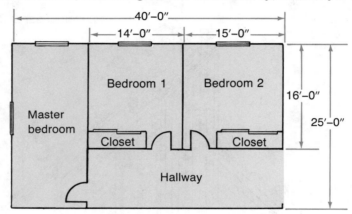

Solution

Step 1:
Multiply 25' − 0" by 40' − 0", or 25.0 ft × 40.0 ft, to find the area in square feet. Since 9 ft² = 1 yd², divide by 9 to find the area in square yards.

 111.11111

The area is 111 yd², to three significant digits.

Step 2:
Find the cost to the nearest hundred dollars.

[1] [1] [1] [×] [1] [8] [=] 1998

The cost is about $2000.

EXERCISES

A Refer to the diagram above.

1. What is the area of the master bedroom floor in square feet? in square yards?

2. What is the area of bedroom 1 in square yards?

B For each floor plan below, do the following.

  a. Make a drawing of the floor on quarter-inch grid paper using the scale $\frac{1}{4}$ in. = 1 ft.
  b. Find the total amount of floor space, to the nearest square yard.
  c. What would be the cost of carpeting the given floor space, to the nearest dollar, at \$15.95/yd²?

3. Master bedroom with closet

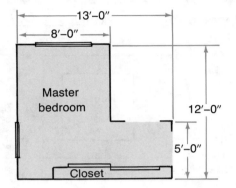

4. Living room without closet and foyer

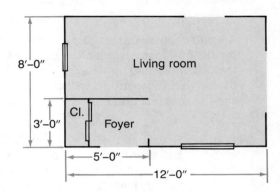

5. Dining room and breakfast nook without kitchen

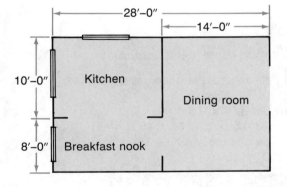

6. Living room and dining room without fireplace

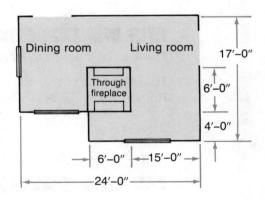

Refer to the floor plan of the first floor of the *contemporary house* on the second page of this chapter.

7. What would be the cost, to the nearest ten dollars, of the ceramic tiling for the kitchen floor at \$1.29/ft²?

8. What would be the cost, to the nearest hundred dollars, of heavy-duty carpeting for all of the family room, conversation room, dining room, living room, and study at \$18.25 /yd² for the carpet and \$2.50/yd² for the carpet padding?

9-7 Landscaping

The landscaping of a piece of property involves grading the earth, planting grass, trees, and shrubs, as well as putting in patios and driveways. A landscape architect studies all aspects of appearance, function, cost, and maintenance when making the landscape plans for a lot.

Example

A landscape architect has determined that it would be most practical to lay grass sod on the lot sketched at the right. If sod costs $1.50/yd², how much will the grass cost, to the nearest ten dollars?

Skill(s) 1, 4, 6, 7

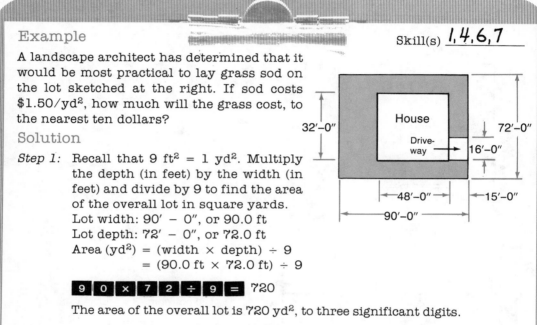

Solution

Step 1: Recall that 9 ft² = 1 yd². Multiply the depth (in feet) by the width (in feet) and divide by 9 to find the area of the overall lot in square yards.

Lot width: 90′ − 0″, or 90.0 ft
Lot depth: 72′ − 0″, or 72.0 ft
Area (yd²) = (width × depth) ÷ 9
$$= (90.0 \text{ ft} \times 72.0 \text{ ft}) \div 9$$

`9 0 × 7 2 ÷ 9 =` 720

The area of the overall lot is 720 yd², to three significant digits.

Step 2: Find the area of the house and driveway in square yards.

House: 48′ − 0″ by 16′ − 0″ + 32′ − 0″ is
48′ − 0″ by 48′ − 0″, or 48.0 ft × 48.0 ft
Driveway: 15′ − 0″ by 16′ − 0″, or 15.0 ft × 16.0 ft
Area (yd²) = (area of house + area of driveway) ÷ 9
$$= (48.0 \times 48.0 + 15.0 \times 16.0) \div 9$$

`4 8 × 4 8 = + 1 5 × 1 6 = ÷ 9 =` 282.66667 *

*For calculators with built-in order of operations.

The area of the house and driveway is 283 yd², to three significant digits.

Step 3: Find the area of the lot, *excluding* the house and driveway, and then find the cost of the grass sod at $1.50/yd², to the nearest ten dollars.

`7 2 0 − 2 8 3 = × 1 · 5 =` 655.5

The grass sod will cost about $660.

EXERCISES

Refer to the *contemporary house* plan in lesson 9-1. Use a calculator to solve the problems.

A **1.** Seven rosebushes will be planted along the terrace of the *contemporary house*. What will be the cost of these rosebushes at $4.75 each?

2. Topsoil will be delivered to the building site for $12.50/yd³. What will it cost to have 45 yd³ of topsoil delivered?

3. The lot for the *contemporary house* requires one birch tree for $45.00, two flowering crab trees for $39.50 each, and one silver maple for $36.75. What will be the total cost of these trees, to the nearest ten dollars?

4. A contractor needs to lay sod for a 50.0 ft by 60.0 ft area. If the sod costs $2.74/yd², how much will the contractor pay for the sod, to the nearest ten dollars?

5. Six inches of topsoil are needed for a 60.0 ft by 90.0 ft lot. At $13.50/yd³, what will the topsoil cost, to the nearest dollar?

6. Find the cost of sod for a 20.0 ft by 45.0 ft backyard at $1.50/yd², to the nearest ten dollars.

7. What is the cost of sod for a 50.0 ft by 60.0 ft lot if sod costs $1.58/yd² and the labor to lay it costs 45¢/yd²?

8. A landscaping service will hydro-seed a lawn for $1.25/yd² plus a $100 fertilizer fee or will sod it for $1.45/yd². How much less will it cost to hydro-seed a lawn that measures 75.0 ft by 90.0 ft? (Round your answer to the nearest dollar.)

B **9.** What is the total cost of the following list of trees and shrubs, to the nearest ten dollars?

Item	Quantity	Unit Price
mountain ash tree	1	$34.00/tree
juniper bush	5	$21.00/bush
rosebush	6	$2.75/bush
seedless ash tree	3	$14.50/tree

10. A tree-moving service will deliver trees at $35.00 each plus 25¢/mi. If five trees are needed and the round-trip distance is 14 mi, what will these trees cost, to the nearest ten dollars?

11. Marble chips cost $3.60/50.0 lb bag and bark chips cost $2.50/40.0 lb bag. A flower bed requires 350.0 lb of marble chips and 500.0 lb of bark chips. What will be the total cost, to the nearest ten dollars?

12. Large concrete tiles are to be laid on the side and back terraces of the *contemporary house*. What would be the cost of the concrete tiles, at $1.85/ft², for the two terraces, to the nearest ten dollars? Refer to the plan in lesson 9-1. The side terrace measures 20′ − 0″ by 11′ − 4″. The back terrace is 27′ − 4″ by 13′ − 4″.

Find the cost of sod for each lot diagram below, to the nearest ten dollars, if sod costs $1.75/yd².

13.

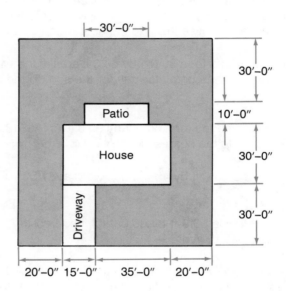

14.

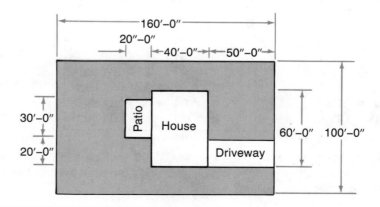

C **15.** Draw a plan for a 100 ft 0 in. by 120 ft 0 in. lot on grid paper. Then draw a house with a driveway and back patio.

   a. Find the cost, to the nearest dollar, of sod for the lot at $1.50/yd².

   b. Find the cost, to the nearest dollar, of a brick patio and driveway at $0.99/ft².

Technology Update

Modular House Construction

The demand for low-cost, flexible housing is being met by modular construction. Houses are put together from factory-assembled and finished modules. Each section of the house is made up of two rectangular modules side by side. All the plumbing, wiring, and duct work is done at the factory. The modules are shipped to the house site on trucks. Once assembled on the foundation, they are ready for use.

Since all the units are rectangles, a variety of room arrangements can be planned. The buyer can choose a suitable floor plan.

Modular housing is low in cost and is quick and easy to assemble. After the foundation is laid, a house can be ready in less than one month.

Check Your Skills

Round to the nearest hundredth. (Skill 1)

   1. 95.4635   **2.** 12.346   **3.** 243.09997   **4.** 356.0049

Round to the nearest whole number.

   5. 35.6001   **6.** 169.298   **7.** 438.9   **8.** 0.4

Find the quotient. (Skill 4)

   9. 25,000 ÷ 4   **10.** 4500 ÷ 18   **11.** 405 ÷ 27   **12.** 5510 ÷ 38

Find the sum or difference. (Skill 6)

13. 2.63 + 9.8453   **14.** 2.435 − 1.33   **15.** 9.12 + 0.887

16. 1000.001 − 4.25   **17.** 8.4352 + 5.7   **18.** 1.24 − 0.0456

19. 7500 − 189 + 56   **20.** 35 − (12.8 + 3.425)   **21.** 6 + (34.123 − 6.7)

Find the product or quotient. (Skill 7 and 8)

22. $54.50 × 620

23. $98.20 / 5

24. 1.69 / 13

25. (23.47)5   **26.** 432 ÷ 0.03   **27.** 11(0.00333)

Self-Analysis Test

1. How many *whole* squares of shingles are needed for a shed roof that is 20.0 ft long and 14.0 ft wide?

2. The roof of the *contemporary house* requires 38 squares of shingles at $24.95/square. What is the cost of these shingles, to the nearest ten dollars?

3. If 2 lb of nails are needed for each square of shingles, how many pounds of nails are needed for a shed roof that is 24 ft long and has 14 ft rafters?

4. The *contemporary house* requires 4800 ft² of siding. If redwood siding is used at 90¢/ft², what is the cost, to the nearest ten dollars?

5. A wall 28.0 ft long and 8.0 ft high is to be covered with vertical tongue-and-groove boards. The wall contains two windows and one door that occupy 64 ft² of space. What is the total area that is to be covered with tongue-and-groove boards?

6. How many gallon cans of paint should you order to paint the walls and ceilings of two 10.0 ft by 12.0 ft bedrooms, ignoring windows and doors? The walls are 8.0 ft high. Assume one gallon can covers 450 ft².

7. How many bricks should be ordered to cover the wall below? Allow 700 bricks per 100 ft².

8. How many square yards of carpeting should be ordered to cover the bedroom floor below without the closet?

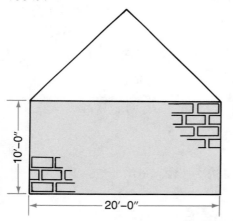

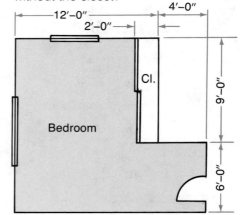

9. What will it cost, to the nearest ten dollars, to buy and install carpeting in the family room of the *contemporary house* at $23.50/yd² with an installation cost of $2.65/yd²? (Use the floor plan on the first page of lesson 9-1 for the room dimensions.)

10. Find the cost of sod for a 20.0 ft by 45.0 ft yard at $1.25/yd², to the nearest dollar.

11. How many cubic yards of topsoil should be ordered for a 50.0 ft by 75.0 ft lot if the desired depth is 0.40 ft?

There are many monthly expenses involved in owning a house. Some of these expenses are *fixed*, or do not change each month. Mortgage payments as well as property tax and house insurance payments are examples of fixed monthly expenses.

The costs of utilities, such as electricity, heat, water, and telephone, are examples of *variable* monthly expenses. These costs change each month.

An electronic spreadsheet, which is available in many different computer software programs, is a very useful device for monitoring all monthly house expenses. Some typical costs have been entered into the spreadsheet at the right.

Notice that only the important digits of the fixed and variable costs have been entered. No dollar signs and decimal points are included.

Formulas asking the electronic spreadsheet to compute totals have been entered in the appropriate cells. For example, in cell B7, "=Sum(B4:B6)" asks the computer to add the values from cells B4 to B6. In cell B16, "=B7+B14" asks the computer to add the values in cells B7 and B14.

The second spreadsheet at the right is the *result* of the computer's calculations on the monthly house expenses data entered.

Note that cells containing numerical values have been formatted as dollar values.

The results of the formula calculations are shown in the cells for totals. For example, the formula "=Sum(C4:C6)" has been entered in cell C7 at the right. The spreadsheet calculated the total as $600.00.

When preparing a spreadsheet, all you have to do is enter labels, numbers, and formulas. The computer will make all the necessary calculations for you. If you change a value, the spreadsheet will automatically recalculate all the totals.

	A	B	C
1	Monthly House Expenses		
2		January	February
3	Fixed costs:		
4	Mortgage	450	450
5	Property tax	138	138
6	Insurance	12	12
7	Total:	=Sum(B4:B6)	=Sum(C4:C6)
8			
9	Variable costs:		
10	Electricity	24	35
11	Heat	112	98
12	Water	15	10
13	Telephone	48	70
14	Total:	=Sum(B10:B13)	=Sum(C10:C13)
15			
16	Monthly total	=B7+B14	=C7+C14

	A	B	C
1	Monthly House Expenses		
2		January	February
3	Fixed costs:		
4	Mortgage	$450.00	$450.00
5	Property tax	$138.00	$138.00
6	Insurance	$12.00	$12.00
7	Total:	$600.00	$600.00
8			
9	Variable costs:		
10	Electricity	$24.00	$35.00
11	Heat	$112.00	$98.00
12	Water	$15.00	$10.00
13	Telephone	$48.00	$70.00
14	Total:	$199.00	$213.00
15			
16	Monthly total	$799.00	$813.00

In recent years, the demand for interior design has increased due to a better understanding of how an environment can affect the way people function best. Today, the home and workplace are being designed with color, fabric, lighting, and decorative accessories that provide harmonious home and work environments.

Job Description

What do interior designers do?

1. An interior designer confers with a client to plan an interior environment that is pleasing to the eye, functional, and within budget. The client's architectural and equipment needs, lifestyle or work habits, and color, lighting, and furniture preferences are determined.

2. The designer then formulates a plan for the client in the form of a drawing that is both practical and aesthetic. Design factors such as space planning, layout and utilization of furnishings, lighting, and color schemes are included in the plan.

3. Following approval of the plan, the designer selects and purchases the furnishings, paint, draperies, carpeting, art works, and accessories. This involves careful data gathering and keeping accurate records.

4. Frequently, the designer subcontracts workers to do any necessary renovations. Subcontractors might also be hired to install carpeting, light fixtures, and draperies.

Qualifications

It is important that someone entering a career in interior design has an eye for color and design, an extensive product knowledge, skill in business management, and a sincere interest in helping people create attractive homes or offices within a certain budget.

Courses and degrees in interior design are offered at most colleges, universities, and trade schools. The coursework could include basic design, interior graphics, design technology, home planning and furnishing, history of interiors, interior lighting, and principles of professional practices.

FOCUS ON INDUSTRY

The retail industry is responsible for distributing goods and services to consumers for their own use. Among retailers there is a great deal of competition. This competition usually takes the form of competitive pricing. Other forms of competition include advertising, the attractiveness of displays, and product diversification. The industry must concern itself with the needs and wants of consumers to be able to serve them more effectively.

Retailers buy goods in large lots and divide them into smaller quantities that are more convenient for consumers. A variety of stock is kept on hand to provide the maximum convenience to customers. Since retailers deal with consumers on a day-to-day basis, they can provide manufacturers with current information on buying habits.

Retailers utilize a great variety of selling methods. Some different types of store retailing are specialty stores, department stores, discount stores, supermarkets, and chain stores. Nonstore retailing includes vending machines, selling by mail order, and selling door-to-door. In recent years, the retailing trend has been towards selling in large shopping malls.

Retailers for Household Supplies*				
Kind of Stores	Number of Stores (1000s)	Store Sales ($1,000,000)	Annual Payroll ($1,000,000)	Paid Employees (1000s)
Furniture	38.3	17,658	2608	214
Home Furnishing	44.3	9435	1320	124
Floor Covering	17.1	5287	721	54
Household Appliance	13.9	5885	697	59
Radio, Television, and Music	35.2	13,813	1662	146

*The statistics are for a recent year.

For the questions below, refer to the table above.

1. Find the total for each column.

2. What percent of total household retail store sales does each kind of store have? Give your answers to the nearest whole percent.

3. What is the average annual salary, to the nearest dollar, for an employee in a furniture store?

Chapter 9 Test

Use a calculator where possible.

1. Estimate the cost to build the house below at a rate of $48/ft^2, to the nearest thousand dollars. (9-1)

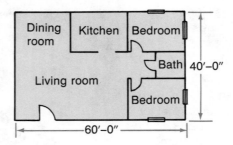

2. A contractor will build a 35.0 ft by 56.0 ft house for $90,720. How much is this per square foot? (9-1)

3. The foundation of the *contemporary house* requires 15 yd^3 of pea gravel at $12.50/yd^3 and 8 yd^3 of mason sand at $11.75/yd^3. What is the total cost of the pea gravel and mason sand? (9-2)

4. One concrete footing in the basement of the *contemporary house* is 28.0 ft long, 16.0 in. wide, and 8.0 in. thick. How many *whole* cubic yards of concrete should be ordered for this footing? (9-2)

5. The *contemporary house* requires 2192 lin ft of wall studs. What is the cost of these studs at $0.26/lin ft, to the nearest ten dollars? (9-3)

6. How many 2.0 in. by 4.0 in. studs are needed for the length of wall frame sketched below? (9-3)

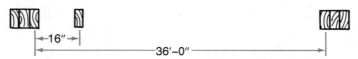

7. A gable roof has 16 ft rafters on each side. The roof is 55 ft long. How many *whole* squares of shingles will be needed for both sides of this roof? (9-4)

8. If 2 lb of nails are needed for each square of shingles, how many *whole* pounds of nails are needed for a shed roof that is 36 ft long and has rafters 18 ft long? (9-4)

What is the area of each wall below to be painted, excluding windows and doors? (9-5)

9.

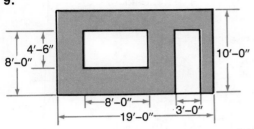

10.

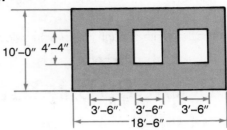

11. The ceilings and walls of the master bedroom of the *contemporary house* contain 779 ft². How many gallon cans of paint will be required to paint this bedroom with two coats if each gallon can will cover 450 ft²? Ignore the windows and doors of the bedroom. (9-5)

What is the cost, to the nearest ten dollars, of carpeting each given floor space at $16.95/yd²? (9-6)

12. Exclude the kitchen.

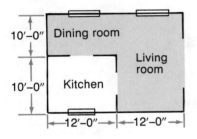

13. Exclude the fireplace.

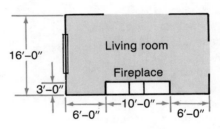

14. How many cubic yards of topsoil should be ordered for a 45 ft by 75 ft lot if the desired depth is 6.0 in.? (9-7)

15. What is the cost of laying grass sod for the lot below, to the nearest ten dollars, if the sod costs $1.65/yd²? (9-7)

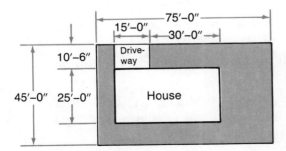

16. A tree-moving service charges $35/tree plus 30¢/mi for delivering trees. Find the price of six trees if the total mileage is 25 mi. (9-7)

Cumulative Review

Use a pica ruler if you have one. Otherwise, copy the pica ruler in the back of the book.

1. What is the width of each line of type below, to the nearest half pica?

   a. The extent of a book is its page length.

   b. Book Publishing

2. A photograph measuring 36 picas wide by 60 picas deep must be reduced to fit a space that is 18 picas deep. What will the reduced width of the photograph be, to the nearest half pica?

3. Find the total cost to the publisher for producing 10,000 *Lawn Care* books.

Preliminary Costing Sheet			
Title: Lawn Care **Extent:** 32 pages **Print Quantity:** 10,000 books			
Non-recurring Costs		**Recurring Costs**	
Editorial work	$1500	Paper	$2800
Graphic design	$1150	Printing	$1950
Typesetting @ $18/page	?	Book cover	$1200
Assembly @ $15/page	?	Book binding	$2500
Film of pages	$256		
Printing plates	$1820	**Total recurring costs:**	?
		Run-on cost per	
Total non-recurring costs:	?	**thousand books:**	$850

4. A printer is printing 16 pages on each side of a sheet of paper for a 320-page book.

   a. If the publisher has ordered 15,000 copies of the book, how many sheets of paper will the printer need?

   b. How many *whole* reams of paper will be required?

5. The bindery cost for the production of a book is 18¢ per book for a print quantity of 1200 books. What is the total bindery cost?

6. What is the cost of building a 2500 ft^2 house at $63/ft^2, to the nearest hundred dollars?

7. A concrete slab foundation is 12.0 ft wide, 18.0 ft long, and 4.0 in. thick. How much concrete is required for the foundation, to the nearest *whole* cubic yard?

8. Ready-mixed concrete will be delivered to a construction site for $77/yd^3. What will be the cost for a concrete slab foundation that is 50.0 ft by 80.0 ft by 6.0 in.? Round your answer to the nearest ten dollars.

9. If 2 by 4 studs are placed 18 in. o.c. in a wall frame, how many are required for a 63 ft 10 in. wall? Allow two corner studs at each end of the wall frame.

10. What is the cost of shingles for the gable roof below with a cost of $27.95/square, to the nearest hundred dollars? Consider the shed dormer roof to be part of the gable roof in your estimate.

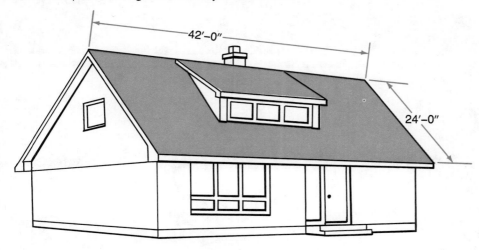

11. What would be the cost of covering the 1500 ft² exterior of a house with brick if each brick costs $0.35? Allow 700 bricks for 100 ft².

12. If one roll of wallpaper will cover about 36 ft², how many rolls of wallpaper are needed for the four walls of a room that is 12.0 ft by 8.0 ft by 8.0 ft? Ignore the windows and doors.

13. What is the cost to carpet the den floor sketched below at $28.95/yd², to the nearest hundred dollars?

14. What is the cost of laying grass sod on the lot below if the sod costs $1.45/yd² plus $0.45/yd² for labor? Round your answer to the nearest ten dollars.

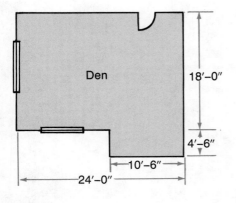

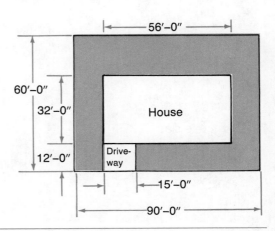

Chapter 10
Right Triangle Trigonometry

After completing this chapter you should be able to:

1. Use the rule of Pythagoras to solve problems involving right triangles.
2. Find the sine, cosine, and tangent ratios of acute angles.
3. Apply the trigonometric ratios to the solution of right triangle problems.

Right Triangles

10-1　The Rule of Pythagoras

Carpenters, designers, and surveyors often have to calculate unknown dimensions of **right triangles** (a triangle containing a right angle). A useful rule, credited to the ancient Greek philosopher Pythagoras, remains helpful today in determining unknown dimensions or distances in right triangles.

In a right triangle, the side opposite the right angle is called the **hypotenuse**. The rule of Pythagoras states that the square of the hypotenuse of a right triangle is equal to the sum of the squares of the other two sides. The diagram at the right illustrates this rule.

The rule of Pythagoras can be summarized in the formula $c^2 = a^2 + b^2$, where a, b, and c are the lengths of the sides of a right triangle.

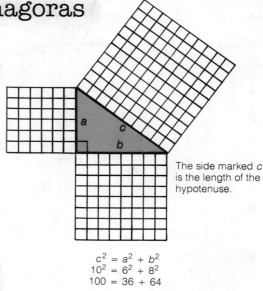

The side marked c is the length of the hypotenuse.

$$c^2 = a^2 + b^2$$
$$10^2 = 6^2 + 8^2$$
$$100 = 36 + 64$$

The example below shows how the rule of Pythagoras is used for finding unknown lengths in a right triangle.

Example

What is the length of the rise for the roof at the right if the run is 11 ft and the rafter length is 14 ft?

Solution

The problem can be solved using the rule of Pythagoras and a calculator, if we assume that the wall studs are at right angles to the top plate of the wall frame.

Skill(s) 6, 7, 29, 30, 32

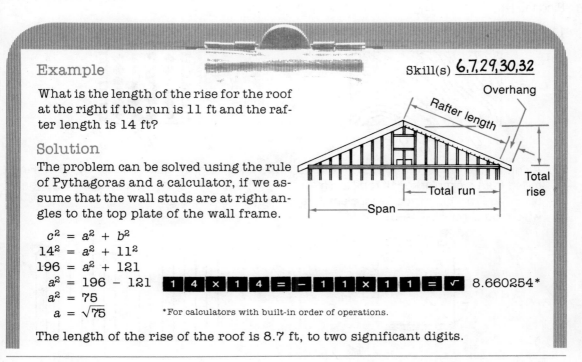

$$c^2 = a^2 + b^2$$
$$14^2 = a^2 + 11^2$$
$$196 = a^2 + 121$$
$$a^2 = 196 - 121$$
$$a^2 = 75$$
$$a = \sqrt{75}$$

[1][4][×][1][4][=][−][1][1][×][1][1][=][√]　8.660254*

*For calculators with built-in order of operations.

The length of the rise of the roof is 8.7 ft, to two significant digits.

EXERCISES

Use a calculator where possible.

A Find the unknown right triangle lengths in the table below.
The hypotenuse is the length of side *c*.

	Length of side *a*	Length of side *b*	Length of side *c*	
1.	3 cm	5 cm +	?	5.8 CM
2.	7.0 m	?	9.0 m −	5.6 M
3.	8 in.	?	10 in. −	6 IN.
4.	2.0 ft	4.0 ft +	?	4.4 FT.
5.	7.3 in. +	5.7 in.	?	9.2 IN.
6.	8.4 m	9.7 m +	?	12.8 M
7.	3.25 m	?	6.75 m −	5.91 M
8.	4.75 in.	?	9.50 in. −	8.22 IN.
9.	?	345 cm	682 cm −	588 CM
10.	?	98.4 cm	179.7 cm −	150.3 CM

A ... *C* ... *B*

B Find the length of the diagonal of the following objects.

11. a 4 ft by 8 ft rectangular piece of plywood **12.** a 4 ft by 4 ft square rug

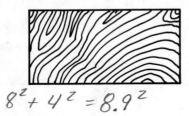

$8^2 + 4^2 = 8.9^2$

$4^2 + 4^2 = 5.6^2$

13. A square bar 1.0 in. on each side is to be milled from a circular rod. If rods are available with 1 in., $1\frac{1}{4}$ in., and $1\frac{1}{2}$ in. diameters, which rod should be used?

14. A rectangular gate is 48 in. by 36 in. Find the length of a corner-to-corner diagonal brace for the gate.

$1^2 + 1^2 = 1.414^2$

$1\frac{1}{2}$ ROD.

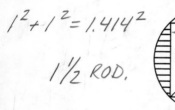

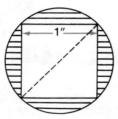

1"

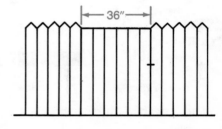

36"

$36^2 + 48^2 = 60^2$

15. The rectangular frame of a billboard is 12 ft by 8.0 ft. Find the length of a diagonal brace for the frame. $12^2 + 8^2 = 14.4^2$

16. The EZ Company makes triangular shelf braces like the one shown below. How many inches of material are needed for each brace? *13.5² −*

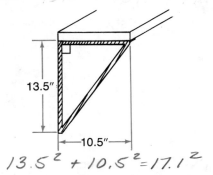

13.5"

10.5"

$13.5^2 + 10.5^2 = 17.1^2$

17. A surveyor has measured the two distances in relation to the pond, shown in the diagram below, to find the distance from point A to point C. What is this distance?

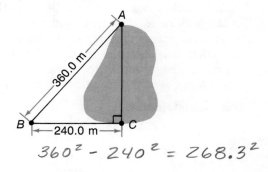

A

360.0 m

B

240.0 m

C

$360^2 - 240^2 = 268.3^2$

18. What is the rafter length indicated in the house frame below?

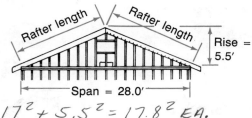

Rafter length   Rafter length

Rise = 5.5'

Span = 28.0'

$17^2 + 5.5^2 = 17.8^2$ EA.

19. An engineer wishes to find the distance AB along some rough terrain. Using the measurements shown, calculate AC, BC, and AB.

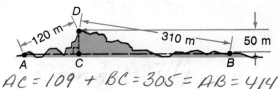

D

120 m   310 m   50 m

A   C   B

$AC = 109 + BC = 305 = AB = 414$

20. An automobile is driven 16.3 mi east and then 18.6 mi north. What is the distance of the car from the starting point? $18.6^2 + 16.3^2 = 24.7^2$

21. A ramp for wheelchairs must span a horizontal distance of 11.3 ft and a vertical distance of 3.1 ft. How long is the ramp? $11.3^2 + 3.1^2 = 11.7^2$

22. When a quilting frame for a 75 in. by 52 in. quilt is set up, a check is made of the diagonals to make sure the frame is rectangular. (The two diagonals are of the same length for a rectangle.) Find the length of the two diagonals. $75^2 + 52^2 = 91.2^2$ EA.

23. A quality-control worker checks 4.0 ft by 8.0 in. plastic panels by measuring the diagonals. Calculate the length of the diagonals. $48^2 + 8^2 = 48.6^2$

24. A doorway is 36 in. wide and 82 in. high. What is the width of the widest piece of plate glass that can be carried diagonally through the doorway? Allow 2.0 in. at each corner for clearance. $82^2 + 36^2 = 89^2 - 4 = 85.5^2$

C 25. A 42 ft tower has three guy wires from the top of the tower and three wires from the midpoint of the tower. If the wires are anchored 32 ft from the base of the tower, how much wire is needed? Allow 7.0 ft of wire for making all of the connections.

$42^2 \times 32^2 = 52.8^2 \times 3 = 158.4$
$21^2 \times 32^2 = 38.2^2 \times 3 = 114.6$ $= 273 + 7 = 280.2$

10-2 The Sine, Cosine, and Tangent Ratios

A right triangle has one 90° angle and two acute angles (angles measuring less than 90°). In a right triangle, the ratio formed by the length of any two sides is called a **trigonometric ratio**. More specifically, such a ratio can be called the **sine**, **cosine**, or **tangent ratio**. The trigonometric ratios can be written in fraction form and in decimal form, usually with four decimal places. The ≈ sign means "approximately equal to." The ratios are defined below for ∠ A of the right triangle ABC at the right.

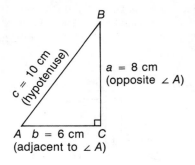

$c = 10$ cm (hypotenuse)

$a = 8$ cm (opposite ∠ A)

$b = 6$ cm (adjacent to ∠ A)

Trigonometric Ratios:

Sine of ∠ A $= \dfrac{\text{side opposite } \angle A}{\text{hypotenuse}}$ → $\sin A = \dfrac{a}{c} = \dfrac{8}{10} = 0.8000$

Cosine of ∠ A $= \dfrac{\text{side adjacent to } \angle A}{\text{hypotenuse}}$ → $\cos A = \dfrac{b}{c} = \dfrac{6}{10} = 0.6000$

Tangent of ∠ A $= \dfrac{\text{side opposite } \angle A}{\text{side adjacent to } \angle A}$ → $\tan A = \dfrac{a}{b} = \dfrac{8}{6} \approx 1.3333$

Example

Skill(s) 6,7,8,18,29,30,32

The television tower at the right is supported by a brace wire anchored at point A. Find the sine, cosine, and tangent ratios of ∠ A rounded to four decimal places.

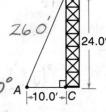

26.0′   24.0′

Solution

Step 1: Use the rule of Pythagoras to find length AB.

$A = 67.38013505 = 90°$
$B = 22.61986495$

10.0′

$$c^2 = a^2 + b^2$$
$$= 24.0^2 + 10.0^2$$
$$= 676$$
$$c = \sqrt{676}$$

26

| 2 | 4 | × | 2 | 4 | = | + | 1 | 0 | × | 1 | 0 | = | √ | 676 * |

*For calculators with built-in order of operations.

The length from A to B is 26.0 ft, to three significant digits.

Step 2: Find the sine, cosine, and tangent ratios for ∠ A.

$\sin A = \dfrac{24 \text{ (opposite)}}{26 \text{ (hypotenuse)}}$
≈ 0.9231

$\cos A = \dfrac{10 \text{ (adjacent)}}{26 \text{ (hypotenuse)}}$
≈ 0.3846

$\tan A = \dfrac{24 \text{ (opposite)}}{10 \text{ (adjacent)}}$
$= 2.4000$

| 2 | 4 | ÷ | 2 | 6 | = |

0.9230769

| 1 | 0 | ÷ | 2 | 6 | = |

0.3846154

| 2 | 4 | ÷ | 1 | 0 | = |

2.4

EXERCISES

Write each trigonometric ratio in decimal form with four decimal places.
Use a calculator where possible.

A Refer to the triangle *ABC* at the right to write each
trigonometric ratio.

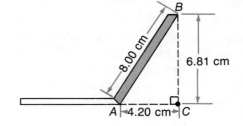

1. sin A .8512
2. tan A .6167
3. cos B .8512
4. sin B 1.9047
5. tan B .6167
6. cos A .525

Write each trigonometric ratio for the roof truss at
the right.

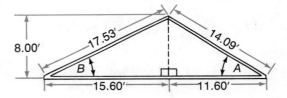

7. sin B .4563
8. sin A .5677
9. cos B .8899
10. cos A .8232
11. tan B .5128
12. tan A .6896

B 13. Metal strap bracing nailed to the 8.0 ft wall
frame of a house is anchored 3.2 ft from the
bottom corner of the wall frame.

   a. Find the tangent of the angle between
the metal strap bracing and the top of
the wall frame. 8 ÷ 3.2 = 2.5
   b. Use the rule of Pythagoras to find the
length of the metal strap. 8.6

14. A sidewalk is constructed diagonally across
a rectangular garden, as shown in the sketch
below.

   a. Use the rule of Pythagoras to find the
distance from point *A* to point *B*. 55.0
   b. Find sin A. .9090
   c. Find cos B. .9090
   d. Find tan A. 2.1739

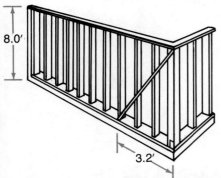

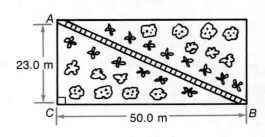

15. A ski lift has a vertical rise of 152 m over a horizontal distance of 914 m.
Find the sine of the angle that the lift makes with the horizontal plane. .1641

10-3 The Isosceles Right Triangle

A triangle with a right angle and two sides of equal length is called an **isosceles right triangle**. One of the triangles in a pair of set squares, as shown at the right, is an example of such a triangle. Each acute angle measures 45°.

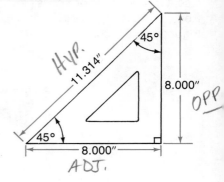

The trigonometric ratios of the sides of an isosceles right triangle in relation to a 45° angle can be expressed in fraction or decimal form. (The decimal form is usually rounded to four decimal places.) The ratios for the 45°– 45°–90° set square at the right are given below.

$$\sin 45° = \frac{8.000}{11.314} \quad \frac{\text{(opposite)}}{\text{(hypotenuse)}} \approx 0.7071 \qquad \cos 45° = \frac{8.000}{11.314} \quad \frac{\text{(adjacent)}}{\text{(hypotenuse)}} \approx 0.7071$$

$$\tan 45° = \frac{8.000}{8.000} \quad \frac{\text{(opposite)}}{\text{(adjacent)}} = 1.0000$$

In general, for any isosceles right triangle, the following trigonometric ratios are true.

$$\sin 45° = \cos 45° = \frac{x}{x\sqrt{2}} = \frac{1}{\sqrt{2}} \approx 0.7071$$

$$\tan 45° = \frac{x}{x} = 1.0000$$

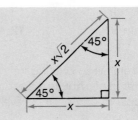

These ratios can be used to find the length of a side in an isosceles right triangle, as shown below.

Example

How long is each side of the shed roof?

Skill(s) **6, 7, 8, 18, 28**

Solution

Either the sine or the cosine ratio can be used to solve the problem. Substitute known values into a cosine (or sine) ratio equation.

$$\cos 45° = \frac{x}{6} \quad \frac{\text{(adjacent)}}{\text{(hypotenuse)}}$$

$$0.7071(6) = x$$

$\boxed{\;\cdot\;}\boxed{7}\boxed{0}\boxed{7}\boxed{1}\boxed{\times}\boxed{6}\boxed{=}$ 4.2426

Each side of the roof is 4.2 ft long, to two significant digits.

EXERCISES

Use the trigonometric ratio tables in the back of the book or a scientific calculator.

A Write each trigonometric ratio in decimal form.

1. sin 45° .7071   **2.** cos 45° .7071   **3.** tan 45° 1

4. sin 28° .4694   **5.** cos 55° .5735   **6.** tan 76° 4.0107

For each equation below, x represents an unknown length, in centimeters, of an isosceles right triangle. Solve for x.

7. $\sin 45° = \dfrac{x \, .7071}{8.0 \text{ cm}}$ **8.** $\sin 45° = \dfrac{25 \text{ cm}}{x \, 35.3}$ **9.** $\sin 45° = \dfrac{x \, 34.08}{48.2 \text{ cm}}$

10. $\cos 45° = \dfrac{18.0 \text{ cm}}{x \, 25.4}$ **11.** $\tan 45° = \dfrac{x \, 6.5}{6.5 \text{ cm}}$ **12.** $\cos 45° = \dfrac{124 \text{ cm}}{x \, 175.3}$

B **13.** A square rug has a diagonal distance of 48.25 in. Calculate the length of each side of the rug.

14. A square piece of land has sides that are 68.5 m long. What is the length of a diagonal across the piece of land? 34.2

15. What is the length of the diagonal of a square floor tile that is 9.0 in. on each side? 12.7

16. Find the depth of the cut in the V-block shown below. The depth is represented by the distance *BD*. Assume *BD* to be perpendicular to *BC*. BD=5.65 BC=5.66

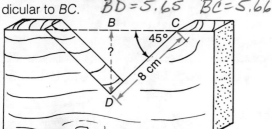

17. What is the length of the diagonal across the quilting square shown below? 48.79

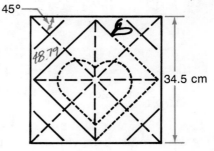

34.5 cm

18. The bases of a baseball diamond are at the corners of a square that measures 90.0 ft on each side. Find the distance between second base and home plate. 127 FT.

19. A surveyor makes the measurements shown at the right.

   a. What is the height of the building? 109.5
   b. What is the distance from the surveyor's eyes to the top of the building? 146.7

C **20.** The diagonal of a square room measures 33.9 ft in length. How much will it cost to lay a vinyl floor at $12.87/yd²? $823.68

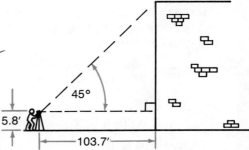

45°

5.8'

103.7'

10-4 The 30°–60°–90° Triangle

A second special right triangle is the 30°–60°–90° triangle, which is *half* of an equilateral triangle, as shown at the right. For this reason, the side opposite the 30° angle (in a 30°–60°–90° triangle) is always *half* the length of the hypotenuse.

equilateral triangle

The trigonometric ratios for the 30° angle of the 30°–60°–90° set square at the right are shown below in both fraction and decimal form.

$$\sin 30° = \frac{7.50}{15.00}$$
$$= 0.5000$$

$$\cos 30° = \frac{12.99}{15.00}$$
$$\approx 0.8660$$

$$\tan 30° = \frac{7.50}{12.99}$$
$$\approx 0.5774$$

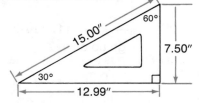

In general, for any 30°–60°–90° triangle, the following ratios are true.

$$\sin 30° = \frac{x}{2x} = \frac{1}{2} = 0.5000 \quad \cos 30° = \frac{x\sqrt{3}}{2x} = \frac{\sqrt{3}}{2} \approx 0.8660$$

$$\tan 30° = \frac{x}{x\sqrt{3}} = \frac{1}{\sqrt{3}} \approx 5.774$$

Example

Skill(s) 6, 7, 8, 18, 28

How high is the total rise of the roof at the right? (Assume the wall studs are perpendicular to the top plate of the wall frame.)

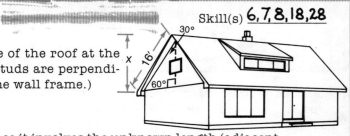

Solution

Choose the cosine ratio since it involves the unknown length (adjacent to the 30° angle) and 16 ft (the hypotenuse). Substitute known values into the cos 30° ratio equation.

$$\cos 30° = \frac{x}{16} \qquad \frac{\text{(opposite)}}{\text{(hypotenuse)}}$$

$$0.8660 = \frac{x}{16}$$

$$x = 0.8660 \times 16 \qquad \boxed{.}\ \boxed{8}\ \boxed{6}\ \boxed{6}\ \boxed{\times}\ \boxed{1}\ \boxed{6}\ \boxed{=}\ 13.856$$

The total rise is 14 ft, to two significant digits.

EXERCISES

Use the trigonometric ratio table in the back of the book or a scientific calculator.

A Find the trigonometric ratios in decimal form rounded to four decimal places.

1. sin 60° .8660   **2.** cos 60° .500   **3.** tan 60° 1.732

For each equation below, x represents an unknown side length, in centimeters, of a 30°–60°–90° triangle. Solve for x.

4. $\cos 60° = \dfrac{x\ \ 19}{38\ cm}$ **5.** $\sin 30° = \dfrac{42\ cm}{x\ 84}$ **6.** $\tan 60° = \dfrac{199\ x}{115\ cm}$

7. $\sin 60° = \dfrac{2.4\ cm}{x\ 2.771}$ **8.** $\tan 30° = \dfrac{x\ 4.33}{7.5\ cm}$ **9.** $\cos 30° = \dfrac{8.6\ cm}{x\ 9.9}$

B 10. A pole has three 8 ft guy wires, which are fastened to the ground at a 60° angle. How high up on the pole are the wires fastened?

11. A hill is inclined at 30° with the horizontal plane. If you walk 85 ft up the hill, how much altitude have you gained?

12. How tall is the tree the surveyors are measuring in the diagram below?

13. What is the distance from point A to point C in the river diagram below?

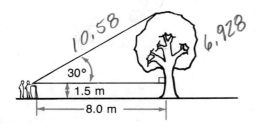

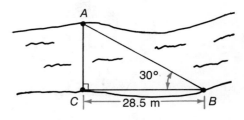

14. A sign has the shape of an equilateral triangle. If each side is 26.3 in. long, what is the height of the sign?

15. A staircase rises at 30° with the horizontal plane. If the rise of the staircase is 7.80 m, what is the overall length of the staircase?

16. What is the total rise on the roof at the right? (Assume the wall studs to be perpendicular to the top plate.)

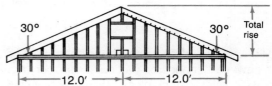

17. A rocket is fired at an angle of 30° and travels in a straight line. When the rocket reaches a height of 5000 ft, how far has it traveled? (Ignore the pull of gravity and the Earth's curvature.)

18. A missile that is traveling in a straight line climbs at an angle of 60°. How high is the missile after it covers a ground distance of 3.7 mi? (Ignore the pull of gravity and the Earth's curvature.)

19. A panel on a geodesic dome is an equilateral triangle with each side 96 in. long. What is the height of the panel?

20. A cross section of a V-shaped thread is shown below. Assume distances *AB*, *BC*, and *CA* to be each 0.05 in. What is the depth of the thread?

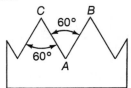

21. A pipe bends away from a wall, as shown below. What is the horizontal distance that the pipe bends?

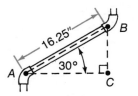

C **22.** Refer to the hex nut shown below. If the distance from *D* to *E* measures 16 mm, what is the distance from *A* to *B*?

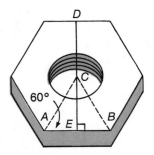

23. Each side of a hexagonal tabletop measures 28.0 in. What is the surface area of the table-top? (Start by tracing the hexagon and dividing it into six equilateral triangles.)

Technology Update

Hydroplaning

Some automobile accidents are caused by a condition called hydroplaning. A vehicle hydroplanes when it slides on a thin film of water on a wet highway rather than having its tires in direct contact with the pavement.

Highway departments now have computers on trucks that locate places on our highways that become particularly slippery during periods of rain. Such a truck travels with a water tank and pulls a one-wheel trailer. Water is sprayed from the truck's tank to the pavement in front of the trailer wheel. The computer then measures the force required to stop the trailer wheel from turning. (When there are hydroplaning conditions on wet pavement, very little force is needed to stop the turning of the wheel.)

If the computer data indicate a particularly slick section of highway, warning signs for motorists are put in place, and that section of roadway is scheduled for remedial work.

Self-Analysis Test

Find the unknown length using the rule of Pythagoras and a calculator.

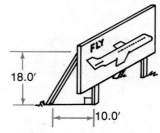

1. The top of a billboard is 18.0 ft above the ground, as shown at the right. It is supported by two equal-sized braces at the back. What is the length of each brace?

18.0′

10.0′

2. The top of a 16.0 ft ladder rests against a building at a point 13.5 ft above the ground. How far is the foot of the ladder from the base of the building?

3. A screen door is 35.4 in. wide and 79.7 in. high. A diagonal wire stabilizes the screen door. Find the length of the wire.

Use the trigonometric ratio table in the back of the book or a scientific calculator. Refer to the quilting pattern at the right for questions 4 to 10.

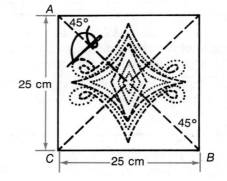

4. Use the rule of Pythagoras to find the length of diagonal *AB*.

Write each trigonometric ratio in decimal form rounded to four decimal places.

25 cm

25 cm

5. sin *A*   **6.** cos *A*

7. tan *A*   **8.** sin *B*

9. cos *B*   **10.** tan *B*

11. A roadway has a vertical rise of 18.3 ft over a horizontal distance of 360.0 ft. Find the tangent of the angle that the roadway makes with the horizontal plane.

12. Find the length of the diagonal of a square storage shed that has sides 13 ft long.

13. A square sheet of aluminum has a diagonal that is 38.4 cm long. Find the length of each side of the square sheet.

14. A cake pan has the shape of an equilateral triangle. If the height of the triangle is 12.3 in., what is the length of each side of the cake pan?

15. The Washington Monument is about 169.0 m high. How far from the base would you have to stand in order that the angle of sight measures 30° to the top of the monument?

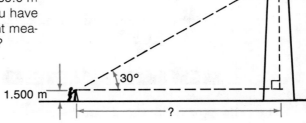

1.500 m

30°

?

Applications

10-5 Angles of Elevation and Depression

There are many applications for right triangle trigonometry that deal with an angle of elevation or an angle of depression. The **angle of elevation** can be described as the angle formed by a horizontal line and that which is *upwards* from it. This is illustrated by a person standing at ground level and observing a helicopter in flight. The angle of elevation extends from the horizontal line at eye level *upwards* to the helicopter.

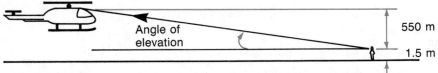

The helicopter pilot's view of the person on the ground is an example of an **angle of depression** (the angle formed by a horizontal line and that which is *downwards* from it).

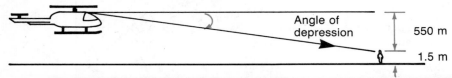

Example

Skill(s) **6, 7, 8, 18, 28**

In the second diagram above, suppose the helicopter pilot measured a 10° angle of depression from eye level downwards to the person on the ground. How far is the helicopter from the person on the ground if the helicopter is at an altitude of 550 m above the person?

Solution

Choose a ratio involving the given altitude, 550 m (opposite the 10° angle), and the unknown distance (hypotenuse).

Step 1: Find the sine ratio of the 10° angle to four decimal places.

$\boxed{1}\ \boxed{0}\ \boxed{\text{SIN}}$ 0.1736482   The sine ratio is about 0.1736.

Step 2: Find the unknown distance, x.

$$\sin 10° = \frac{550}{x} \quad \frac{\text{(opposite)}}{\text{(hypotenuse)}} \qquad 0.1736 = \frac{550}{x}$$

$\boxed{5}\ \boxed{5}\ \boxed{0}\ \boxed{\div}\ \boxed{\cdot}\ \boxed{1}\ \boxed{7}\ \boxed{3}\ \boxed{6}\ \boxed{=}$ 3168.2028 $x = \dfrac{550}{0.1736}$

The helicopter is 3200 m from the person, to two significant digits.

The trigonometric ratios can also be used to find an unknown angle measure in a right-angled triangle. The following example shows how the **INV** key on a scientific calculator helps you do this.

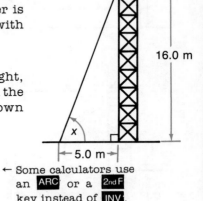

Example 2

What is the angle of elevation formed by a wire brace and the ground if the radio tower at the right is 16.0 m tall and the distance from the stake to the tower is 5.0 m? Assume the tower forms a right angle with the ground.

Solution

Choose a ratio involving the given tower height, 16.0 m (opposite the angle of unknown size), and the distance, 5.0 m (adjacent to the angle of unknown size).

$$\tan x = \frac{16}{5} \quad \frac{\text{(opposite)}}{\text{(adjacent)}}$$

1 **6** **÷** **5** **=** 3.2   **INV** **TAN** 72.645975   ← Some calculators use an **ARC** or a **2nd F** key instead of **INV**.

The angle of elevation is 73°, to two significant digits.

EXERCISES

Use the trigonometric ratio table in the back of the book or a scientific calculator.

A Use the given equation to solve for the unknown distance, x.

1. $\cos 48° = \dfrac{195 \text{ m}}{x}$

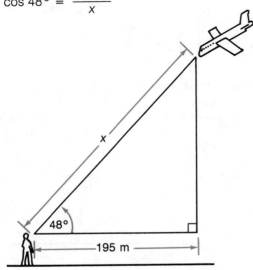

2. $\sin 74° = \dfrac{640 \text{ m}}{x}$

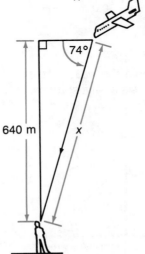

Write an equation with a trigonometric ratio. Then solve for the unknown dimension, x.

3.

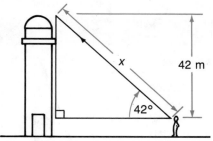

42 m

42°

4.

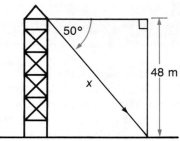

50°

48 m

x

5.

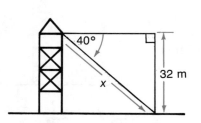

40°

32 m

x

6.

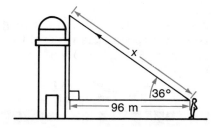

x

36°

96 m

B Write an equation with a trigonometric ratio. Then solve for the unknown angle measure.

7.

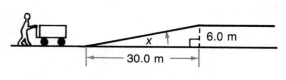

x 6.0 m

30.0 m

8.

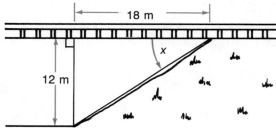

18 m

x

12 m

9. When the angle of elevation of the sun is 42°, a flagpole casts a 42 ft shadow. How tall is the flagpole?

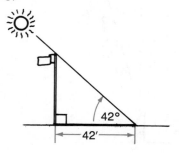

42°

42′

10. An airplane pilot measures a 25° angle of depression from the cockpit down to a skyscraper 100 m tall. How far is the pilot from the skyscraper if the plane has an altitude of 3500 m?

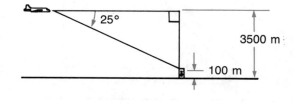

25°

3500 m

100 m

11. A surveyor with a transit 1.5 m off the ground sights a 44° angle to the top of a building. If the transit is 56 m from the base of the building, what is the height of the building?

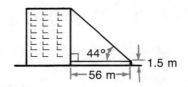

12. A mine tunnel is dug at an angle of 26°. How long must the tunnel be to reach a depth of 75 ft?

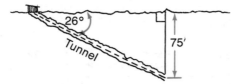

13. A railroad track is 14,400 m long with a rise of 432 m. Find the angle of elevation for the section of track.

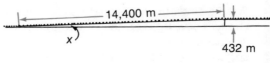

C **14.** The angle of depression from an airplane to a fixed point on the ground is 27°. The airplane is directly over a location 1250 m from the fixed point. What is the altitude of the airplane?

15. A roadbed rises at an angle of 8°. Find the amount of rise in the roadbed over a horizontal distance of 1250 yd.

16. A plane takes off and climbs at an angle of 7°. How much ground distance must the plane travel to reach an altitude of 3500 ft?

17. The angle of depression from a ranger's tower to a fire in a valley is 14°. If the tower is 650 ft above the valley, calculate the distance from the tower to the fire.

18. A ladder on a fire truck can be extended to 85 ft. The maximum angle of elevation is 73°. How high will the ladder reach on a building if the base of the ladder is 5.5 ft off the ground?

Tricks of the Trade

A Transit

The transit is widely used in surveying and engineering work for the measurement of field distances, angles, and elevations. It is used for such things as laying out property boundaries, making maps, and building roads.

A transit (sometimes called a theodolite) consists of a tripod with a telescope mounted on it. The telescope is aimed at some point by means of cross hairs. The direction of the point can be read on a horizontal vernier scale, and the angle of elevation or depression of the point can be read on a vertical vernier scale.

10-6 Navigation

A navigator frequently uses the trigonometric ratios to calculate the distance his or her vessel is from a visible landmark on shore.

Example 1

Skill(s) **6,7,8,18,28**

With a sextant, a tugboat navigator measures the angle of elevation to the top of a lighthouse in the distance to be 4°.
What distance, x, is the tugboat from the base of the lighthouse? The top of the lighthouse is 32 m above sea level. Ignore the height above sea level the viewer is.

Solution

Choose a ratio that involves the height of the lighthouse (opposite to 4° angle) and the unknown distance, x (adjacent to 4° angle).

Step 1: Find the tangent ratio of 4° to four decimal places.

> **4** **TAN**   0.0699268   The tangent ratio is about 0.0699.

Step 2: Find the distance from the tugboat to the lighthouse.

$$\tan 4° = \frac{32}{x} \qquad 0.0699 = \frac{32}{x}$$

> **3 2 ÷ · 0 6 9 9 =**
> 457.79685

The tugboat is about 460 m from the lighthouse, to two significant digits.

Example 2

The diagram at the right shows the course of a ship from A to B on a bearing of 025° (that is, 25° *clockwise* from North). The right-angled triangle illustrates how far east the ship has traveled (AC) and how far north it has traveled (CB). Find the distance the ship traveled east.

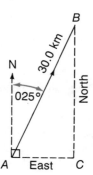

Solution

Step 1: Since 90° − 25° is 65°, ∠ *BAC* is 65°.

Step 2: Find the distance traveled east.

$$\cos 65° = \frac{x}{30.0} \qquad 0.4226 = \frac{x}{30.0}$$

> **3 0 × · 4 2 2 6 =**
> 12.678

The ship traveled 12.7 km east, to three significant digits.

EXERCISES

Use the trigonometric ratio table in the back of the book or a scientific calculator. For this text, ignore the effects of side drift due to current and/or wind as well as the curvature of the Earth.

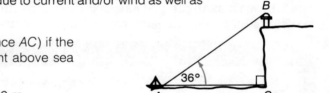

A 1. How far is the boat from the cliff (distance *AC*) if the top of the lighthouse is the given height above sea level?

a. 980 m

b. 1400 m

B 2. A ship travels 16 km in a northeasterly direction on a bearing of 050°. Find the distance traveled north.

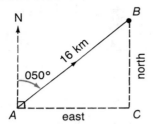

3. A barge travels 22 km in a southeasterly direction on a bearing of 132°. How far east has the barge traveled?

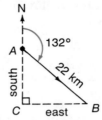

4. A ship sails due west for 9.3 km and then due north for 11.3 km. How far is the ship from its starting point?

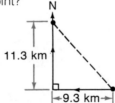

5. A boat sails 7.8 mi along a course that is 10.0° off the correct course (indicated by the dashed line below.) How far is the boat from the correct course?

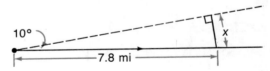

6. A sailboat's navigator measures the angle of elevation to the top of a lighthouse to be 7°. If the top is 275 ft above sea level, how far is the sailboat from the base of the lighthouse?

7. A buoy must be placed 620 m from the shore. What must the angle of depression be from the top of a 32 m cliff above the shoreline to place the buoy in the correct position?

8. The diagram at the right shows where a fishing boat traveled from starting point *A*.

a. Find the measure of ∠*BAC*.

b. What was the bearing of the fishing boat as it went from *A* to *B*?

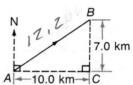

9. How many miles east were covered for each situation below?

a. The bearing is 015° and the distance traveled is 24 mi.

b. The bearing is 060° and the distance traveled is 45 mi.

10-7 More Applications

The following example and exercise questions illustrate the variety of ways in which trigonometry can be applied to real-life problems.

Example

Skill(s) **6,7,8,28**

The diagram at the right shows a staircase that joins the upper and lower floors of a house. What is the angle of elevation made by the staircase and the lower floor?

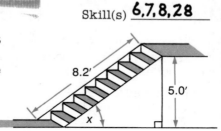

Solution

Choose a ratio that involves 5.0 ft (opposite to the unknown angle) and 8.2 ft (the hypotenuse).

Let x be the unknown angle size. Use the sine ratio to solve the problem.

$$\sin x = \frac{5.0}{8.2} \quad \frac{\text{(opposite)}}{\text{(hypotenuse)}}$$

`5 ÷ 8 · 2 = INV SIN` 37.571869

The angle of elevation is 38°, to two significant digits.

EXERCISES

Use a calculator where possible.

A 1. A kite is flying at the end of a string 80 m long, making a 53° angle with the ground. How high is the kite above the ground? (Ignore the sag of the string.)

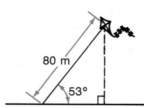

2. A wire stay, 26 m long, supports a pole and has one end fastened in the ground. The other end is fastened to a point 24 m up the pole. What angle does the stay make with the ground?

3. A hill has an angle of elevation measuring 5° from a fixed point *M* in the diagram at the right. The elevation at point *N* is 16.4 ft. How many feet does a car travel in moving from point *M* to point *N*?

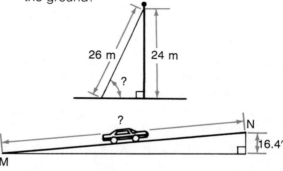

B **4.** A ladder 8.0 m long is placed against a wall with its foot 2.0 m away from the wall. Find the angle the ladder makes with the ground.

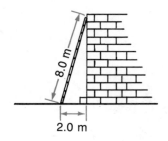

5. A square field has a footpath running diagonally from one corner to the opposite corner. If the length of the footpath is 150 m, what is the length of a side of the field?

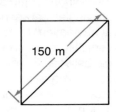

6. A boat is moored to the side of a quay by a rope 6.0 m long. When the rope is tight, it makes a 37° angle with the horizontal plane. How far is the boat from the side of the quay?

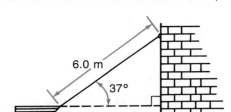

7. The two levels of an outdoor parking garage are 21.5 ft apart. The ramps between levels are 43.5 yd long. Find the angle of elevation of the ramps.

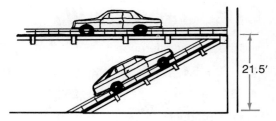

8. A rectangular snack tray is 48 cm by 38 cm. What is the length of the diagonals of the tray?

C **9.** The sloping roof of a tent is 2.0 m long and is inclined to the horizontal plane at an angle of 56°. What is the width of the tent?

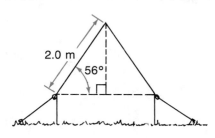

10. After take-off, a plane climbs for 10.0 km at an angle of 20° with the horizontal plane.

a. What is the altitude reached in meters?
b. What is the ground distance traveled in kilometers?

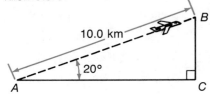

11. The Leaning Tower of Pisa is about 4° off being perpendicular to the horizontal. If an object is dropped from the top, about how far will it hit from the base? The height of the tower is about 182 ft.

Tricks of the Trade

Sextants have been used by navigators since 1730 to locate the position of a ship, using the positions of stars or by using measurements to known landmarks. A sextant can be used vertically to measure angles between heavenly bodies and the horizon in celestial navigation. In coastal navigation, the sextant can be used vertically to obtain the distance from shore by measuring the angle between an object on shore and sea level. The sextant is used horizontally to fix positions between three or more shore features. The sextant is used daily on ships at sea to find the latitude at noon and to fix the position of the ship.

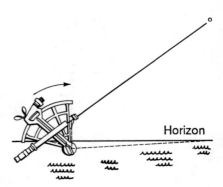

Horizon

The modern sextant has a micrometer head along with a vernier, scale which makes reading an angle in degrees, minutes ($\frac{1}{60}$ of a degree), and tenths of a minute relatively easy.

Check Your Skills

Find the sum or difference. (Skill 6)

1. $4.3 + 5.6$

2. $9.875 + 4.37$

3. $9.8 - 0.91$

4. $6.23 - 5.4$

5. $253.9 + 837.54$

6. $427.981 - 23.6$

Find the product. (Skill 7)

7. 43.6×9.824

8. $19.2\,(42.65)$

9. $(16.352)(25.4)$

Divide. Round to the nearest hundredth if necessary. (Skill 8)

10. $3672 \div 3.2$

11. $2433 \div 12.5$

12. $19.65 \div 24$

A house has 826 wall studs, 220 rafters, and 24 stair treads. Write each ratio in simplest form. (Skill 18)

13. studs to rafters

14. treads to rafters

15. treads to studs

Solve each equation.

16. $\frac{x}{43} = 28.6$

17. $19.2y = 333.12$

18. $0.364z = 5460$ (Skill 28)

19. $\frac{x}{3.4} + 2.6 = 11.44$

20. $5.6q + 2.33q = 793$

21. $\frac{w}{2.4} - 3.63 = 19.2$ (Skill 29)

Simplify. (Skill 30)

22. $(0.87)^2$

23. 11^3

24. $(0.15)^2$

Find the square root. (Skill 32)

25. $\sqrt{841}$

26. $\sqrt{5.29}$

27. $\sqrt{1369}$

Self-Analysis Test

Use the trigonometric ratio table in the back of the book or a scientific calculator.

1. A section of road rises 26.7 ft over a horizontal distance of 943 ft. Find the angle of elevation.

2. A balloon is attached to 150.0 ft of rope. The angle between the rope and the ground is 82°. What is the altitude of the balloon?

3. Find the angle of elevation of a cliff that is 82.5 ft high from a point that is 38.6 ft from the base of the cliff.

4. A bridge is being built across a canyon that is 650 ft wide. If one side of the canyon is 25 ft higher than the other side, find the angle of elevation of the bridge.

5. A weather balloon is sighted at an angle of elevation of 36° from a ship. Radar determines the distance between the ship and the balloon to be 3450 m. What is the height of the balloon?

6. A submarine dives from the surface at an angle of 11°. It travels diagonally 720 ft. How deep is the submarine below the surface?

7. A ship travels due south for 21.3 mi and then due east for 24.5 mi. How far is the ship from its starting point?

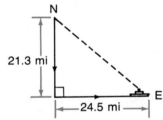

8. A fishing boat traveled 35 km on a bearing of 075°. How far east has the fishing boat gone?

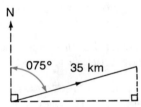

9. A roof has the dimensions shown below. Find the measure of the angle between the roof and the ceiling top plate.

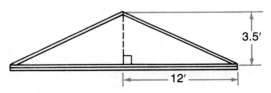

10. A sloping approach to an overpass is 58 m long, as shown below. Find the angle of elevation between the approach and the horizontal plane.

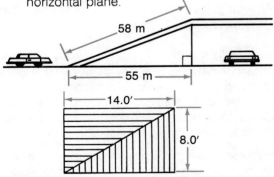

11. A wall 14.0 ft long and 8.0 ft high is to be covered with tongue-and-groove boards, as shown at the right. What is the measure of the angle between the diagonal seam and the floor?

For all kinds of construction, whether it is the building of bridges, roads, houses, shopping malls, or skyscrapers, it is important to consider the topographical features of the land. Such features include the basic contour of the land as well as the positions of ridges, valleys, rivers, and vegetation. It is the work of the surveyor to determine as exactly as possible these topographical features.

Job Description

What does a surveyor do?

1. A surveyor uses various instruments (levels, transits, theodolites) to pinpoint as closely as possible the topographical features of an area.

2. Once the topographical measurements have been made, a surveyor prepares a map that graphically represents all of the important physical features. Such maps may be used for geological surveys, highway and building planners, or mining surveys.

3. A surveyor might be asked to write a description of a piece of land for deeds, leases, and other legal documents.

4. The official land and/or water boundaries for parcels of land can be established by a surveyor.

5. A surveyor is required to convey topographical information to the construction engineers and architects who will design and construct highways, bridges, buildings, and mining tunnels.

6. Marine surveyors survey harbors, rivers, and lakes to measure shorelines, water depths, and other features.

7. A large manufacturing plant might ask a surveyor to check the positioning of machines.

Qualifications

A good surveyor is familiar with the use of all surveying instruments and has the ability to visualize objects, distances, and sizes. It is important to be able to make accurate mathematical calculations and pay close attention to details. Courses in algebra, geometry, trigonometry, drafting, mechanical drawing, and computer science are beneficial.

Many community colleges offer courses in surveying technology. In some cases, coursework can be combined with on-the-job training.

FOCUS ON INDUSTRY

Since its first commercial use in the 1850s, petroleum has become a major energy source for the U.S. and the rest of the industrialized world. Petroleum is the mixture of complex hydrogen and carbon compounds known as hydrocarbons, found in the Earth's crust. Petroleum can occur in liquid and gaseous forms known as crude oil and natural gas. Natural gas is often found along with oil deposits in sedimentary rock and, until recently, had been thought of as a waste product of oil drilling. Since we are becoming more conscious of conserving energy, natural gas is not wasted as much today.

Oil and gas are found in huge sedimentary basins. The U.S. possesses many such basins on land as well as in the water, especially in the Gulf of Mexico and off the coast of California. Since the number of basins that are actually producing oil or gas on land is diminishing, the greatest potential for future petroleum sources lies offshore.

Estimates of the undiscovered reserves in offshore basins vary considerably. A recent survey yielded these high and low estimates of potential resources.

Estimates of Offshore Oil and Gas Reserves					
	Oil in billions of barrels			Gas in trillions of cubic feet	
Region	Low		High	Low		High
Alaskan Seas	4.6	to	24.2	33.3	to	109.6
Pacific Coast	1.7	to	7.9	3.7	to	13.6
Gulf of Mexico	3.1	to	11.1	41.7	to	114.2
Atlantic Coast	1.1	to	12.9	9.2	to	42.8

The U.S. per-capita demands for oil and natural gas have been among the highest in the world. This results largely from our high standard of living. The automobiles we use, for example, are at present completely fueled by petroleum products.

Refer to the table above to answer the following questions.

1. Write out the estimated amounts of energy reserves given below.

    a. the high estimate of reserve oil in the Alaskan Seas
    b. the low estimate of gas reserves in the Gulf of Mexico
    c. the high estimate of reserve oil off the Pacific Coast

2. What area appears to have the greatest potential for high-level gas production?

Chapter 10 Test

Use the trigonometric ratio table in the back of the book or a scientific calculator.

1. The size of a television screen is usually given by the length of its diagonal. If a television screen is 21.0 in. by 15.5 in., could the set be advertised as a "26 inch TV set?" Ignore the rounded corners of most TV sets. (10-1)

2. A rectangular football field is 120.0 yd long from goalpost to goalpost and 50.0 yd wide from sideline to sideline. Find the distance between opposite corners of the field. (10-1)

3. A ramp for loading heavy equipment on trucks is 22.1 ft long. If the horizontal distance covered by the ramp is 21.8 ft, find the amount of vertical rise. (10-1)

4. An 8.0 ft ladder is leaning against a wall with the base of the ladder 3.2 ft from the base of the wall. Find the tangent of the angle between the ladder and the ground. (10-2)

Refer to the sketch at the right measuring cloud height to write each trigonometric ratio.

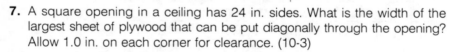

5. Find sin ∠ A; cos ∠ A; tan ∠ A. (10-2)

6. Find sin ∠ B; cos ∠ B; tan ∠ B. (10-2)

7. A square opening in a ceiling has 24 in. sides. What is the width of the largest sheet of plywood that can be put diagonally through the opening? Allow 1.0 in. on each corner for clearance. (10-3)

8. A square window opening has sides that are 31.5 in. long. Calculate the length of the diagonal of the square opening. (10-3)

9. When the angle of elevation of the sun is 30°, a tower casts a shadow that is 28.3 ft long. How tall is the tower? (10-4)

10. A company logo that is put on large equipment has the shape of an equilateral triangle. If each side of the logo is 38 cm, what is the height of the logo? (10-4)

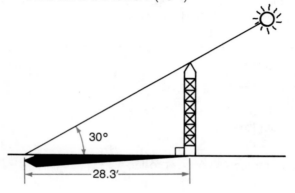

11. A helicopter is hovering directly over an intersection. The angle of depression to an accident 1.2 mi away from the intersection is 26°. How far is the helicopter from the accident? (10-5)

12. The angle of elevation of a ski slope averages 38°. If the slope is 2600 m long, what is the vertical drop? (10-5)

13. A loading ramp has a vertical rise of 4.25 ft over a horizontal distance of 12.5 ft. Find the angle of elevation. (10-5)

14. A warning light is located at the top of a cliff 52.1 m above the shoreline. The navigator of a ship measures the angle of elevation to the light at 16°. How far is the ship from the base of the cliff? (10-6)

15. A ship sails 15.2 km east, makes a 90° turn, and sails south an additional 11.7 km. How far is the ship from the starting point? (10-6)

16. A sailboat has traveled 28 km on a bearing of 056°. How far east has the sailboat gone? (10-6)

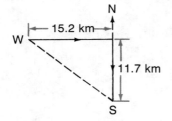

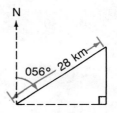

17. From the top of a lighthouse that is 155 ft above sea level, the angle of depression to a ship is 31°. Find the distance from the ship to the base of the lighthouse. (10-6)

18. An airplane is 5400 ft above a body of water. The navigator of the airplane measures the angle of depression to a ship on the water below to be 25°. How far is the boat from the airplane at the time of the sighting? (10-7)

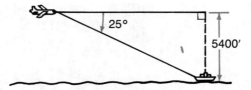

19. A guy wire from a 64 ft pole is to be fastened at a distance of 28 ft from the base of the pole. What will be the measure of the angle the guy wire makes with the ground? (10-7)

20. The angle of elevation from horizontal ground level to the top of a hill is 8°. How far does a car travel on this hill if its final position is 8.5 ft *above* its initial position, which was at horizontal ground level? (10-7)

Chapter 11
Statistics

After completing this chapter, you should be able to:

1. Organize data into a tally and frequency table.

2. Find the mean, median, and mode.

3. Determine the best measure of central tendency for a set of data.

4. Read and interpret stem-and-leaf plots, box plots, bar graphs, histograms, broken-line graphs, pictographs, and circle graphs.

5. Recognize the misuse of graphs in the representation of data.

6. Graph the relationships between two sets of data on a coordinate grid and recognize these relationships as direct, inverse, joint, or combined variation.

Data Analysis

11-1   Collecting and Organizing Data

Numerical facts called **data** are found in many places in everyday life. We often need to analyze and draw conclusions from data, making it necessary to organize the facts into a table, chart, or graph. This organizing, analyzing, and interpreting of data is part of the branch of mathematics called **statistics**.

When collecting data, it is not always possible to obtain a total count from the entire population. By using a **sample** count that is **representative** of all characteristics of the total population, conclusions from the data can be more reliably estimated.

The example below shows how data can be organized in a **tally and frequency table**.

Example

Skill(s) *2, 20, 22, 24*

A restaurant manager collected data on the items chosen by customers in one day. The menu choices, quarter chicken (Q), half chicken (H), two chicken legs (T), chicken and ribs (R), and chicken fingers (F), are shown below.

Menu choices

Q, R, F, T, T, H, H, Q, F, R, T, H, F, R, H, F, Q, Q, H, T, F, R, H, Q, H, H, T,
R, R, H, H, H, Q, Q, T, Q, F, Q, H, T, R, F, H, Q, F, H, R, R, T, Q, H, F, H, Q,
H, T, T, F, Q, Q, F, F, H, R, H, R, T, H, Q, H, T, Q, H, R, Q, F, H, Q, F, Q, R,
F, H, R, H, F, F, T, R, T, T, F, H, H, T, Q, Q, Q, F, F, H, H, R, T, H, F, Q, F.

What menu item was selected most frequently that day? What percent of all the choices was that choice? Round your answer to the nearest whole percent.

Solution

Make a tally and frequency table of the data collected.

Menu Item	Tally	Frequency
Quarter chicken	卌 卌 卌 卌 III	23
Half chicken	卌 卌 卌 卌 卌 卌	30
Two chicken legs	卌 卌 卌 II	17
Chicken and ribs	卌 卌 卌 I	16
Chicken fingers	卌 卌 卌 卌 II	22
Total		108

The menu item most frequently selected was the half chicken dinner. It was chosen 30 out of 108 times.

$$3 \;\; 0 \;\; \div \;\; 1 \;\; 0 \;\; 8 \;\; = \;\; 0.2777778$$

Of the choices made, 28% were for the half chicken dinner.

EXERCISES

A An environmentalist sighted the following numbers of deer at dusk at the edge of a forest. The data collected is shown in the table below.

Deer Sighted	Tally	Frequency
Adult males	ꟷꟷꟷꟷꟷꟷ	?
Adult females	ꟷꟷꟷꟷꟷꟷꟷꟷ III	?
Fawns	ꟷꟷꟷ II	?
Total		?

1. In each category, how many deer were sighted?

2. What percent of the deer sighted were fawns? Answer to the nearest whole percent.

3. Ordinarily, 36 adult males are sighted at the same time and place. How many fewer deer were sighted on this day? What percent is this of the total number of deer sighted?

B Quality-control inspectors on a production line recorded the following numbers of defective computer disks found in a *representative sample* of all disks manufactured for each working day of one week in the table below.

Day	Tally	Frequency
Monday	ꟷꟷꟷꟷ	?
Tuesday	ꟷꟷꟷ	?
Wednesday	ꟷ II	?
Thursday	ꟷꟷ I	?
Friday	ꟷꟷꟷ III	?
Total		?

4. How many defective computer disks were found each day?

5. Which single day produced as many defective computer disks as two other days combined?

6. If we assume that only 0.2% of the computer disks manufactured on Wednesday were defective, what is the total number of computer disks that were produced on Wednesday?

A survey was taken at an intersection during rush hour one day. The data in the table below show which directions vehicles passing through the intersection traveled, right (R), left (L), or straight (S).

Directions Traveled Through an Intersection

R, L, R, S, S, L, R, L, R, S, S, R, S, R, R, R, S, S, L, L, R, R, S, S, S, R, L, S,
S, S, S, R, R, R, R, S, S, S, S, L, S, S, L, R, S, S, R, R, S, S, S, S, L, L, R, S,
S, S, R, L, L, L, R, R, S, S, S, S, R, R, S, L, L, L, L, R, S, L, R, L, R, L, S, S.

7. Make a tally and frequency table for the traffic data above.

8. Which direction was most frequently traveled?

9. What percent of the time was either a right or a left turn made? Answer to the nearest whole percent.

A survey was taken of the frequency of pedestrian traffic passing by three different locations in a city during the evening hours. The data collected is shown below.

Hour (P.M.)	Location 1	Percent	Location 2	Percent	Location 3	Percent
4:00–5:00	252	?	73	?	99	?
5:00–6:00	188	?	152	?	167	?
6:00–7:00	97	?	296	?	163	?
7:00–8:00	180	?	89	?	167	?
8:00–9:00	212	?	132	?	148	?
9:00–10:00	94	?	135	?	155	?
Total	?	100.0%	?	100.0%	?	100.0%

10. What is the total number of pedestrians seen at each location at each time?

11. Calculate each percent of the total for each time for all locations to the nearest tenth of a percent.

12. Which location would be best for a shoe store?

13. Which is the best location for a meat market that is open until 6 P.M.?

C Inspectors at a ball-bearing manufacturing plant took measurements of the diameters of a *representative sample* of the bearings produced one day. The data below range in size from 1.00 cm to 1.20 cm.

Ball-bearing diameters in centimeters
1.07, 1.00, 1.07, 1.08, 1.05, 1.10, 1.07, 1.07, 1.09, 1.08,
1.07, 1.04, 1.07, 1.06, 1.19, 1.07, 1.07, 1.08, 1.09, 1.10,
1.12, 1.07, 1.07, 1.20, 1.07, 1.06, 1.02, 1.07, 1.15, 1.08.

14. Group the data in *intervals* as you make a tally and frequency table. Use intervals of 1.00 cm–1.04 cm, 1.05 cm–1.09 cm, 1.10 cm–1.14 cm, 1.15 cm–1.19 cm, and 1.20 cm–1.24 cm.

15. If an acceptable diameter is 1.07 cm $\pm$ 0.02 cm, what percent of the ball bearings tested are acceptable?

16. If 500 ball bearings are made per hour and the plant is in operation for eight hours, how many bearings, to the nearest hundred, can be expected to be produced within acceptable limits?

Technology Update

The first United States census was taken in 1790. Since then, the census has been taken every ten years in an effort to create a statistical portrait of the nation. The 1980 census cost the government over one billion dollars, used 275,000 temporary workers in addition to Census Bureau employees, and took over three years to complete. The information gathered for the 1980 census represents the most accurate and complete collection of data ever collected about a society.

11-2 Mean, Median, and Mode

A forester measured the heights of a representative sample of trees after one year of growth. The data collected are shown below.

Heights (m) of trees
1.8, 1.6, 1.5, 1.9, 1.9, 1.3, 1.3, 1.5, 1.4, 2.0,
1.8, 1.9, 1.6, 1.8, 1.8, 1.5, 1.4, 2.0, 1.8, 1.2.

If this set of data were to be represented by a single statistic, it would be most appropriate to use a measure of **central tendency**. There are three different measures of central tendency: **mean**, **median**, and **mode**. The example below explains each measure of central tendency and shows how each would be calculated for the set of data above.

Example

Skill(s) *2, 4, 24*

Find the mean, median, and mode of the heights of the trees.

Solution

1. The **mean** is the sum of all values divided by the number of values in the set. In this case, the 20 heights of the trees are added and their sum is divided by 20.

$$\text{mean} = \frac{\text{sum of all heights of trees}}{20} = \frac{33}{20} \qquad \boxed{3}\ \boxed{3}\ \boxed{\div}\ \boxed{2}\ \boxed{0}\ \boxed{=}\ 1.65$$

The mean height is 1.7 m, to two significant digits.

2. The **median** is the middle value when the heights are arranged in order.

1.2, 1.3, 1.3, 1.4, 1.4, 1.5, 1.5, 1.5, 1.6, 1.6, | 1.8, 1.8, 1.8, 1.8, 1.8, 1.9, 1.9, 1.9, 2.0, 2.0
1.7

The median height is *halfway* between 1.6 m and 1.8 m, or 1.7 m.

3. The **mode** is the value that occurs most frequently in the overall tally. There can be no mode, one mode, or two modes. If there are more than two modes, we generally say there is no mode, since there is no clear mode.

For the trees, the most frequently occurring height is 1.8 m. Thus, the mode is 1.8 m.

EXERCISES

A The table below shows the amount of money an individual was able to save each month for one year.

> **Amounts of money saved each month**
> $100, $180, $140, $40, $60, $90, $55, $60, $128, $120, $125, $110.

Answer *true* or *false*.

1. The median amount of money saved is $100.

2. There is no mode of the savings data.

3. The mean amount of money saved in a month, to the nearest dollar, is $101.

B A naturalist studying the quality of water determined the number of mayfly larvae per 125 mL jar in several parts of a stream. The presence of mayfly larvae is an indication of good quality water. The data found is shown in the table below.

> **Mayfly larvae found per 125 mL jar**
> 47, 49, 53, 55, 54, 32, 42, 48, 46, 44, 42,
> 47, 40, 51, 57, 31, 28, 38, 34, 46, 29, 38,
> 50, 38, 26.

4. Arrange the data in order from lowest to highest.

5. What is the median number of mayfly larvae per 125 mL jar?

6. What is the mean for these data?

7. Do these data have a mode? If so, what is it?

A tire company wear tested twelve tires for 50,000 mi each. The data below show the amount of tread wear for the tires after 50,000 mi.

> **Amount of tire tread wear (in.)**
> 0.231, 0.342, 0.197, 0.292, 0.137, 0.179,
> 0.341, 0.315, 0.264, 0.281, 0.244, 0.215.

8. Arrange the data in order from lowest to highest.

9. What is the mean, median, and mode for these data?

10. What percent of the twelve tires had less than 0.200 in. of tread wear after 50,000 mi?

11. If the tires had 0.500 in. of tread when the test was started, what was the mean amount of tread remaining after the test, to the nearest thousandth of an inch?

11-3 Best Measure of Central Tendency

The mean, median, and mode are each statistics that represent the central value in a set of data. The appropriateness of each measure depends upon the data represented. For example, the mean is not always the statistic that will best summarize a set of data.

Example 1

Skill(s) _2.4_

Environmentalists tagged 21 birds to study their health and migration patterns from region to region. As the birds were tagged, their masses were measured. The data obtained one day are shown below.

Masses (kg) of 21 birds
2.4, 1.2, 2.3, 2.2, 1.2, 2.2, 2.2, 2.4, 2.1, 1.1, 2.2, 1.4, 2.2, 1.3, 2.1, 2.1, 2.3, 1.2, 1.3, 2.1, 2.1.

Find the mean, median, and mode of the data. Then decide which is the best measure of central tendency to describe this particular set of data.

Solution

Step 1: Find the mean mass.

$$\text{mean} = \frac{\text{sum of all masses}}{21} = \frac{39.6}{21}$$

$$\boxed{3}\;\boxed{9}\;\boxed{.}\;\boxed{6}\;\boxed{\div}\;\boxed{2}\;\boxed{1}\;\boxed{=}\;1.8857143$$

The *mean mass* is 1.9 kg, to two significant digits.

Step 2: Find the median by first arranging the masses in order from lowest to highest.

1.1, 1.2, 1.2, 1.2, 1.3, 1.3, 1.4, 2.1, 2.1, 2.1, 2.1, 2.1, 2.2, 2.2, 2.2, 2.2, 2.2, 2.3, 2.3, 2.4, 2.4.

The *median* mass is the eleventh value, or 2.1 kg.

Step 3: There are *two modes* for this set of data, 2.1 kg and 2.2 kg.

Fourteen of the 21 masses are over the mean, 1.9 kg. In this case, the mean is not a good *central value*. The mode is also not an appropriate measure of central tendency for this data since there are two modes.

For this set of data, the median is the best measure of central tendency. In general, the median is more descriptive of data than the mean if there is any possibility of having even a few unusually large or small values.

Example 2

A meteorologist checked the amounts of rainfall in millimeters during a storm in 16 locations. The data collected is shown at the right.

> **Rainfall (mm)**
> 8, 5, 4, 3, 7, 22, 3, 4,
> 6, 7, 4, 19, 4, 6, 4, 5.

Which measure of central tendency best supports the claim that the amount of rainfall in the 16 regions *averaged* 7 mm?

Solution

Find the mean, median, and mode of the data.

Step 1: The mean is the sum of all rainfall amounts, divided by 16.

$$\text{mean} = \frac{111}{16} \qquad \boxed{1}\,\boxed{1}\,\boxed{1}\,\boxed{\div}\,\boxed{1}\,\boxed{6}\,\boxed{=}\ 6.9375$$

The mean amount of rainfall is 7 mm, to one significant digit.

Step 2: Arrange the values in numerical order to find the median.

3, 3, 4, 4, 4, 4, 4, 5 | 5, 6, 6, 7, 7, 8, 19, 22

The median amount of rainfall is 5 mm.

Step 3: The mode is 4 mm. This amount of rainfall occurred five times.

The mean, 7 mm, is closest to the claim that the amounts of rainfall *averaged* 7 mm.

EXERCISES

A 1. Five fishing boats came ashore with the following numbers of salmon: 1345, 1496, 1417, 2982, and 3015. One fisherman claimed the five boats averaged more than 2000 salmon caught. A second fisherman claimed more boats caught fewer than 1500 salmon. Find a measure of central tendency that supports each claim.

B 2. A subdivision has houses priced at $90,000, $70,000, $70,000, $65,000, $70,000, $130,000, and $70,000. One real estate agent claims the average house price is $80,000. A second agent claims most houses are priced no higher than $70,000. Find a measure of central tendency supporting both claims.

3. One officer in a small company earns $70,000 per year. Four other officers each earn $30,000 per year and one officer earns $20,000 per year. Which measure of central tendency best describes the salaries of the six officers?

4. Below are fifteen measurements for the strength of steel specimens. Which measure of central tendency would best describe the data?

> **Strength of steel measurements (pounds per square inch)**
> 40,250, 60,400, 47,150, 65,100, 44,300, 73,200, 49,400, 26,550, 35,800, 22,400, 31,200, 45,350, 38,400, 55,600, 52,300.

5. The annual earnings per share of a corporation for the past seven years are shown below.

> **Annual earnings per share**
> $0.12, $0.14, $0.28, $0.14, $0.27, $0.28, $0.15

   a. Which measure of central tendency would you use if you wanted to promote the sale of this stock?
   b. If you wanted to discredit the value of this stock, which measure of central tendency would be best to use?

6. The amounts of money spent on advertising by a small company during its first eight years of operation are listed in the table at the right.
   a. Find the mean cost of advertising for the first five years.
   b. Find the mean cost of advertising for the last five years.
   c. Considering what the cost of advertising might be in the 9th year, which of the means (from questions 6a and 6b) best describes the data?

Year	Cost of Advertising
1st	$1000
2nd	$2000
3rd	$4000
4th	$8000
5th	$16,000
6th	$32,000
7th	$64,000
8th	$128,000

Tricks of the Trade

The Spreadsheet Average Function

Most computer spreadsheet programs contain several useful functions that allow the user to perform calculations *without* the aid of a calculator. One of these functions is called AVERAGE, which computes the mean of a set of data.

The grids at the right show how the AVERAGE function works. In spreadsheet grid 1, a set of lengths have been entered into cells A2, A3, A4, A5, A6, A7, and A8. In order to obtain the average (mean) of these seven lengths, the formula, = Average(A2:A8), is typed in cell A9. This formula tells the computer to place the average of cells A2 to A8 into cell A9, as shown in spreadsheet grid 2. Here, the average length, 2.52 cm, is shown.

1.

	A
1	Lengths (m)
2	2.25
3	2.86
4	2.58
5	2.44
6	2.48
7	2.14
8	2.90
9	=Average(A2:A8)

2.

	A
1	Lengths (m)
2	2.25
3	2.86
4	2.58
5	2.44
6	2.48
7	2.14
8	2.90
9	2.52

Self-Analysis Test

Use a calculator where possible.

A market researcher surveyed a *representative sample* of 170 homes to find the number of magazines subscribed to each month. The table below shows the data collected.

Number of Magazines	Tally	Frequency
0 or 1	~~THL THL THL THL~~ III	?
2 or 3	~~THL THL THL THL THL THL THL~~	?
4 or 5	~~THL THL THL THL THL THL THL THL THL THL~~ II	?
6 or 7	~~THL THL THL THL THL THL~~ I	?
8 or 9	~~THL THL THL~~ II	?
10 or 11	~~THL THL~~	?
12 or over	III	?

1. How many homes subscribed to the following numbers of magazines?

   a. 0 or 1   **b.** 2 or 3   **c.** 4 or 5   **d.** 6 or 7

   e. 8 or 9   **f.** 10 or 11   **g.** 12 or over

2. What percent, to the nearest whole number, of the 170 homes surveyed subscribed to 6 or 7 magazines each month?

3. If the entire population from which the *representative sample* was taken is 400,000 homes, how many of these homes might be expected to subscribe to 6 or 7 magazines each month?

The prices paid for apples each week for a 12-week period are listed in cents per pound in the table below.

Prices received for apples (cents per pound)
37.9, 38.1, 37.2, 36.8, 38.5, 39.0, 40.2, 41.5, 40.8, 39.7, 39.4, 40.1.

4. What is the mean price per pound paid for apples over the 12-week period?

5. What is the median price paid?

6. What is the mode for this set of data?

The Sunday circulations for seven newspapers are shown in the table below.

Sunday newspaper circulations
1,645,000, 1,630,000, 1,126,000, 740,000, 624,000, 508,000, 118,000.

7. Which measure of central tendency would you use to support the claim that the seven newspapers average a Sunday circulation just under 1,000,000?

8. If you wanted to point out that most Sunday circulations for the seven newspapers are 740,000 or under, which measure of central tendency would you use?

Data Graphs

11-4   Stem-and-Leaf Plots and Box Plots

Twenty-five single, working, adults living in the same community were surveyed to find the amount of money they spend on food in one month. The data collected is shown below.

Money spent on food in one month by single adults
$295, $348, $305, $358, $315, $340, $350, $318, $335, $350, $284, $310, $296, $342, $322, $285, $316, $328, $360, $336, $318, $312, $345, $292, $321.

The data can be quickly organized and displayed in a **stem-and-leaf plot**, as shown in the example below.

Example 1   Skill(s) __2__

Organize the data on the money spent on food into a stem-and-leaf plot. What is the median amount of money spent on food in one month?

Solution

Step 1:   Determine the spread of the scores. The highest amount is $360 and the lowest amount is $284. The **range**, or the difference between the highest and the lowest amount, is $360 − $284, or $76.

Step 2:   Think of each dollar amount as a *stem* and a *leaf*. For example, $305 would be the following.

$$\text{stem} \mid \text{leaf}$$
$$30 \mid 5$$

Write all of the stems vertically with a line to their right so the full range of dollar amounts can be included, as shown at the right. This gives a quick visual display of the way the amounts of money spent are clustered.

Food Money For One Month ($)	
28	4 5
29	2 5 6
30	5
31	0 2 5 6 8 8
32	①2 8 ◄— The median is $321.
33	5 6
34	0 2 5 8
35	0 0 8
36	0

28 | 4 represents $284.

Step 3:   Plot the leaves to the right of the stems in numerical order.

The median amount of money spent on food is found by locating the thirteenth dollar amount in the stem-and-leaf plot. This is easy to do since the amounts are arranged in numerical order. The median amount is $321.

In order to highlight the median, the middle half of the data, and the data extremes, a 50% **box plot** can be used. The example below shows how this is done with the same data that was used for the stem-and-leaf plot in Example 1.

Example 2

Organize the data on the money spent on food by single adults in one month into a box plot showing the upper and lower extremes, the median, and the medians of the upper and lower halves of the amounts.

Solution

Step 1: Draw a horizontal number line that includes the full range of dollar amounts.

Step 2: Use the stem-and-leaf plot for the data to locate the median amount and then mark it with a dot under the number line.

There are twelve amounts above and below the median. Locate the medians of each half of the data and mark them with dots on the number line. The medians of each half of the data are also known as **hinges**.

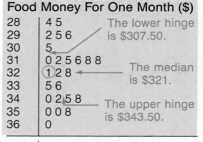

Food Money For One Month ($)

28	4 5
29	2 5 6
30	5
31	0 2 5 6 8 8
32	①2 8
33	5 6
34	0 2 5 8
35	0 0 8
36	0

The lower hinge is $307.50.

The median is $321.

The upper hinge is $343.50.

28 | 4 represents $284.

Mark the upper and lower extremes with dots.

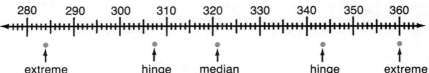

Step 3: Draw a box between the hinges or the middle half of the data. Draw a line through the box at the median. Then draw the two *whiskers* from the hinges to the extremes. Notice that the median and the two hinges divide the data into four quarters.

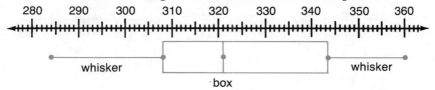

EXERCISES

A What is the median and the range for each set of data displayed below on a stem-and-leaf plot?

1. Heights (cm)

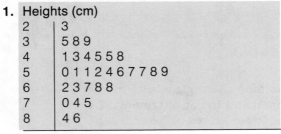

2	3
3	5 8 9
4	1 3 4 5 5 8
5	0 1 1 2 4 6 7 7 8 9
6	2 3 7 8 8
7	0 4 5
8	4 6

2 | 3 represents 23 cm.

2. Time (s)

1	8 9
2	5
3	4 7 8 8
4	0 5
5	2 6 7 9 9
6	1 4 5 5 8
7	2 3
8	0 4 6
9	5 5 6 7 8 8 9
10	0 0 1 3 3 4 4 6 6
11	0 2 2 3 4 5 7 7 8 9
12	2 2 2 3 4 6 9
13	0 0 0 1 2 5 7 8 9
14	0 1 2 3 8 9
15	0 0 1 2 7
16	1 1 2 4 8 8 9 9
17	5
18	3 7 8
19	7
20	0 1

7 | 2 represents 7.2 s.

3. Costs ($)

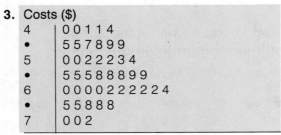

4	0 0 1 1 4
•	5 5 7 8 9 9
5	0 0 2 2 2 3 4
•	5 5 5 8 8 8 9 9
6	0 0 0 0 2 2 2 2 2 4
•	5 5 8 8 8
7	0 0 2

5 | 2 represents $52.

Identify the median, the upper hinge, and the lower hinge for each set of data displayed below in a box plot.

4.

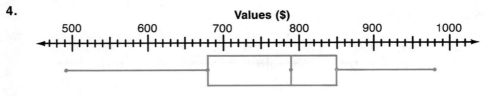

Values ($)

5.

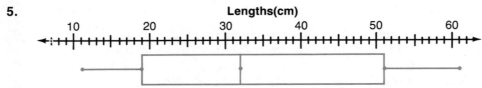

Lengths(cm)

B The data in the table at the right show the number of invoices processed in one day for several days.

6. Make a stem-and-leaf plot for the data.

7. Identify the range and median.

Invoices processed

450, 385, 510, 392, 444, 421, 392, 412, 399, 381, 362, 377, 428, 386, 369, 431, 437, 449, 403, 421, 378, 366, 384, 398, 409, 452, 422, 438.

Use the data given in each stem-and-leaf plot to make a box plot.

8. New Houses

12	1 2 2 5 7 7 8
13	0 1 2 3
14	1 2 3 5 6 7 8 9
15	2 5 5
16	0 1 3
17	1

12 | 5 represents 125 new houses.

9. Books Sold

8	1
9	2 3 5 5 7
10	5 5 6 8
11	2 3 3
12	2 4 5 6 6 8 9
13	0 5
14	0 1 2
15	2 5 8 9

8 | 1 represents 8,100,000 books sold.

The data in the table below shows some prices of cars.

Car prices ($1000)
7.5, 8.3, 9.0, 9.6, 9.8, 10.9, 11.0, 11.6, 12.0, 13.0, 13.6, 15.2, 17.9.

10. Find the median and the lower and upper hinges of the data.

11. Make a box plot for the data.

The data in the table below shows the heights of various seedling plants.

Heights of seedling plants (cm)
18.2, 28.4, 37.5, 24.6, 19.9, 25.0, 39.2, 41.0, 38.6, 35.0, 18.2, 28.4, 37.5, 24.6, 19.9, 25.0, 39.2, 41.0, 28.6, 35.0, 33.7, 26.1, 32.3, 18.9, 29.0, 21.5, 37.0, 42.9, 26.5, 38.7.

12. Identify the range, the upper and lower hinges, and the median of the data.

13. Make a box plot of the data.

14. Since the data are split into four quarters by the median and the two hinges, which quarter has the greatest range?

C The box plot below displays how 25 single, working, adults living in the same community spend their money on food, shelter, and savings in one month.

15. Which quarter of the savings data for one month has the greatest range?

16. For most of the single adults surveyed, about how much is spent individually on food, shelter, and savings per month, to the nearest ten dollars?

Money Spent by Single Adults ($)

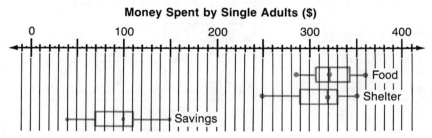

11-5 Bar Graphs and Histograms

Bar graphs and histograms use bars to provide a visual representation of data on an evenly divided scale. **Bar graphs** are used for data that fall into separate and distinct categories. They may contain vertical or horizontal bars to display data. In a bar graph, the bars do *not* touch.

Example 1

Skill(s) _24_

The loudness of the sounds made in our environment is measured in decibels. The table below summarizes some data.

Example	Loudness in Decibels
jackhammer	120
jet engine	140
quiet garden	30
metals foundry	110
a whisper	15
vacuum cleaner	80

Make a bar graph of the data.

Solution

Notice that the bars do *not* touch on a bar graph as each bar represents a separate category. The *horizontal axis* shows an evenly divided scale ranging from 0 to 160 decibels of loudness.

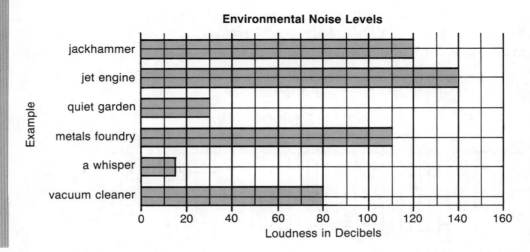

Environmental Noise Levels

Histograms are used to display *frequencies* of data on an evenly divided scale. A histogram is a vertical bar graph with *no space* between the bars.

Example 2

A study was made of the usage in successive weeks after the introduction of a bank machine in an average-size city. The number of bank customers for the first four weeks was recorded each week in the tally and frequency table shown below. Make a histogram of the data.

Bank Machine	Tally	Frequency
1st week	卌 卌 卌 卌 卌 I	26
2nd week	卌 卌 卌 卌 卌 卌 卌 卌 卌 卌 卌 卌 卌 III	68
3rd week	卌 卌 卌 卌 卌 卌 卌 卌 卌 卌 卌 卌 卌 卌 卌 卌 I	81
4th week	卌 卌 卌 卌 卌 卌 卌 卌 卌 卌 卌 卌 卌 卌 卌 卌 卌 卌 卌 II	97

Solution

Notice that the bars of a histogram are *touching* and of equal width so their heights and areas are proportional to their frequencies. The *vertical axis* shows the frequency with an evenly divided scale ranging from 0 to 100 persons using the bank machine.

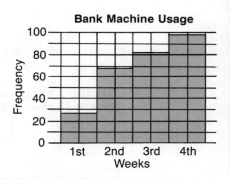

EXERCISES

A Refer to the bar graph at the right.

1. What information is given along the horizontal axis of the graph?

2. What amount does the smallest division of the scale along the vertical axis of the graph represent?

3. In which year were appliance sales the greatest?

4. What is the amount of sales, in dollars, for 1988?

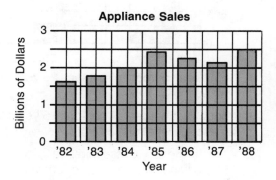

B The number of passengers at the ten busiest international airports in the United States in a recent year are shown in the histogram below.

Number of Passengers

5. How many passengers passed through Chicago?

6. How many more passengers were handled by the Atlanta airport than La Guardia?

7. To the nearest ten million, how many passengers were handled by all ten airports?

Below is a *double bar graph* showing the high school and college education of people 25 years old or more.

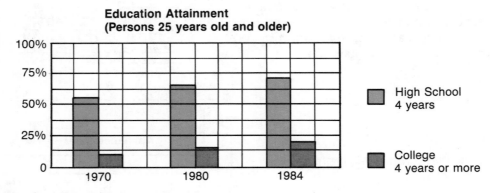

**Education Attainment
(Persons 25 years old and older)**

8. About what percent of persons 25 years and older attended four years of high school in 1980?

9. About what percent of persons 25 years and older attended four or more years of college in 1984?

10. What is the percent difference between 1970 and 1984 for persons 25 years and older attending four years of high school?

The bars on the graph are colored to indicate each of three age groups in the United States population.

Population Age Distribution

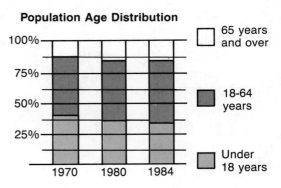

11. About what percent of the population was under 18 years in 1984?

12. About what percent of the population was between 18 and 64 years in 1980?

13. Which age group has decreased in percent of the population from 1970 to 1984?

14. The table below shows the high wind speeds in miles per hour for several weather stations for a recent year. Make a vertical bar graph of the data.

Station	High (mi/h)
Atlanta, Georgia	70
Bismark, North Dakota	72
Cape Hatteras, North Carolina	110
Fort Smith, Arkansas	58
Galveston, Texas	100

15. Make a histogram of the menu choices frequency table in the example in lesson 11-1.

16. Make a bar graph of the data in the ball-bearing diameters table found in the exercise questions for lesson 11-1.

C The masses of codfish caught in one day are recorded in the table below.

Masses Intervals (kg)	Tally	Frequency
2.00–2.39	卌 卌 ll	12
2.40–2.79	卌 卌 卌 卌 卌 lll	28
2.80–3.19	卌 卌 卌 卌 l	21
3.20–3.59	卌 卌 卌 卌 卌 卌 卌 l	36
3.60–3.99	卌 卌 卌 卌 卌 卌 卌 卌 l	41
4.00–4.39	卌 卌 卌 卌 卌 卌 卌 卌 卌 卌 卌 ll	57
4.40–4.79	卌 卌 卌 卌 卌 卌 卌 卌 卌 ll	47
4.80–5.19	卌 卌 卌 卌 卌 lll	28
5.20–5.59	卌 卌 卌 lll	18

17. Make a histogram of the data.

18. What percent of the cod caught had a mass of 4.00 kg or more, to the nearest whole percent?

19. What percent of the cod caught had a mass of from 2.00 kg to 2.79 kg, to the nearest whole percent?

20. In what interval is the median mass of the cod?

11-6 Broken-Line Graphs and Pictographs

A **broken-line graph** is used to display *continuous data* that change gradually over time, such as temperature and weight, and are recorded only at specific times. This kind of graph is made by joining successive plotted points with straight lines.

Example 1

Skill(s) *2, 4, 18, 24*

The temperatures of a patient in the hospital taken every morning and evening are shown in the table below. These temperatures are an indication of the patient's state of well-being.

Make a broken-line graph of the data.

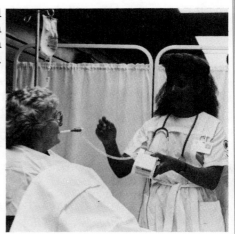

Date	Temperature (°F)			
June 21	morning	99.5	evening	100.0
June 22	morning	101.0	evening	100.5
June 23	morning	100.5	evening	100.0
June 24	morning	99.5	evening	100.0
June 25	morning	99.0	evening	99.5
June 26	morning	98.6	evening	98.0
June 27	morning	98.0	evening	98.6
June 28	morning	98.6		

Solution

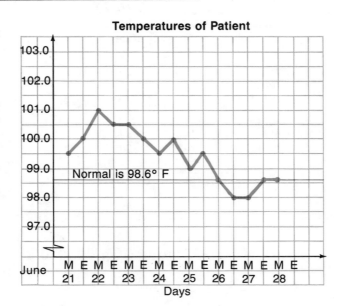

A **pictograph** uses pictures or symbols to represent *large* values.

Example 2

The table below shows the number of television sets sold by a chain of appliance stores.

Year	Television Sets Sold
1982	1,800,000
1983	2,400,000
1984	2,250,000
1985	2,100,000
1986	2,700,000
1987	3,150,000
1988	3,450,000

Make a pictograph of the data.

Solution

Number of Television Sets Sold

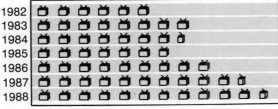

= 300,000 television sets

EXERCISES

A Refer to the television set data above.

1. How many television sets were sold in 1982? in 1986?

2. About what might you expect the total sales for 1989 to be?

3. What was the average number of television sets sold for the seven-year period?

Refer to the temperature data in Example 1.

4. When was the patient's temperature normal?

5. What was the patient's highest temperature? When did it occur?

6. What was the patient's temperature on:

a. the morning of June 24?   **b.** the evening of June 27?

7. What was the patient's mean morning temperature?

B The amount of salt contained in five leading brands of mineral water is given in the pictograph below.

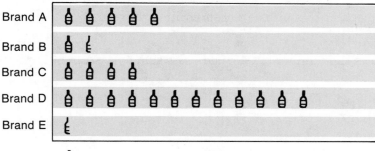

Salt Contained in Mineral Water

$\stackrel{\text{\i}}{\text{\j}}$ = 100 ppm (parts per million)

8. How much salt is in brand B?

9. Any beverage that contains 50 or fewer parts per million salt can be considered salt-free. Identify the salt-free brands shown in the pictograph.

10. a. If 1 ppm is the same as 1 mg/L, how many milligrams of salt are contained in a 1L bottle of brand D?
  b. How many milligrams of salt are contained in a 1L bottle of brand B?

The broken-line graph below shows the pulse rate of a patient in the hospital. The pulse rate tells how fast the heart is beating per minute.

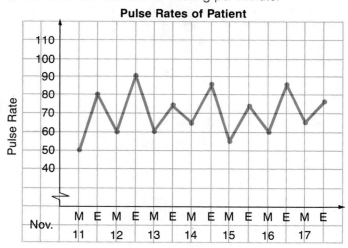

Pulse Rates of Patient

11. What was the pulse rate on the morning of November 14?

12. At what times did the patient's resting pulse rate, measured in the morning, fall within the normal range of between 72 and 80?

13. What was the mean evening pulse rate, to the nearest whole number?

The national average price for a gallon of gasoline for each of the first six months of a recent year is listed in the table below.

Average Gasoline Prices						
Month	January	February	March	April	May	June
Price	$1.15	$1.14	$1.16	$1.21	$1.23	$1.24

14. Display the data on a broken-line graph.

15. What might you expect the price of gasoline to have been in July?

A test of the lives of 400 light bulbs resulted in the data shown below.

Light Bulb Test	
Number of Hours Lasted	Number of Bulbs
600–649	28
650–699	48
700–749	64
750–799	102
800–849	72
850–899	50
900–949	36

16. Display the data on a pictograph.

17. The average life of a light bulb is 750 h. What percent of the bulbs tested lasted fewer than 750 h, to the nearest whole percent?

18. What percent of the bulbs lasted more than 849 h, to the nearest whole percent?

19. Find the ratio of the number of bulbs that lasted fewer than 750 h to the number of bulbs that lasted more than 799 h.

The average precipitation for Bismark, North Dakota, for each of the twelve months of the year is given below.

Average Precipitation for Bismark, North Dakota						
Month	Jan.	Feb.	Mar.	Apr.	May	June
Precipitation (in.)	0.5	0.5	0.7	1.5	2.2	3.0
Month	July	Aug.	Sept.	Oct.	Nov.	Dec.
Precipitation (in.)	2.0	1.7	1.4	0.8	0.5	0.5

20. Display these data on a broken-line graph.

21. What is the total average precipitation for one year?

22. What percent of the total average precipitation falls during the growing season, from April through September, to the nearest whole percent?

23. What percent of the total average precipitation falls during the months of November, December, January, and February, to the nearest whole percent?

11-7 Circle Graphs

A circle graph is used to display the relationship between parts of a set of data and the whole set. It also displays the relationships of the parts to each other. This is accomplished by changing each part of a set of data to a percent of the whole set and then representing this percent as a central angle of a 360° circle.

Example

Skill(s) *24, 25*

The table at the right shows the costs involved in the production of one $12 LP record.

Display the data on a circle graph.

Cost of Producing a $12 LP Record	
Expense Items	Cost
Distribution and selling	$4.80
Manufacturing	$1.80
Record company profit	$3.00
Songwriter, singer, musicians	$2.40
Total:	$12.00

Solution

For each expense, find its percent of the total cost to the nearest whole number. Then find the central angle size needed on a circle graph by multiplying each percent by 360°.

Expense	Money Spent	Percent	Central Angle
Distribution and selling	$4.80	40% (4.8 ÷ 12)	144° (40% × 360°)
Manufacturing	$1.80	15% (1.8 ÷ 12)	54° (15% × 360°)
Record company profit	$3.00	25% (3 ÷ 12)	90° (25% × 360°)
Songwriter, singer, musicians	$2.40	20% (2.4 ÷ 12)	72° (20% × 360°)
Total:	$12.00	100%	360°

To make the circle graph, use a compass to draw a circle and mark off the required central angles with the aid of a protractor.

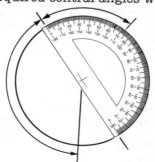

Expenses for a $12 LP Record

EXERCISES

A Refer to the cost data for a $12 LP record on the previous page.

1. What percent of the total cost of making a record is for manufacturing, distribution, selling, and profit?

2. Does the cost of manufacturing and the profit equal the cost of distribution and selling?

Refer to the United States Coastlines graph at the right. The United States has 12,383 mi of coastline. To the nearest whole mile, how much of the total is:

3. the Atlantic Ocean coastline?

4. the Pacific Ocean coastline?

5. the Gulf of Mexico coastline?

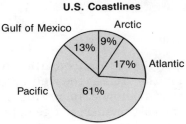

U.S. Coastlines

B For questions 6 to 8, draw a circle with a compass. Then mark off and shade the given percent of the circle. What size central angle is shaded for the given percent?

6. 25%   **7.** 45%   **8.** 12.5%

Below is a table showing how one person spends money each month.

Monthly Expenditures		
Expense	Percent	Central Angle
Apartment (rent, heat, electricity, telephone)	40%	?
Transportation	10%	?
Living (food, clothes, recreation)	15%	?
Loans/Savings	30%	?
Other	5%	?

9. Give the required central angle size for each expense item.

10. Make a circle graph of the monthly expenditures data.

11. How much of a net income of $1500 per month is spent on rent?

12. For a net income of $1950 per month, how much is spent in paying off loans or in savings?

Make a circle graph of the data in each market survey table below.

13.

Favorite Courier Service	
Company	Percent
A	45%
B	12.5%
C	37.5%
D	5%

14.

Favorite Business Magazine	
Magazine	Percent
A	24%
B	28%
C	22%
D	26%

11-8 Misuse of Graphs

Data graphs in newspapers or magazines often display information that is *distorted* in such a way as to create a false impression. The example below shows one way this is done.

Example

A company's profits from 1981 to 1988 are shown in the table at the right. Both broken-line graphs below display this data. Which graph suggests a sharper rise in profits? How was this impression created?

Company Profits	
1981	$150,000
1982	$160,000
1983	$170,000
1984	$172,000
1985	$175,000
1986	$180,000
1987	$190,000
1988	$200,000

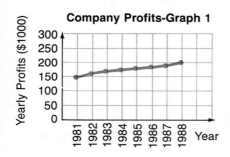

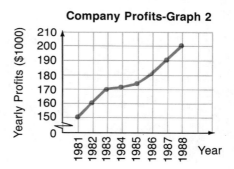

Solution

Graph 2 suggests a sharper rise in profits. This impression was created by using a break in the vertical axis.

EXERCISES

A A bar graph of part of the Company Profits data above is shown at the right.

1. How are the company profits data represented differently in the bar graph than in graphs 1 and 2 above?

2. What impression is made by not showing all of the data?

3. By giving the bars depth, what impression is created?

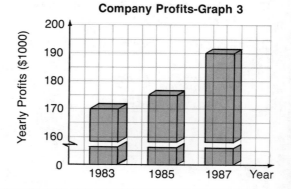

B Below is a bar graph showing the number of new homes built in a certain area during a four-year time period.

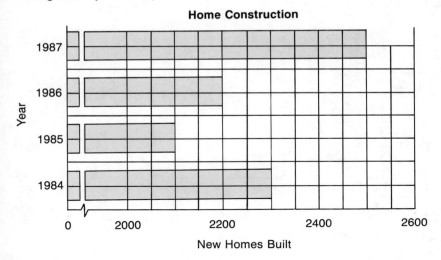

Home Construction

4. What impression is created by the break in the bar graph when the number of new homes built in 1986 and 1987 are compared?

5. Redraw the bar graph without a break in the horizontal axis, to more accurately represent the data.

The broken-line graph below represents the amount of sales, in thousands of dollars, made by a furniture salesperson.

Furniture Sales

6. Would this graph lead you to expect the sales figures to increase sharply for the coming January?

7. Redraw the broken-line graph without a break in the vertical axis, to show only a slight change in sales over the same period.

Refer to the bar graph below used in a soap advertisement.

"Sudzie Soap outsells them all."

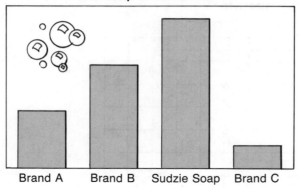

Brand A   Brand B   Sudzie Soap   Brand C

8. What impression is created?

9. Why is a graph such as this of no real value?

The graph below indicates the amount of money spent by a successful company in the last two years on salaries.

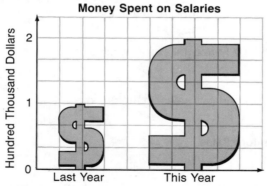

10. How many times larger is the overall size of this year's dollar symbol as compared to the overall size of last year's symbol?

11. In actual fact, how many times larger are this year's salary figures than last year's figures?

12. On a separate sheet of graph paper, redraw the above graph using the $ symbol to represent $25,000.

C The numbers of traffic accidents on a certain highway from April through September are 39, 40, 43, 43, 44, and 45.

13. Draw a bar graph to convince the local authorities to install more signs to lower the speed limit.

14. From the standpoint of the local authorities, draw a bar graph to show that the increase in traffic accidents is not that great and perhaps lower speed limit signs are not needed.

Self-Analysis Test

1. Find the median and hinge points for a box plot of the data for salmon released from fish hatcheries over a nine-year period.

> **Salmon (millions)**
> 9.5, 18.9, 23.7, 27.9, 28.0, 25.5, 30.8, 33.0, 33.0.

2. The total amount of sales a small company has made in a year to seven clients is shown in the table below. Make a bar graph of these data.

Client	A	B	C	D	E	F	G
Total Sales	$7200	$10,900	$10,500	$8400	$8900	$11,900	$12,400

3. Would a bar graph or a histogram be better for representing the seasonal precipitation for seven different cities?

4. The line graph below shows times of sunrise.

  a. What does one small division represent on the scale of dates? times?
  b. What is the time of sunrise on January 14?
  c. Approximate the time of sunrise on January 18.

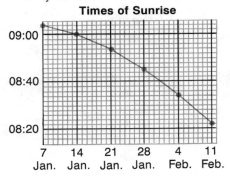

Times of Sunrise

5. The pictograph below shows the live weight in billions of pounds of fish caught by various countries in a recent year.

  a. Which two countries catch half as many pounds of fish as two other countries?
  b. The combined catch of which two countries is closest to that of Japan?

Fish Caught in a Year

Canada, Japan, Peru, Norway, U.S.

1 fish = 1,000,000,000 lb

6. A consultant in the investment industry recommends that clients distribute their investments as follows: Stocks 55%, Bonds 20%, Gold 5%, and Cash 20%. Make a circle graph of these data.
A survey of the productivity of three work units was taken at a plant that manufactures automotive parts. The data collected is recorded in the table.

Productivity of Three Work Units	
Work Unit	Pieces Produced per Shift
A	550
B	320
C	425

7. Make a bar graph that would create the impression that work unit A was twice as productive as work unit C.

8. Redraw the bar graph to show more accurately the productivity of the three work units.

Graphing Data Relationships

11-9 Direct Variation

We often need to determine the pattern or relationship between two sets of data. For example, the table at the right shows how many quarts there are in different numbers of cases of motor oil.

Cases (x)	Quarts (y)
1	12.0
2	24.0
3	36.0
4	48.0
5	60.0

The relationship between the two sets of data is graphed on a *coordinate grid* at the right. The graph starts at the *origin* (0,0) and shows a constant growth between the number of cases and the number of quarts.

> As the number of cases increases, the number of quarts also increases uniformly.

The number of quarts is *directly related* to the number of cases. The number of quarts is always 12 times the number of cases. This is an example of **direct variation**.

The relationship between the two sets of data can be expressed by the formula, $y = 12x$. The number of quarts per case, 12, is constant and is called the **constant of variation**.

Contents of Cases of Motor Oil

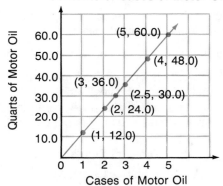

Example   Skill(s) __28__

How many quarts of motor oil are in 2.5 cases. Use the data above.

Solution

The solution can be seen on the graph above. The ordered pair (2.5, 30.0) indicates that for 2.5 cases you have 30.0 qt of oil.

The solution can also be found using the relationship formula.

$$y = 12x$$
$$y = 12(2.5)$$

$\boxed{1}\ \boxed{2}\ \boxed{\times}\ \boxed{2}\ \boxed{\cdot}\ \boxed{5}\ \boxed{=}$ 30

For 2.5 cases, there are 30.0 qt of motor oil.

> For a direct variation, the quotient (ratio) of two variables (sets of data) is constant. Direct variation can be defined by the equation $\frac{y}{x} = k$ or $y = kx$, where k is a nonzero constant and x and y are corresponding elements from different sets of data.

EXERCISES

Is the set of data an example of direct variation? If *yes*, state the constant of variation.

1.

Number of cars (x)	1	2	3	4	5
Number of spark plugs (y)	4	8	12	16	20

2.

Cost of spark plugs (y)	$5.40	$10.80	$16.20	$21.60
Number of spark plugs (x)	4	8	12	16

B **3.** Write a formula with a constant of variation that describes the relationship in each graph below.

a.

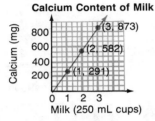

Calcium Content of Milk

b.

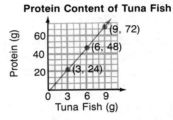

Protein Content of Tuna Fish

4. Draw a graph of the relationship described in each table below.

a.

Hours worked (x)	4	8	12	16	20	24
Salary earned (y)	$54	$108	$162	$216	$270	$324

b.

Concrete blocks (y)	16	32	48	64	80	96
Wall area (m²) (x)	1.0	2.0	3.0	4.0	5.0	6.0

5. How do you know each graph you have just made shows an example of direct variation?

6. What is the constant of variation for each graph in question 4?

First write a formula with a constant of variation. Then solve each problem.

7. At $12.95/yd, what is the cost of 5.0 yd of drapery fabric? For the same price rate, what is the cost of 12 yd?

8. A truck is carrying a load of bags of lawn fertilizer. If each bag is 40.0 lb and the load contains 36 bags, how heavy is the load? How heavy is a load of 45 bags?

First identify the constant of variation. Then solve each problem.

9. Three pounds of nails cost $5.40. How much will 5.0 lb cost? What is the cost of 8.0 lb?

10. A truck hauls 96 m³ of dirt with six loads. How much dirt could be hauled with 20 loads? with 28 loads?

11-10 Inverse Variation

A library needs $800 to buy some new books. The table below shows the various ways the $800 can be raised through charitable donations.

Ways to get $800								
Number of Donations (x)	10	20	40	80	100	200	400	800
Size of Donation (y)	$80	$40	$20	$10	$8	$4	$2	$1

The graph of this data shows an *inverse* relationship.

As the number of donations increases, the size of the donation decreases.

This is an example of **inverse variation**. It can be described by the formula, $xy = 800$. The product of the pairs of data remains *constant* at 800.

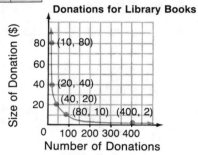

Donations for Library Books

Example
Skill(s) __28__

Graph the times for traveling a distance of 72 mi at the different speeds given in the table below. What speed would allow the 72 mi distance to be covered in one half hour?

Time (x)	2.0 h	3.0 h	4.0 h	6.0 h	8.0 h	9.0 h
Speed (y)	36 mi/h	24 mi/h	18 mi/h	12 mi/h	9 mi/h	8 mi/h

Solution

The graph shows that as the time increases, the speed decreases. The time and speed are related by the formula $xy = 72$.

Substitute into the relationship formula to find the speed for one half hour.

$xy = 72$
$0.5y = 72$

 144

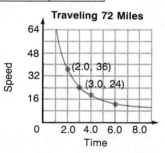

Traveling 72 Miles

To cover 72 mi in one half hour, a speed of 140 mi/h is needed, to two significant digits.

> For inverse variation, the product of two variables (sets of data) is constant. Inverse variation can be defined by the equation $xy = k$ or $y = \frac{k}{x}$, where k is a nonzero constant and x and y are corresponding elements from different sets of data.

EXERCISES

A Is each of the following situations an example of direct or inverse variation?

1.

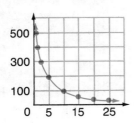

2.

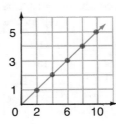

3. As the pressure increases, the volume of a gas decreases.

4. As the temperature increases, the volume of a gas increases.

5. As the speed increases, the time to cover a given distance decreases.

6. The amount of fuel used is related to the distance traveled.

B Write a sentence describing the relationship in each graph below. Then write a formula describing the relationship.

7.

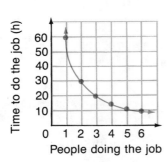

People doing the job

8.

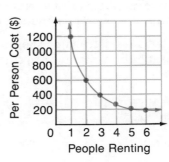

People Renting

Find the missing values in each table below. Then draw a graph of the relationship.

9.

Age of Automobile (years)	1	2	3	4	5
Value of Automobile ($)	15,000	7500	5000	?	?

10.

Interest rate (%)	3	4	6	8	9	12	18	24
Years Needed for Money to Double	24	18	12	9	?	?	?	?

C Solve each problem by first writing a formula.

11. A designer would like to use a rectangular plate with an area of 48 cm² in a piece of equipment. List five possible pairs of whole number dimensions that could be suitable for the plate.

12. A non-stop flight takes 3.0 h at 600.0 mi/h to cover a distance of 1800 mi. How long would it take if the plane traveled at 720 mi/h? 400.0 mi/h?

11-11 Joint and Combined Variation

The tables below show some possible distances that can be covered by a car when it travels at different speeds and times.

Distance, z (mi)		Speed, x (mi/h)		Time, y (h)
6 =		30.0	×	0.2
12 =		60.0	×	0.2
24 =		120.0	×	0.2
18 =		90.0	×	0.2
36 =		90.0	×	0.4
72 =		90.0	×	0.8
10 =		50.0	×	0.2
60 =		100.0	×	0.6

← When the speed is doubled, the distance is doubled.

← When the time is doubled, the distance is doubled.

← When the speed is doubled and the time is tripled, the distance is six times greater.

We say that the distance traveled by the car *varies directly* as the *product* of its speed and time. A relationship between two sets of data like this, speed and time, is called **joint variation**. This relationship is expressed in the formula $z = xy$. The constant of variation in this case is 1.

A joint variation is defined by the equation $z = kxy$, where k is a non-zero constant and x, y, and z are corresponding members of different sets of data.

Example

Skill(s) __28__

The area of a triangle varies directly as one half the product of the base and height. Thus, the relationship is an example of joint variation. The table below gives some possible areas obtained from different dimensions for the base and the height.

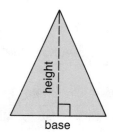

height

base

Area, z (cm²)		Base, x (cm)	Height, y (cm)
6.0	0.5 ×	4.0 ×	3.0
12	0.5 ×	8.0 ×	3.0
18	0.5 ×	12 ×	3.0

Write a formula describing the relationship. Let x represent the base, y the height, and A the area. What is the constant of variation?

Solution

The joint variation can be expressed by the formula $z = 0.5xy$. The constant of variation is 0.5.

The tables below show the possible time in which different numbers of foresters can plant various numbers of trees, knowing that one forester can plant 40 trees in 1h. In other words, one forester can plant one tree in $\frac{1}{40}$ of an hour.

Time, z		No. of Trees, x	No. of Foresters, y
3 h	$= \frac{1}{40}$	× 240	÷ 2
6 h	$= \frac{1}{40}$	× 480	÷ 2
12 h	$= \frac{1}{40}$	× 960	÷ 2

← As the number of trees doubles, the time doubles. This is *direct* variation.

6 h	$= \frac{1}{40}$	× 480	÷ 2
3 h	$= \frac{1}{40}$	× 480	÷ 4
1.5 h	$= \frac{1}{40}$	× 480	÷ 8

← As the number of foresters doubles, the time is cut in half, This is *inverse* variation.

The time, z, it takes foresters to plant trees varies *directly* with the number of trees to be planted, x, and *inversely* with the number of foresters, y. Such a relationship is called **combined variation**. This relationship can be expressed by the formula $z = \frac{1}{40}\left(\frac{n}{y}\right)$.

A combined variation can be defined by the equation $z = k\left(\frac{x}{y}\right)$, where k is a constant and x, y, and z are corresponding elements from different sets of data.

Example 2

If it takes two bakers 3 h to bake 60 dozen doughnuts, how long should it take four bakers to bake 200 dozen doughnuts?

Solution

The time, z, varies *directly* as the number of doughnuts, x, to be baked and *inversely* as the number of bakers, y. Express this combined variation relationship with the formula $z = k\left(\frac{x}{y}\right)$.

Step 1: Find the time it takes one baker to bake one dozen doughnuts first; or, find the constant of variation, k.

$$z = k\left(\frac{x}{y}\right)$$
$$3 = k\left(\frac{60}{2}\right)$$
$$3 = 30k$$
$$k = \frac{1}{10}$$

It takes one baker $\frac{1}{10}$ h to bake one dozen doughnuts.

Step 2: Find how long it should take four bakers to bake 200 dozen doughnuts.

$$z = k\left(\frac{x}{y}\right)$$
$$= \frac{1}{10}\left(\frac{200}{4}\right)$$
$$= 5$$

It should take 5 h for four bakers to bake 200 dozen doughnuts.

EXERCISES

A **1.** As the speed of a nail-making machine is doubled while the amount of time remains the same, what change occurs in the number of nails produced?

2. If the speed of a bottle-capping machine is doubled and the number of hours the machine is in production is doubled, what effect does this have on the number of bottles capped?

3. When the speed of laying bricks is doubled and the number of hours worked is halved, what is the change in the number of bricks laid?

4. If the length of the base of a triangle is doubled and its height is tripled, what effect does this have on the area?

5. As the number of house painters is tripled and the amount of time is doubled, how is the amount of painted wall surface changed?

B **6.** The weight of a block of metal varies jointly as the length, width, and height. If a block of metal has a weight of 12 kg, what is the weight of a block of the same metal with the same height, twice the length, and three times the width?

The simple interest earned on a sum of money varies jointly as the interest rate and the number of years. Find the missing values in the table below for an investment of $1000.

	Simple Interest	Investment ($)	Interest Rate	Time (years)
	$40 =	1000 x	4% x	1
7.	?	1000 x	8% x	1
8.	?	1000 x	12% x	1
9.	?	1000 x	9% x	1
10.	?	1000 x	9% x	2
11.	?	1000 x	9% x	4

12. What is the change in the amount of simple interest if the interest rate doubles and the time is cut in half?

The time for a printer to print out a report varies directly as the number of words in the report and inversely as the speed of the printer in words per minute. Find the missing values in the table below. The constant of variation is 1.

	Time (min)	Speed (words/min)	Words Printed
13.	2.5	400	?
14.	5.0	400	?
15.	7.5	400	?
16.	3.0	300	?
17.	1.5	600	?
18.	0.75	1200	?

19. If the printer took 2.5 min to print out a 3000-word report at a printing speed of 1200 words/min, how long should it take the printer to print out an 8000 word report, to the nearest minute?

The surface area, *SA*, of the paper wrapping around the side of a can varies jointly as the radius, *r*, and the height, *h*, of the can. The relationship is expressed in the formula $SA = 6.28rh$, with the constant of variation being 6.28. Find the missing values in the table below.

	Radius (cm)	Height (cm)		Surface Area (cm²)
20.	4.0	5.0	× 6.28	?
21.	8.0	5.0	× 6.28	?
22.	12	5.0	× 6.28	?
23.	6.0	8.0	× 6.28	?
24.	6.0	16	× 6.28	?
25.	6.0	24	× 6.28	?

C **26.** The time to manufacture a number of screwdrivers varies directly as the number to be produced and inversely as the number of machines for production.

  a. If two machines can produce 600 screwdrivers in 5 h, how long, to the nearest hour, should it take to produce 15,000 screwdrivers with six machines?

  b. At the same rate, how long, to the nearest hour, should it take to produce 15,000 screwdrivers with 12 machines?

Check Your Skills

Find the sum or difference. (Skill 2)

1. 234 − 145 **2.** 92 + 959 **3.** 1284 − 976

Divide. Express each remainder as a fraction. (Skill 4)

4. $17\overline{)1648}$ **5.** 896 ÷ 3 **6.** $\frac{3654}{36}$

Express each ratio in simplest form. (Skill 18)

7. 17 : 34 **8.** 18 : 96 **9.** 36 : 412

Write each decimal as a percent. Write each percent as a decimal.

10. 0.0532 **11.** 27.821 **12.** 84.3% **13.** 0.023% (Skills 20 and 22)

Find the percent or the part. Round to the nearest tenth where needed. (Skills 24 and 25)

14. 40 out of 65 = _?_ % **15.** 7 out of 36 = _?_ % **16.** _?_ out of 350 = 75%

Solve the equation.

17. $1.05x = 1.89$ **18.** $3.4x = 0.068$ **19.** $\frac{46}{9.2}x = 1$ (Skill 28)

Self-Analysis Test

Write a formula with a constant of variation that describes the relationship in each graph.

1.

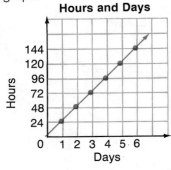

Hours and Days

2.

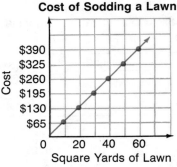

Cost of Sodding a Lawn

3. As the number of quarts of oil purchased increases, the cost of the oil increases. Find the missing values in the table below. Then draw a graph of the relationship.

Quarts of Oil	1.0	2.0	3.0	4.0	5.0	6.0	7.0	8.0
Cost ($)	?	?	6.75	?	?	?	?	?

Write a sentence describing the relationship in each graph below. Then write a formula describing the relationship.

4.

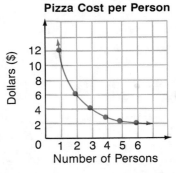

Pizza Cost per Person

5.

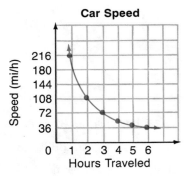

Car Speed

6. If two pickers can harvest 60 kg of berries in 3 h, how many kilograms of berries can be picked by 10 pickers in 6 h?

Identify the type of variation implied by each statement. Then decide whether or not the variation suggested makes sense.

7. If Tom has a mass of 25 kg at age 8, he will have a mass of 50 kg at age 16.

8. If it took 20 min to get an order of two scoops of ice cream, it would take 10 min to get an order of one scoop.

9. If the average life of a 60 W bulb is 750 h, the average life of a 100 W bulb should be 380 h.

Large amounts of information can be easily handled with the use of a computer and a data base management system. Data base management systems can store records, sort information, and retrieve only the information needed by the user.

Data bases are used for many different purposes, such as telephone directories, bank records, and baseball statistics. Travel agencies have computers that are connected to data bases. They can retrieve all the necessary information on airline flights.

Suppose that you are a travel agent and the following data base of flights between New York and Toronto appears on your computer screen.

From	To	Depart Time	Arrival Time	Carrier	Flight	Class of Service
Toronto	New York	7:00a	8:17a	AA	547	FYBQM
Toronto	New York	9:17a	10:34a	AA	357	FYBQM
Toronto	New York	11:20a	12:38p	AC	706	YBV
Toronto	New York	12:50p	2:53p	AL	446	YBQMK
Toronto	New York	1:50p	3:00p	AC	708	FJYBV
Toronto	New York	2:45p	3:53p	AC	710	YBV
Toronto	New York	3:30p	4:43p	AA	468	FYBQM
Toronto	New York	5:00p	6:35p	OC	609	YB
Toronto	New York	6:20p	7:41p	AC	714	YBV
Toronto	New York	7:10p	8:45p	OU	611	YB
Toronto	New York	8:10p	9:26p	AC	718	YBV
New York	Toronto	6:59a	8:24a	AA	171	FYBQM
New York	Toronto	7:10a	8:44a	AC	701	YBV
New York	Toronto	8:50a	10:19a	AC	731	YBV
New York	Toronto	9:55a	11:30a	OU	604	YB
New York	Toronto	11:25a	12:47p	AC	707	YBV
New York	Toronto	12:00p	1:35p	OU	606	YB
New York	Toronto	1:22p	2:45p	AA	285	FYBQM
New York	Toronto	3:50p	5:17p	AC	712	FJYBV
New York	Toronto	4:55p	6:22p	AC	713	YBV
New York	Toronto	7:00p	8:29p	AC	737	YBV
New York	Toronto	8:00p	9:25p	AA	122	FYBQM

List the possible departure and arrival times, carrier, flight, and class of service for each client's needs.

1. Ms. Sauvais is a fashion coordinator who flies to New York each season to see the latest fashions. She collects bonus points for traveling with the same airline. Ms. Sauvais would like to return to Toronto in the afternoon between 4:00 P.M. and 6:00 P.M. She wants to fly with Air Canada (AC under Carrier).

2. Gina Garland is a singer with the New York Metropolitan Opera. She has a special appearance at Roy Thompson Hall in Toronto. Her rehearsals with the New York Symphony finish at noon. Gina would like to leave for Toronto after her rehearsal, but before 2:00 P.M. She wants to travel in first class (letter F under Class of Service).

3. Mr. Hastings is a New York businessman who wants to go to Toronto and back the same day. He needs to be in Toronto for a 10:00 A.M. to 4:00 P.M. meeting. Mr. Hastings wants to return to New York before 9:00 P.M. He insists on traveling business class (letter B under Class of Service).

Most computer systems require specialized personnel to enter data and instructions, operate the system, and print and transmit the results. In a large
company, data processors usually enter information
into a large network system of computers, while the
computer operators control the workings of the
computer system at the various work stations. In
smaller companies, a computer operator may be
involved with both data entry and the monitoring of
the computer system.

Although computer operators are employed in almost every industry, most of them work in manufacturing firms, banks, insurance companies, colleges,
universities, and firms that provide data processing
services.

The growth in recent years of the use of computers, especially of large network
systems, has increased the demand for computer operators.

Job Description

What do computer operators do?

1. Computer operators monitor the entire computer system. They operate all
equipment involved in the computer system, such as the terminal, the Central
Processing Unit (CPU), and the printer.

2. On a daily basis, computer operators see that programs run properly and
solve problems as they occur.

3. Computer operators open and close the computer system each day. The
closing procedure takes more time than the opening, since the computer
operator usually makes duplicates, called backups, on disks or tapes of the
work done by the computer.

4. Each day, computer operators organize the work to be done by the computer.
For example, if the work for the day involves running programs to invoice
customers, prepare orders, and issue credits, a good computer operator
would set up a time schedule indicating the time each program is to be run
through the system.

Qualifications

Employers usually require a high school education, and many prefer to hire
computer personnel who have some community college training. Skills in data
processing and the operation of various kinds of computer equipment are usually
required. Larger companies, which have a large network system, may even
require programming skills. The ability to reason logically and the ability to work
quickly and accurately are also helpful.

The fishing industry plays an important role in the economy of the United States. In a recent year, 230,000 Americans were employed as fishermen, accompanied by 110,000 shore workers.

In addition to fishing as an industry, tourism is enhanced by thousands of visitors who are attracted by both saltwater and freshwater fishing. In fact, sports fishermen and commercial fishermen can compete for this valuable natural resource.

Statistics plays an important role in the fishing industry. The National Marine and Fisheries Service (NMFS) and the National Oceanic and Atmospheric Administration (NOAA) issue annual data on the commercial fishing and food processing industries. In addition, statistics are used to estimate the number of fish available and the number caught in order to ensure that stocks are maintained and that adequate numbers of fish will be available for the future.

The table below gives the supply of fish products in the United States in eight successive recent years. The imported fish column in the table shows that we do not depend entirely on the fish we can catch.

Supply of Fish Products in the United States		
Year Reported	Domestic Catch (1,000,000 lb)	Imported Fish (1,000,000 lb)
1st	6028	5481
2nd	6267	5564
3rd	5482	4875
4th	5977	5376
5th	6367	5644
6th	6439	5913
7th	6438	6114
8th	6258	8803

1. Draw a bar graph to display the data for the domestic catch.

2. Draw a pictograph to display the data for the imported fish.

3. Why would a broken-line graph be a poor choice for presenting these data?

4. The amount of canned fish produced in the United States, in millions of pounds, for an eight-year period is 1382, 1439, 1516, 1476, 1294, 1391, 1411, and 1161. Find the mean and median of these data.

Chapter 11 Test

1. A lawn and garden center tabulated the number of orders it had for asters (A), daisies (D), marigolds (M), petunias (P), and roses (R). The amounts of each flower tabulated are given in the table below. Make a tally and frequency table for this data.(11-1)

Flower Orders

R, A, P, M, A, P, R, M, A, A, R, P, P, R, M, R, D, D, R, P, D, D, R, M, A, M, R, M, A, A, R, R, R, P, P, M, D, A, M, R, P, P, R, R, A, R, D, A, A, P, M, A.

2. The amount of profit for a small business in each of the last eight years is $20,100, $18,500, $25,800, $26,000, $28,400, $32,100, $23,400, and $35,700. Find the mean, median, and mode of this data. (11-2)

3. The pulse rates for 50 workers at an assembly plant are listed below. Which measure of central tendency would best support the claim that the pulse rates of the workers fall within a normal range of 72 to 80? (11-3)

Pulse Rates

89, 84, 82, 77, 82, 93, 79, 86, 83, 79, 68, 63, 81, 90, 88, 74, 77, 71, 81, 79, 76, 96, 90, 73, 94, 77, 70, 71, 78, 66, 83, 85, 72, 80, 78, 76, 80, 62, 81, 74, 87, 80, 78, 75, 72, 65, 80, 87, 80, 82.

4. The data in the table below shows the number of customers in a cafeteria at lunchtime for the past three weeks. (11-4)

   a. Make a stem-and-leaf plot for the data.
   b. Identify the range and the median.
   c. Use the stem-and-leaf plot to make a box plot.

Cafeteria Customers

312, 320, 335, 305, 280, 322, 319, 332, 308, 327, 295, 286, 319, 328, 314, 338, 334, 300, 282, 323, 334.

5. Make a bar graph of the data below. (11-5).

Home Heating Fuels Used by Homeowners	
Oil	18%
Natural Gas	45%
Electricity	22%
Wood	3%
Propane	0.5%

6. Make a histogram of the data below. (11-5)

Ice-Cream Cone Sales	
Vanilla	250
Butterscotch	160
Chocolate	290
Strawberry	190
Banana	70
Maple nut	180

7. The closing price of a certain stock was recorded for eight successive days in the table below. Make a broken-line graph of the data. (11-6)

Stock Price (Per Share)

$17.50, $17.25, $17.74, $18.50, $19.25, $19.75, $16.00, $16.25.

8. The number of passengers per week on a commuter train service for a six-week period are shown in the table below. Make a pictograph of the data. (11-6)

Commuter Train Passengers
8000, 11,500, 12,000, 6000, 3600, 5500.

9. A survey of waterfowl tagged for scientific study in a state park included the percents shown in the table below.

a. What size central angle, to the nearest whole number, would be required to make a circle graph of the data?

b. Make a circle graph of the information. (11-7)

Waterfowl	Percent Found
Ducks	60%
Geese	28%
Swans	12%

10. The circle graph below shows percents of the total market for different brands of computers. What false impression is created by this graph? (11-8)

Computer Brands Used

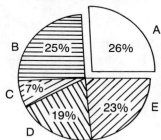

11. The cost of carpet per square yard varies directly with the number of square yards purchased. Find the missing values in the table below. Then draw a graph of the data. (11-9)

Carpet (yd²)	4.0	8.0	12.0	16.0	20.0
Cost	?	$144	?	?	?

12. Is the situation an example of *direct* or *inverse* variation? (11-10)

a. The time to complete the bricklaying job decreases as the number of bricklayers increases.

b. As the running time increases, the pulse increases.

c. The number of stones required to make a patio decreases as the size of the patio stone used increases.

13. The number of people who can safely take an elevator varies inversely as the average mass of the persons. If twelve people with the average mass of 75 kg can safely ride an elevator, how many people with an average mass of 60 kg can take the elevator? (11-10)

14. If it takes two florists 3 h to prepare six floral arrangements, how long should it take four florists to prepare 24 arrangements? (11-11)

Cumulative Review

1. Use the rule of Pythagoras to find the length of the diagonal for the rectangular wooden panel shown below.

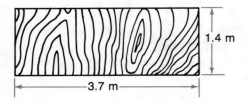

2. Write each trigonometric ratio for the roof truss below.

    a. cos P **b.** tan Q **c.** sin P

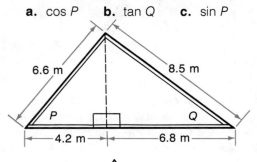

3. What is the height of the plastic panel shown at the right? The panel has the shape of an equilateral triangle.

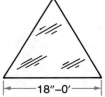

4. The ramp below is inclined at an angle of 15° to the ground. How much height is gained by using the ramp?

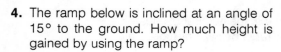

5. A square sheet of aluminum has a diagonal 135.7 in. long. Find the lengths of each side of the sheet.

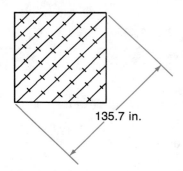

6. Find the distance from the ground to the top of the ladder in the diagram below.

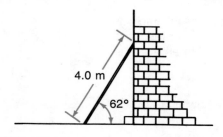

7. What is the distance from the ship to the base of the lighthouse in the diagram below?

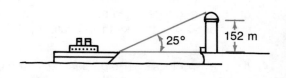

8. An ecologist measured the wing spans of a *representative sample* of 50 birds to study their life span. The data found is listed in the table below.
   a. Make a tally and frequency table for these data.
   b. What is the mean length of the birds?
   c. Find the median of these data.
   d. Make a stem-and-leaf plot of the data.
   e. Draw a box plot of the data.

> **Wingspans of Birds in Centimeters**
> 74, 84, 79, 76, 82, 83, 75, 77, 75, 84, 81, 78, 79, 80, 80, 81, 79, 77,
> 78, 76, 78, 79, 79, 82, 81, 80, 80, 81, 80, 82, 76, 80, 78, 80, 83, 82,
> 80, 79, 77, 78, 80, 78, 82, 81, 80, 78, 83, 81, 81, 80.

9. Use the broken-line graph below to answer each question.
   a. A surplus of cement occurs when supply is greater than demand. How much was the surplus in 1983?
   b. A shortage of cement occurs when demand is greater than supply. How much was the shortage in 1986?
   c. In 1988, was there a shortage or a surplus? How much?

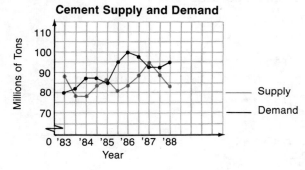

10. Of all the money spent by skiers at a certain resort in a recent year, 45% was spent on accommodations, 25% on lift tickets, 14% on ski gear, and 16% on food and drink. Draw a circle graph of this data.

11. As a sewing machine salesperson, your sales from May through December are 12, 13, 13, 14, 15, 15, 16, 17.
   a. Draw a broken-line graph to convince your boss you need a raise.
   b. Draw a graph from your boss's standpoint indicating that you do not deserve a raise.

12. Identify the relationship in each statement as an example of direct or inverse variation.
   a. The pressure increases as the volume decreases.
   b. The cost increases as the number purchased increases.
   c. The velocity of a body falling from rest increases as its time increases.
   d. The number of oscillations per second of a pendulum decreases as the length of the pendulum increases.

SKILLS FILE

SKILL 1 Rounding

When rounding a whole number to a certain place, look at the digit in the place immediately to the right. If it is 5 or higher, *round up* the number. If it is less then 5, *round down*.

To the nearest ten, 235 *rounds up* to 240.

To the nearest hundred, 5732 *rounds down* to 5700.

To the nearest thousand, 4368 *rounds down* to 4000.

Round to the nearest ten.

1. 26   **2.** 215   **3.** 428   **4.** 602   **5.** 1875   **6.** 5046

Round to the nearest hundred.

7. 745   **8.** 555   **9.** 999   **10.** 7168   **11.** 10,608   **12.** 39,726

Round to the nearest thousand.

13. 6415   **14.** 498   **15.** 26,086   **16.** 10,548   **17.** 649   **18.** 308,916

When rounding numbers to a certain decimal place, *round up* if the digit in the place immediately to the right is 5 or greater. *Round down* if the digit is less than 5.

To the nearest tenth, 3.1573 *rounds up* to 3.2.

To the nearest thousandth, 0.003126 *rounds down* to 0.003.

To the nearest whole number, 7.564 *rounds up* to 8.

To the nearest cent, $0.5736 *rounds down* to $0.57.

Round to the nearest tenth.

19. 12.22   **20.** 1.28   **21.** 0.269   **22.** 0.55   **23.** 5.455   **24.** 85.267

Round to the nearest thousandth.

25. 0.1234   **26.** 0.4952   **27.** 0.49999   **28.** 1.2557   **29.** 0.9013   **30.** 1.2134

Round to the nearest whole number.

31. 10.2   **32.** 16.58   **33.** 50.5   **34.** 62.98   **35.** 68.4   **36.** 2.002

Round to the nearest cent.

37. $1.6905   **38.** $25.992   **39.** $75.752   **40.** $1.986   **41.** $0.2596   **42.** $9.856

Round to the nearest dollar.

43. $6.21   **44.** $8.53   **45.** $23.47   **46.** $0.75   **47.** $122.50   **48.** $0.49

Skills

SKILL 2 Adding and Subtracting Whole Numbers

To add whole numbers, be sure the digits with the same place value are lined up.

Add 6407 and 376.

$$\begin{array}{r} \overset{1}{6407} \\ +376 \\ \hline 6783 \end{array}$$

Find the sum.

1. 4783 + 356

2. 4950 + 1625

3. 2588 + 92

4. 15,494 + 893

5. 37,381 + 21,407

6. 49,641 + 6830

7. 7649 + 350 + 22

8. 12 + 987 + 15

9. 74,125 + 21,456

10. $958 + $6240

11. $12,355 + $9638

12. $12 + $25 + $458

Find the sum if m = 295.

13. 64 + m

14. m + 1125 + 1815

15. 19 + m + 824

16. 1269 + m + m

17. m + 2369

18. m + 4125 + m

Before subtracting whole numbers, make certain the digits with the same place value are lined up.

Subtract 946 from 62,834.

$$\begin{array}{r} {}^{1\,17\,12\,14}62,\!834 \\ -946 \\ \hline 61,\!888 \end{array}$$

Subtract.

19. 625 – 21

20. 1345 – 255

21. 2624 – 259

22. 97,526 – 1255

23. 47,452 – 5224

24. 58,577 – 7187

25. 35,795 – 14,326

26. 45,673 – 10,267 – 15

27. 97,126 – 8123 – 1563

28. $198 – $56

29. $1587 – $594

30. $4560 – $1298 – $58

31. $12,599 – $525

32. $225,800 – $52,300

33. $523 – $25 – $129

Find the difference if p = 568.

34. p – 49

35. p – 395

36. 840 – p

37. 5132 – p – p

38. 5982 – 1249 – p

39. p – 129 – 25

SKILL 3 Multiplying Whole Numbers

Multiplication can be indicated by a times sign or by parentheses as shown below.

Find the product. 24×53, or $24(53)$, or $(24)(53)$

```
     53
 ×   24
    212  = 4 × 53      ← The digits with the same place
   1060  = 20 × 53       value must be lined up.
   1272
```

Algebraic expressions can indicate the multiplication of whole numbers.

Find the value of the algebraic expression $25a$, if a is 475.

```
     475
   × 25
   2 375  = 5 × 475
   9 500  = 20 × 475
  11,875
```

Find the product.

1. 51×76

2. 68×36

3. 32×2

4. $27(35)$

5. $58(71)$

6. $30(122)$

7. $(54)(27)$

8. $(47)(35)$

9. $(58)(61)$

10. $(736)(72)$

11. $(47)(6501)$

12. $(412)(5327)$

13. 27×5743

14. 31×1942

15. 9010×99

16. $16,822 \times 35$

17. $11,400 \times 15$

18. $21,492 \times 61$

19. $37,642 \times 123$

20. $14,231 \times 315$

21. $49,237 \times 432$

22. $783 \times 4 \times 12$

23. $9 \times 56 \times 943$

24. $(70)(54)(32)$

25. $\$23 \times 6$

26. $\$156 \times 9$

27. $\$843 \times 16$

28. $\$1986(12)$

29. $(26)(\$126)$

30. $813 \times \$985$

Find the value of each algebraic expression if $n = 48$.

31. $10n$

32. $6n$

33. $14n$

34. $92n$

35. $128n$

36. $652n$

SKILL 4 Dividing Whole Numbers

A division with whole numbers can be written in three different ways, as shown below. The calculation for the problem at the right shows that the problem does not come out evenly. When this happens, the remainder can be expressed as a fraction, remainder over divisor.

$$15\tfrac{7}{44}$$
$$44\overline{)667}$$
$$\underline{44}$$
$$227$$
$$\underline{220}$$
$$7$$

$$\frac{667}{44} \quad \text{or} \quad 667 \div 44 \quad \text{or} \quad 44\overline{)667}$$

Divide. Express the remainders that occur as fractions.

1. $43\overline{)84}$ 2. $111 \div 37$ 3. $\dfrac{61}{28}$ 4. $56\overline{)616}$

5. $754 \div 59$ 6. $\dfrac{695}{61}$ 7. $33\overline{)1781}$ 8. $4230 \div 45$

9. $2900 \div 53$ 10. $\dfrac{3956}{83}$ 11. $\dfrac{1369}{43}$ 12. $24\overline{)2184}$

13. $5205 \div 8$ 14. $\dfrac{3186}{25}$ 15. $\dfrac{193}{15}$ 16. $55\overline{)1210}$

Sometimes zeros are annexed to the dividend. In the example below, the quotient came out even after two zeros were annexed.

$$13.25$$
$$24\overline{)318.00}$$
$$\underline{24}$$
$$78$$
$$\underline{72}$$
$$60$$
$$\underline{48}$$
$$120$$
$$\underline{120}$$
$$0$$

or $\frac{318}{24} = 13.25$

$= 13.3$

(rounded to tenths)

Divide. Annex zeros until there is no remainder.

17. $28\overline{)378}$ 18. $\dfrac{1662}{48}$

19. $923 \div 26$ 20. $\dfrac{2737}{34}$

21. $2288 \div 65$ 22. $95\overline{)8151}$

Divide. When the quotient does not come out evenly, annex enough zeros so that the result can be rounded to the nearest tenth.

23. $67\overline{)701}$ 24. $858 \div 17$ 25. $400 \div 71$ 26. $59\overline{)852}$

27. $\dfrac{347}{26}$ 28. $308 \div 14$ 29. $83\overline{)168}$ 30. $\dfrac{9682}{47}$

If $c = 112$, find the value of the algebraic expression.

31. $\dfrac{504}{c}$ 32. $\dfrac{c}{56}$ 33. $\dfrac{1064}{c}$ 34. $\dfrac{c}{80}$

35. $\dfrac{c}{2}$ 36. $\dfrac{1456}{c}$ 37. $\dfrac{5936}{c}$ 38. $\dfrac{c}{28}$

SKILL 5 Using Order of Operations

When more than one operation is required in an expression, do multiplication and division first, as they occur from left to right; then do addition and subtraction, as they occur from left to right.

$$3 \times 12 + 5 \times 3 = 36 + 15 \qquad 14 \times 7 - 6 \div 1 = 98 - 6$$
$$= 51 \qquad\qquad\qquad\qquad = 92$$

Simplify.

1. $6 \times 8 + 5 \times 6$

2. $4 \times 9 - 2 \times 5$

3. $10 \div 2 + 9 \times 6$

4. $10 \div 5 \times 3$

5. $11 + 8 \times 3$

6. $25 \times 17 - 5$

7. $2 \times 8 + 27 \div 9$

8. $34 \div 17 - 2$

9. $120 \div 5 \times 2$

10. $18 \div 3 + 14 \div 2$

11. $64 \div 8 + 3$

12. $7 - 4 \div 2$

13. $7 + 14 \times 2 \div 4 - 3$

14. $6 \times 9 \div 3 + 4 - 2$

15. $18 \div 6 + 7 \times 2 - 3$

Find the value of each algebraic expression if $a = 6$.

16. $42 - 3a + 2$

17. $38 - \frac{a}{2} \times 10$

18. $65a + 12a \times 5$

The work inside parentheses should be done first. Compare the values of the two expressions below.

$$12 \times 4 + 3 = 48 + 3 \qquad 12 \times (4 + 3) = 12 \times 7$$
$$= 51 \qquad\qquad\qquad = 84$$

Simplify.

19. $10 \times (51 + 17)$

20. $21 \times (83 - 55)$

21. $(65 + 91) \div 2$

22. $15 + (23 + 62) \times 10$

23. $24 \times (78 - 42) + 567$

24. $(41 - 19) \div 11 + 32$

25. $(12 \times 24 + 60) \div 4$

26. $(45 + 67) \times (83 - 64)$

27. $(18 + 16 \times 2) \div 5$

Find the value of each algebraic expression if $b = 14$.

28. $(12b + 4) \div 2$

29. $90 - (\frac{b}{7} + 6) \times 8$

30. $(8 + 4 \times 3b) \times 5$

Long-division bars act like grouping symbols. Do all the work on one side of the bar before doing the work on the other side. Then do the division.

$$\frac{12 \times 2 + 6}{2 \times 9 - 3} = \frac{30}{15} = 2$$

Simplify.

31. $\dfrac{12 + 9 \times 2}{15 - 4 \times 3}$

32. $\dfrac{15 \times (14 - 11)}{27 \div 3 - 6}$

33. $\dfrac{(17 + 23) \div 2}{(18 + 11 \times 2) \div 2}$

SKILL 6 Adding and Subtracting Decimals

It is important to line up the decimal points when adding or subtracting decimal numbers. This automatically lines up digits with the same
place value.

Add 22.5, 2.75, and 7.   Subtract 17.37 from 25.52.   Subtract 15.25 from 27.

```
  1 1                    1 15 4 12                6  9 10
 22.5                     25.52                   27.00   ← Zeros can
  2.75                  − 17.37                 − 15.25      be annexed.
 +7                       8.15                    11.75
 ─────
 32.25
```

Find the sum.

1. 12 + 6.3

2. 52.9 + 21.53

3. 73.21 + 41.385

4. 15.2 + 7.25

5. 142.31 + 156.46

6. 0.263 + 1.5934

7. 0.165 + 0.238 + 1.893

8. 12.46 + 789.9846 + 123.497

Find the difference.

9. 89.6 − 7.75

10. 85.1 − 8.87

11. 100.2 − 16.847

12. 15.8 − 4.91

13. 46 − 15.4

14. 32 − 12.86

15. 126.1 − 89.542

16. 881.9 − 364.72

17. 5674 − 862.135

18. 123.26 − 12.593 − 69.563

19. 12.985 − 4.323 − 0.705

Find the sum or difference.

20. 50.148 + 15.532

21. 216.05 − 118.38

22. 527.9 + 64.275

23. 6.972 − 4.21687

24. 46.45 − 31.735

25. 946.463 + 498.56

26. 104.79 − 26.3 − 62.621

27. 327.915 + 1.07 + 154.002

Find the value of each numerical expression.

28. 6.2 + (8 − 0.9)

29. 7.25 − (1.8 + 4.3) + 6

30. 18.6 + 9.5 − (8.84 − 3.52)

31. 43 + 1.25 − 2.56

32. 4.67 − 2.9 + 3.6

33. 12.23 − 4.73 + 5.78

Find the value of each algebraic expression if n = 8.2.

34. 4.5 + n − 0.6

35. 16.36 + 1.5 − n

36. 50 − (n + 6.48)

37. n + n − 0.67

38. 11.26 − n + 3.6

39. n + 1.6 − 2.4

40. n − 5.3 + n

41. 15.3 − n + 2.6

42. n + n + 1.3

SKILL 7 Multiplying Decimals

When multiplying two decimals, the number of decimal places in the product is equal to the total number of decimal places in the two factors.

Multiply 5.75 by 3.5.

$$
\begin{array}{r}
5.75 \leftarrow 2 \text{ decimal places} \\
\times\ 3.5 \leftarrow 1 \text{ decimal place} \\
\hline
2\ 875 \\
17\ 250 \\
\hline
20.125 \leftarrow 3 \text{ decimal places}
\end{array}
$$

Find the product.

1. 3.5
   ×1.8

2. 7.36
   × 1.5

3. 10.2
   ×7.62

4. 63($14.36)

5. ($26.07)583

6. 7.5(5.432)

7. (0.123)(2.4)

8. (3.89)(0.02)

9. (6.715)(70.3)

10. 0.58
    ×0.04

11. 3.12
    ×0.07

12. 0.05
    ×0.06

13. 100($42.61)

14. 2.4(237.8)

15. 13.8(6231.87)

16. (0.1543)(0.5162)

17. (555.55)(0.0002)

18. (0.007)(1000)

19. 15.924
    × 0.75

20. 85.00
    ×0.875

21. 4.9321
    × 4.63

22. $1.98(12)

23. $5.68 × 16

24. 16 × ($97.50)

25. (12.96)(8.52)

26. (85.5)74

27. 5.4(76.4)

28. (46.45)(78)

29. (15.57)(763.3)

30. (17.36)(89.5)

Find the value of each numerical expression.

31. 5.8(6.2) + 8.65

32. (49.85)(0.5)(150)

33. 64.3 − (0.2)(8.4)

34. 6.7(2.6)(2.9) − 10

35. 4.9 + (1.9)(2.9)

36. (74)(1.5) − (26)(0.5)

37. 1.6(7.9)(4.3)

38. (42.3 + 12)(12)

39. (15.6 + 1.3)(4.3 − 2.1)

Find the value of each algebraic expression if $x = 7.8$.

40. $10.5x - 0.3x$

41. $8.25x(50)$

42. $394x + (126x)4.2$

43. $12x + 16x$

44. $12x + 5x$

45. $x(2.6)(7.9)$

SKILL 8 Dividing Decimals

When dividing by a decimal number, multiply the divisor and dividend by the same power of ten (10, 100, 1000, etc.) so the divisor becomes a whole number.

Divide 9.625 by 2.75.

$$\text{dividend} \rightarrow \frac{9.625}{2.75} \quad \frac{9.625 \boxed{\times 100}}{2.75 \boxed{\times 100}} = \frac{962.5}{275} \leftarrow \text{The divisor is a whole number.} \qquad 2.75\overline{)9.625}^{\;3.5}$$

Divide. Round answers to the nearest hundredth where needed.

1. $11.5 \div 3.5$ **2.** $4.56 \div 2.8$ **3.** $18.272 \div 4.2$

4. $6.5\overline{)27.823}$ **5.** $3.15\overline{)43.55}$ **6.** $12.24\overline{)359.036}$

7. $\dfrac{36.52}{3.08}$ **8.** $\dfrac{10.57}{2.03}$ **9.** $\dfrac{8.24}{0.24}$

10. $0.51\overline{)4.569}$ **11.** $21.2\overline{)4.386}$ **12.** $0.37\overline{)8.49}$

13. $43.32 \div 0.123$ **14.** $8.378 \div 6.49$ **15.** $0.3814 \div 1.32$

16. $\dfrac{130.2}{0.12}$ **17.** $\dfrac{15.64}{0.065}$ **18.** $\dfrac{7.614}{3.84}$

19. $\dfrac{51.42}{0.36}$ **20.** $\dfrac{5.392}{9.21}$ **21.** $\dfrac{24.513}{7.01}$

22. $1.1\overline{)39.9}$ **23.** $14.93\overline{)27.391}$ **24.** $3.02\overline{)21.429}$

25. $0.693 \div 0.006$ **26.** $52.421 \div 1.05$ **27.** $0.062 \div 4.5$

28. $\dfrac{96}{5.06}$ **29.** $\dfrac{0.009}{0.086}$ **30.** $\dfrac{29}{0.05}$

Find the value of each numerical expression.

31. $35.4 \div 0.2 + 1.8$ **32.** $16.8(0.4) \div 1.2$ **33.** $\dfrac{48}{0.6}(3.5)$

34. $4.5 + 2.5 \div 1.25$ **35.** $6.4 \div 0.4 \div 0.2$ **36.** $(1.2)(12) \div 0.72$

Find the value of each algebraic expression if $g = 10.5$.

37. $\dfrac{g}{2.2}(8.6)$ **38.** $4g \div 0.06$ **39.** $7.5g \div 2g$

40. $15.5 \div g + 2.5$ **41.** $12.2g \div 2 \div 2.6$ **42.** $g \div 3.6 \times 5.6$

Divide. Round your answers to the nearest cent where needed.

43. $\$1.53 \div 3$ **44.** $\$127.20 \div 12$ **45.** $\$269 \div 24$

46. $\$30.00 \div 8$ **47.** $\$5425 \div 64$ **48.** $\$695.99 \div 4$

SKILL 9 Changing Mixed Numbers and Fractions

A mixed number can be changed to a fraction by multiplying the whole number by the denominator, adding the numerator, and putting the result over the denominator.

$$5\frac{3}{4} = \frac{(5 \times 4) + 3}{4} = \frac{23}{4}$$

Write each mixed number as a fraction.

1. $5\frac{1}{4}$ 2. $2\frac{3}{4}$ 3. $9\frac{3}{8}$ 4. $7\frac{1}{2}$ 5. $7\frac{2}{3}$

6. $4\frac{5}{6}$ 7. $8\frac{4}{5}$ 8. $3\frac{1}{8}$ 9. $3\frac{3}{10}$ 10. $2\frac{1}{16}$

11. $5\frac{17}{32}$ 12. $3\frac{1}{5}$ 13. $4\frac{1}{8}$ 14. $3\frac{1}{6}$ 15. $1\frac{9}{10}$

16. $2\frac{1}{10}$ 17. $2\frac{5}{16}$ 18. $9\frac{5}{7}$ 19. $4\frac{5}{32}$ 20. $12\frac{1}{3}$

21. $6\frac{9}{32}$ 22. $9\frac{23}{32}$ 23. $12\frac{1}{2}$ 24. $6\frac{3}{8}$ 25. $2\frac{5}{64}$

A fraction can be changed to a mixed number by dividing the numerator by the denominator and expressing the remainder as a fraction. When the division has no fraction remainder, the fraction is expressed as a whole number.

$$\frac{18}{5} = 5\overline{)18}^{\,3\frac{3}{5}} \quad\quad \frac{15}{3}$$

$$\frac{51}{3} = 3\overline{)51}^{\,17}$$

Write each fraction as a mixed number or as a whole number.

26. $\frac{11}{2}$ 27. $\frac{16}{4}$ 28. $\frac{44}{8}$ 29. $\frac{17}{16}$ 30. $\frac{35}{10}$

31. $\frac{32}{7}$ 32. $\frac{23}{5}$ 33. $\frac{30}{8}$ 34. $\frac{43}{16}$ 35. $\frac{47}{15}$

36. $\frac{137}{64}$ 37. $\frac{25}{3}$ 38. $\frac{36}{6}$ 39. $\frac{47}{7}$ 40. $\frac{57}{18}$

41. $\frac{47}{8}$ 42. $\frac{49}{8}$ 43. $\frac{61}{2}$ 44. $\frac{49}{7}$ 45. $\frac{69}{4}$

46. $\frac{84}{3}$ 47. $\frac{93}{9}$ 48. $\frac{27}{2}$ 49. $\frac{156}{11}$ 50. $\frac{53}{16}$

SKILL 10 Equivalent and Simplest Form Fractions

A fraction equivalent to a given fraction can be obtained by *multiplying* both numerator and denominator by the same number.

$$\frac{2}{3} = \frac{2 \times 4}{3 \times 4} = \frac{8}{12}$$

$\llcorner$ equivalent fraction $\lrcorner$

Find an equivalent fraction by multiplying both the numerator and denominator of each by 5; by 6.

1. $\frac{2}{3}$ 2. $\frac{1}{2}$ 3. $\frac{4}{9}$ 4. $\frac{7}{8}$ 5. $\frac{3}{10}$

Write the value of x that would make the equation true.

6. $\frac{3}{8} = \frac{x}{32}$ 7. $\frac{9}{10} = \frac{45}{x}$ 8. $\frac{5}{16} = \frac{x}{64}$ 9. $\frac{9}{4} = \frac{x}{32}$ 10. $\frac{15}{16} = \frac{90}{x}$

11. $\frac{7}{18} = \frac{x}{36}$ 12. $\frac{x}{13} = \frac{33}{39}$ 13. $\frac{7}{24} = \frac{x}{72}$ 14. $\frac{9}{x} = \frac{36}{64}$ 15. $\frac{46}{x} = \frac{230}{255}$

A fraction can be expressed in simplest form by dividing both numerator and denominator by the same number.

$$\frac{25}{40} = \frac{25 \div 5}{40 \div 5} = \frac{5}{8} \leftarrow \text{simplest form fraction}$$

$\lfloor$ equivalent fractions $\rfloor$

Write an equivalent fraction that is the simplest form of the given fraction.

16. $\frac{8}{16}$ 17. $\frac{3}{15}$ 18. $\frac{6}{16}$ 19. $\frac{14}{20}$ 20. $\frac{8}{64}$

21. $\frac{32}{36}$ 22. $\frac{9}{45}$ 23. $\frac{24}{64}$ 24. $\frac{45}{25}$ 25. $\frac{16}{64}$

26. $\frac{24}{36}$ 27. $\frac{21}{27}$ 28. $\frac{20}{35}$ 29. $\frac{28}{64}$ 30. $\frac{112}{128}$

31. $\frac{65}{80}$ 32. $\frac{90}{270}$ 33. $\frac{184}{512}$ 34. $\frac{72}{468}$ 35. $\frac{168}{256}$

36. $\frac{232}{1450}$ 37. $\frac{209}{228}$ 38. $\frac{988}{1024}$ 39. $\frac{183}{366}$ 40. $\frac{1256}{2560}$

41. $\frac{75}{160}$ 42. $\frac{171}{228}$ 43. $\frac{768}{4096}$ 44. $\frac{21}{112}$ 45. $\frac{45}{240}$

SKILL 11 Changing Fractions and Decimals

A fraction can be changed to a decimal by dividing the numerator by the denominator.

$$\frac{7}{2} = 2\overline{)7.0}^{\,3.5} \quad \leftarrow \text{Annex as many zeros as needed.}$$

Write each as a decimal. Round answers to the nearest hundredth where needed.

1. $\dfrac{3}{5}$ 2. $\dfrac{7}{10}$ 3. $1\dfrac{3}{8}$ 4. $\dfrac{3}{4}$ 5. $3\dfrac{2}{16}$

6. $\dfrac{6}{32}$ 7. $1\dfrac{11}{2}$ 8. $\dfrac{3}{5}$ 9. $\dfrac{7}{10}$ 10. $2\dfrac{6}{25}$

11. $3\dfrac{3}{8}$ 12. $\dfrac{9}{10}$ 13. $5\dfrac{1}{12}$ 14. $\dfrac{2}{15}$ 15. $\dfrac{1}{30}$

16. $4\dfrac{7}{10}$ 17. $7\dfrac{1}{8}$ 18. $6\dfrac{5}{8}$ 19. $2\dfrac{7}{32}$ 20. $1\dfrac{7}{8}$

21. $\dfrac{65}{3}$ 22. $\dfrac{48}{7}$ 23. $\dfrac{124}{64}$ 24. $\dfrac{568}{4}$ 25. $\dfrac{879}{5}$

A decimal can be changed to a fraction by writing the decimal place value as the denominator and then simplifying when needed.

$$0.56 \;=\; \frac{56}{100} \;=\; \frac{14}{25} \qquad\qquad 3.25 \;=\; 3\frac{25}{100} \;=\; 3\frac{1}{4}$$

"fifty-six hundredths" simplest form "three and twenty-five hundredths" simplest form

Write each as a fraction in simplest form.

26. 0.3 27. 0.03 28. 0.003 29. 0.26 30. 2.75

31. 1.125 32. 2.25 33. 5.375 34. 3.5 35. 5.625

36. 0.532 37. 0.42 38. 2.875 39. 6.5 40. 1.1875

41. 3.125 42. 0.04 43. 1.411 44. 1.05 45. 0.105

46. 1.6432 47. 1.029 48. 0.1482 49. 5.55 50. 7.8125

51. 1.6 52. 2.93 53. 6.125 54. 0.0625 55. 1.3125

56. 0.9375 57. 15.25 58. 16.25 59. 1.946 60. 0.693

SKILL 12 Comparing Fractions

When comparing two fractions, one fraction is either greater than ($>$) the other, less than ($<$) the other, or equal to ($=$) the other. Two fractions can be compared by writing equivalent fractions with the same denominators.

Write $>$, $<$, or $=$ for the square (■).

$\frac{3}{4}$ ■ $\frac{5}{8}$ Write a fraction equivalent to $\frac{3}{4}$ in eighths, since eighths is a *common denominator* for fourths and halves.

$$\frac{3}{4}\boxed{\begin{array}{c}\times\ 2\\ \times\ 2\end{array}} = \frac{6}{8}$$

Compare the fractions.

$\frac{6}{8} > \frac{5}{8}$ Therefore, $\frac{3}{4} > \frac{5}{8}$

Compare each pair of fractions. Write $>$, $<$, or $=$ for each ■.

1. $\frac{3}{8}$ ■ $\frac{1}{2}$ 3. $\frac{3}{4}$ ■ $\frac{7}{8}$ 3. $\frac{4}{8}$ ■ $\frac{8}{16}$ 4. $\frac{3}{5}$ ■ $\frac{12}{20}$

5. $\frac{7}{9}$ ■ $\frac{2}{3}$ 6. $\frac{1}{2}$ ■ $\frac{17}{32}$ 7. $\frac{5}{16}$ ■ $\frac{3}{8}$ 8. $\frac{17}{32}$ ■ $\frac{9}{16}$

9. $\frac{2}{3}$ ■ $\frac{17}{27}$ 10. $\frac{3}{8}$ ■ $\frac{12}{32}$ 11. $\frac{15}{25}$ ■ $\frac{65}{100}$ 12. $\frac{2}{3}$ ■ $\frac{10}{15}$

13. $\frac{13}{2}$ ■ $\frac{18}{4}$ 14. $\frac{31}{32}$ ■ $\frac{157}{160}$ 15. $\frac{4}{5}$ ■ $\frac{41}{50}$ 16. $\frac{57}{18}$ ■ $\frac{37}{9}$

Write $>$, $<$, or $=$ for the ■.

$\frac{1}{6}$ ■ $\frac{1}{9}$ Write fractions equivalent to $\frac{1}{6}$ and $\frac{1}{9}$ with the lowest common denominator (LCD), 18.

$$\frac{1}{6} = \frac{3}{18} \qquad\qquad \frac{1}{9} = \frac{2}{18}$$

The LCD is 18.

Compare the fractions.

$\frac{3}{18} > \frac{2}{18}$ Therefore, $\frac{1}{6} > \frac{1}{9}$

Compare each pair of fractions. Write $>$, $<$, or $=$ for each ■.

17. $\frac{5}{8}$ ■ $\frac{4}{5}$ 18. $\frac{5}{6}$ ■ $\frac{4}{9}$ 19. $\frac{7}{10}$ ■ $\frac{11}{12}$ 20. $\frac{3}{4}$ ■ $\frac{13}{16}$

21. $\frac{5}{8}$ ■ $\frac{7}{10}$ 22. $\frac{1}{3}$ ■ $\frac{1}{4}$ 23. $\frac{9}{11}$ ■ $\frac{5}{6}$ 24. $\frac{2}{5}$ ■ $\frac{1}{3}$

25. $\frac{6}{12}$ ■ $\frac{50}{100}$ 26. $\frac{5}{2}$ ■ $\frac{7}{3}$ 27. $\frac{17}{4}$ ■ $\frac{13}{3}$ 28. $\frac{31}{64}$ ■ $\frac{123}{256}$

SKILL 13 Adding Fractions

To add fractions with **like denominators**, add the numerators, place the sum over the denominator, and simplify the result.

$$\frac{7}{16} + \frac{5}{16} = \frac{7+5}{16} = \frac{12}{16} \text{ or } \frac{3}{4} \quad \leftarrow \text{simplest form}$$

Express the sum in simplest form.

1. $\frac{1}{5} + \frac{3}{5}$ 2. $\frac{1}{4} + \frac{3}{4}$ 3. $\frac{1}{8} + \frac{3}{8}$ 4. $\frac{4}{9} + \frac{8}{9}$

5. $\frac{3}{10} + \frac{5}{10}$ 6. $\frac{17}{32} + \frac{19}{32}$ 7. $\frac{1}{3} + \frac{2}{3}$ 8. $\frac{5}{16} + \frac{7}{16}$

To add fractions with **unlike denominators**, write equivalent fractions to make both denominators the same.

$$\frac{3}{4} = \frac{15}{20} \quad \leftarrow \text{The LCD is 20.}$$
$$+ \frac{2}{5} = \frac{8}{20}$$
$$\overline{\qquad\qquad}$$
$$\frac{23}{20} \text{ or } 1\frac{3}{20} \quad \leftarrow \text{simplest form}$$

Express the sum as a mixed number in simplest form.

9. $\frac{1}{4} + \frac{1}{8}$ 10. $\frac{3}{8} + \frac{1}{16}$ 11. $\frac{13}{16} + \frac{5}{8}$ 12. $\frac{2}{5} + \frac{27}{30}$

13. $\frac{1}{2} + \frac{3}{16}$ 14. $\frac{9}{11} + \frac{3}{10}$ 15. $\frac{7}{16} + \frac{17}{64}$ 16. $\frac{17}{4} + \frac{2}{3}$

The sum of mixed numbers is found by adding the whole numbers and fractions separately.

$$8\frac{2}{3} = 8\frac{8}{12} \quad \leftarrow \text{The LCD is 12.}$$
$$+ 3\frac{3}{4} = 3\frac{9}{12}$$
$$\overline{\qquad\qquad}$$
$$11\frac{17}{12} = 11 + 1\frac{5}{12} = 12\frac{5}{12} \quad \leftarrow \text{simplest form}$$

Express the sum as a mixed number in simplest form.

17. $7\frac{1}{2} + 3\frac{3}{8}$ 18. $5\frac{3}{8} + 7\frac{3}{16}$ 19. $4\frac{5}{64} + 11\frac{1}{2}$

20. $4\frac{5}{10} + 8\frac{15}{16}$ 21. $6\frac{7}{3} + 15\frac{2}{11}$ 22. $14\frac{7}{8} + 10\frac{7}{16}$

23. $2\frac{7}{32} + \frac{13}{16}$ 24. $11\frac{4}{9} + 13\frac{3}{8}$ 25. $15\frac{5}{8} + 10\frac{3}{10}$

SKILL 14 Subtracting Fractions with Like Denominators

When finding the difference between fractions with like denominators, subtract the numerators and place the difference over the like denominator.

$$\frac{15}{16} - \frac{3}{16} = \frac{15-3}{16} = \frac{12}{16} \quad \text{or} \quad \frac{3}{4} \leftarrow \text{simplest form}$$

Express the difference in simplest form.

1. $\frac{5}{8} - \frac{3}{8}$ 2. $\frac{11}{4} - \frac{2}{4}$ 3. $\frac{9}{16} - \frac{5}{16}$ 4. $\frac{5}{7} - \frac{1}{7}$

5. $\frac{3}{4} - \frac{1}{4}$ 6. $\frac{2}{3} - \frac{1}{3}$ 7. $\frac{7}{15} - \frac{1}{15}$ 8. $\frac{11}{64} - \frac{7}{64}$

9. $\frac{25}{16} - \frac{17}{16}$ 10. $\frac{7}{2} - \frac{3}{2}$ 11. $\frac{28}{5} - \frac{24}{5}$ 12. $\frac{47}{50} - \frac{13}{50}$

Finding the difference of mixed numbers with like denominators can involve regrouping.

$$\begin{array}{rcl} 5 &=& 4\frac{3}{3} \\ -2\frac{2}{3} &=& -2\frac{2}{3} \\ \hline && 2\frac{1}{3} \leftarrow \text{simplest form} \end{array}$$ Change 5 to $4\frac{3}{3}$, then subtract.

$$\begin{array}{rcl} 6\frac{1}{8} &=& 5\frac{9}{8} \\ -2\frac{5}{8} &=& -2\frac{5}{8} \\ \hline && 3\frac{4}{8} \text{ or } 3\frac{1}{2} \leftarrow \text{simplest form} \end{array}$$ Change $6\frac{1}{8}$ to $5\frac{9}{8}$, then subtract.

Express the difference as a mixed number in simplest form.

13. $8\frac{3}{4} - 5\frac{1}{4}$ 14. $12\frac{4}{5} - 7\frac{1}{5}$ 15. $11\frac{5}{8} - 7\frac{3}{8}$

16. $9\frac{6}{7} - 2\frac{5}{7}$ 17. $6\frac{1}{16} - 2\frac{5}{16}$ 18. $8\frac{7}{10} - 5\frac{8}{10}$

19. $15\frac{9}{20} - 10\frac{17}{20}$ 20. $14 - 2\frac{9}{32}$ 21. $17 - 5\frac{15}{16}$

22. $12 - 11\frac{9}{10}$ 23. $19 - \frac{7}{32}$ 24. $8\frac{4}{7} - 5\frac{5}{7}$

25. $14\frac{1}{8} - 12\frac{3}{8}$ 26. $6\frac{3}{64} - 5\frac{11}{64}$ 27. $7\frac{1}{10} - 2\frac{7}{10}$

28. $7 - 4\frac{6}{35}$ 29. $23 - \frac{7}{8}$ 30. $56 - 23\frac{3}{8}$

31. $12\frac{5}{8} - 9\frac{3}{8}$ 32. $24 - 23\frac{2}{5}$ 33. $93\frac{1}{16} - \frac{1}{16}$

SKILL 15 Subtracting Fractions with Unlike Denominators

To subtract fractions with unlike denominators, write equivalent fractions to make both denominators the same.

$$\frac{6}{7} = \frac{18}{21}$$ ← The LCD is 21.
$$-\frac{1}{3} = \frac{7}{21}$$
$$\frac{11}{21}$$ ← simplest form

Express the difference in simplest form.

1. $\dfrac{3}{4} - \dfrac{1}{2}$ 2. $\dfrac{3}{4} - \dfrac{1}{8}$ 3. $\dfrac{2}{3} - \dfrac{1}{6}$ 4. $\dfrac{3}{8} - \dfrac{1}{16}$

5. $\dfrac{5}{8} - \dfrac{1}{4}$ 6. $\dfrac{7}{32} - \dfrac{13}{64}$ 7. $\dfrac{9}{11} - \dfrac{3}{10}$ 8. $\dfrac{9}{8} - \dfrac{13}{16}$

9. $\dfrac{27}{30} - \dfrac{7}{10}$ 10. $\dfrac{19}{24} - \dfrac{5}{18}$ 11. $\dfrac{6}{7} - \dfrac{10}{21}$ 12. $\dfrac{1}{2} - \dfrac{2}{5}$

The difference of mixed numbers with unlike denominators is found by first writing equivalent fractions and then subtracting the whole numbers and fractions separately. This might involve borrowing.

$$12\tfrac{2}{3} = 12\tfrac{16}{24} \overset{11 \ \ 40}{}$$
$$-\ 4\tfrac{7}{8} = 4\tfrac{21}{24}$$
$$7\tfrac{19}{24}$$ ← simplest form

Express the difference as a mixed number in simplest form.

13. $6\dfrac{1}{2} - 4\dfrac{4}{9}$ 14. $9\dfrac{1}{2} - 3\dfrac{5}{8}$ 15. $12\dfrac{7}{8} - 5\dfrac{3}{4}$

16. $32\dfrac{2}{16} - 11\dfrac{3}{4}$ 17. $19\dfrac{4}{5} - 9\dfrac{11}{12}$ 18. $52\dfrac{4}{7} - 37\dfrac{5}{9}$

19. $28\dfrac{3}{7} - 26\dfrac{9}{10}$ 20. $30\dfrac{19}{30} - 29\dfrac{3}{4}$ 21. $6\dfrac{3}{4} - 3\dfrac{7}{8}$

22. $15\dfrac{1}{2} - 7\dfrac{3}{10}$ 23. $15\dfrac{5}{8} - 6\dfrac{3}{4}$ 24. $3 - 1\dfrac{1}{2}$

25. $21\dfrac{5}{6} - 19\dfrac{5}{8}$ 26. $9\dfrac{11}{12} - 7\dfrac{3}{8}$ 27. $21\dfrac{1}{2} - 20\dfrac{1}{8}$

SKILL 16 Multiplying Fractions

To find the product of fractions, multiply the numerators and then multiply the denominators.

$$\frac{2}{3} \times \frac{4}{5} \quad = \quad \frac{2 \times 4}{3 \times 5} \quad = \quad \frac{8}{15} \quad \leftarrow \text{simplest form}$$

Express the product in simplest form.

1. $\frac{1}{2} \times \frac{3}{4}$ **3.** $\frac{2}{3} \times \frac{1}{4}$ **3.** $\frac{5}{8} \times \frac{2}{16}$ **4.** $\frac{3}{4} \times \frac{8}{9}$

5. $\frac{11}{12} \times \frac{5}{11}$ **6.** $\frac{5}{16} \times \frac{1}{2}$ **7.** $4 \times \frac{7}{32}$ **8.** $12 \times \frac{2}{5}$

9. $\frac{1}{5} \times \frac{1}{3}$ **10.** $\frac{8}{9} \times \frac{7}{8}$ **11.** $\frac{5}{32} \times \frac{3}{4}$ **12.** $12 \times \frac{5}{64}$

To find the product of mixed numbers, write the mixed numbers as fractions first.

$$3\frac{1}{4} \times 4\frac{1}{2} \quad = \quad \frac{13}{4} \times \frac{9}{2} \quad = \quad \frac{13 \times 9}{4 \times 2} \quad = \quad \frac{117}{8} \quad = \quad 14\frac{5}{8} \quad \leftarrow \text{simplest form}$$

$$6 \times 7\frac{1}{3} \quad = \quad \frac{6}{1} \times \frac{22}{3} \quad = \quad \frac{6 \times 22}{1 \times 3} \quad = \quad \frac{132}{3} \quad = \quad 44 \quad \leftarrow \text{simplest form}$$

Express the product in simplest form.

13. $2\frac{1}{2} \times 3\frac{1}{3}$ **14.** $3\frac{2}{5} \times 4\frac{2}{3}$ **15.** $\frac{1}{8} \times 3\frac{4}{5}$

16. $9 \times 8\frac{1}{3}$ **17.** $12\frac{1}{2} \times 3\frac{1}{2}$ **18.** $13\frac{1}{2} \times 3\frac{1}{4}$

19. $3\frac{1}{16} \times 1\frac{5}{8}$ **20.** $2\frac{3}{32} \times 2\frac{1}{4}$ **21.** $5 \times 1\frac{3}{8}$

22. $7\frac{4}{9} \times 3\frac{3}{8}$ **23.** $12\frac{1}{2} \times 2\frac{2}{3}$ **24.** $1\frac{3}{4} \times 12$

25. $7\frac{1}{2} \times \frac{5}{32}$ **26.** $4\frac{1}{3} \times 3\frac{2}{5}$ **27.** $5\frac{3}{5} \times 1\frac{2}{6}$

28. $2\frac{5}{9} \times \frac{7}{10}$ **29.** $\frac{2}{3} \times 3\frac{2}{3}$ **30.** $7 \times 8\frac{1}{2}$

31. $6 \times 3\frac{1}{8}$ **32.** $5 \times 15\frac{3}{32}$ **33.** $\frac{15}{32} \times 2\frac{2}{15}$

SKILL 17 Dividing Fractions

To divide fractions, multiply the dividend by the reciprocal of the divisor.

$$\frac{3}{4} \div \frac{2}{3}$$ The divisor is $\frac{2}{3}$. The reciprocal of the divisor is $\frac{3}{2}$.

dividend   divisor

$$\frac{3}{4} \div \frac{2}{3} = \frac{3}{4} \times \frac{3}{2} = \frac{3 \times 3}{4 \times 2} = \frac{9}{8} \text{ or } 1\frac{1}{8} \leftarrow \text{simplest form}$$

Express the quotient in simplest form.

1. $\frac{1}{4} \div \frac{3}{8}$ 2. $\frac{1}{8} \div \frac{1}{4}$ 3. $\frac{3}{4} \div \frac{5}{6}$ 4. $5 \div \frac{2}{5}$

5. $7 \div \frac{7}{11}$ 6. $8 \div \frac{8}{15}$ 7. $\frac{3}{4} \div \frac{1}{8}$ 8. $\frac{2}{3} \div \frac{5}{6}$

9. $\frac{4}{9} \div \frac{1}{27}$ 10. $\frac{7}{12} \div \frac{1}{6}$ 11. $\frac{3}{4} \div 4$ 12. $\frac{3}{4} \div \frac{1}{4}$

13. $\frac{3}{4} \div \frac{2}{4}$ 14. $\frac{4}{3} \div \frac{4}{5}$ 15. $\frac{4}{3} \div \frac{1}{4}$ 16. $\frac{9}{10} \div \frac{3}{5}$

To divide mixed numbers, follow the same procedure as above except first change the mixed numbers to fractions.

$$3\frac{1}{6} \div 1\frac{1}{4} = \frac{19}{6} \div \frac{5}{4} = \frac{19}{6} \times \frac{4}{5} = \frac{19 \times 4}{6 \times 5} = \frac{76}{30} = 2\frac{16}{30} \text{ or } 2\frac{8}{15} \leftarrow \text{simplest form}$$

Express the quotient in simplest form.

17. $4\frac{2}{3} \div 1\frac{1}{2}$ 18. $7\frac{5}{6} \div 1\frac{1}{5}$ 19. $1\frac{1}{6} \div 2\frac{1}{3}$

20. $5\frac{2}{3} \div 1\frac{3}{5}$ 21. $\frac{3}{4} \div 2\frac{1}{2}$ 22. $\frac{1}{8} \div 3\frac{1}{2}$

23. $15 \div 1\frac{1}{5}$ 24. $12\frac{1}{4} \div 2\frac{1}{5}$ 25. $13 \div 2\frac{13}{17}$

26. $14\frac{1}{7} \div 3\frac{1}{6}$ 27. $7\frac{1}{2} \div 1\frac{1}{4}$ 28. $3\frac{5}{7} \div \frac{13}{21}$

29. $\frac{175}{4} \div \frac{526}{120}$ 30. $\frac{256}{8} \div \frac{128}{16}$ 31. $\frac{810}{90} \div \frac{728}{81}$

32. $\frac{15}{16} \div \frac{5}{32}$ 33. $2\frac{1}{16} \div 11$ 34. $19\frac{1}{2} \div 19\frac{1}{2}$

SKILL 18 Writing Ratios and Rates

A **ratio** is a comparison of numbers. If 16 white, 20 blue, and 12 red spools of thread were on a table, the ratio of the number of white spools to the number of blue spools to the number of red spools would be written as 16 to 20 to 12, or 16 : 20 : 12. The *terms* of the ratio are 16, 20, and 12. The *simplest form* of the ratio is 4 to 5 to 3, or 4 : 5 : 3. The simplest form is found by dividing each term by the greatest common factor, 4.

Write each comparison of numbers of screws in the table below as a ratio in simplest colon form.

1. wood screws to tapping screws
2. tapping screws to thumb screws
3. set screws to tapping screws
4. tapping screws to wood screws to set screws
5. thumb screws to wood screws to tapping screws to set screws

SUPPLY INVENTORY			
Wood Screws (boxes)	Tapping Screws (boxes)	Set Screws (boxes)	Thumb Screws (boxes)
22	6	8	12

A *corresponding fraction* can be written for a two-term ratio.

The number of eyes to the number of toes is: 2 to 10, 2 : 10, $\frac{2}{10}$, or $\frac{1}{5}$ ← simplest form

Write each comparison of numbers as a simplest form fraction.

6. The ratio of casement windows to double-hung windows in a building is 30 to 45.

7. The ratio of wall phones to desk phones is 3000 to 1800.

8. The ratio of yellow light bulbs to blue light bulbs is 32 to 16.

9. The width-to-length ratio for a toolbox is 28 cm to 64 cm.

A **rate** is a comparison of quantities, such as kilometers and hours. A **unit rate** is a comparison of quantities in which the second term is 1.

rate: 510 km/6h $= \dfrac{510 \div 6}{6 \div 6} = \dfrac{85}{1}$ unit rate: 85 km/h

Write each rate in fraction form. Then express the rate as a unit rate using the slash mark.

10. 4480 km in 8 h

11. 200 students for 8 instructors

12. 300 m² for 12 lathes

13. 96 m in 16 s

14. $60,000 in 12 years

15. $1280 in 40 h

16. 72 L in 48 bottles

17. 500,000 people in 200 km²

18. $28.50 for 3 kg

19. 32 L for 400 km

20. 4200 books 20 h

21. 4950 words in 9 min

SKILL 19 Writing Proportions

A **proportion** is a statement of equality between two ratios or rates. To check if a proportion is true, see if the *cross products* are the same.

Is the rate 6 kg for $8 the same as 3 kg for $4?

$$\frac{\text{kilograms:}}{\text{dollars:}} \quad \frac{6}{8} = \frac{3}{4} \qquad \frac{6}{8} \times \frac{3}{4} \qquad \begin{matrix} 3 \times 8 = 6 \times 4 \\ 24 = 24 \end{matrix}$$

← The cross products are equal, so the proportion is true.

Consider each of the following as a proportion. Would the proportion be true? Write *Yes* or *No*.

1. $\frac{3}{7}$ and $\frac{9}{21}$ 2. $\frac{6}{13}$ and $\frac{36}{76}$ 3. $\frac{8}{9}$ and $\frac{64}{72}$ 4. $\frac{18}{24}$ and $\frac{9}{12}$

5. $\frac{13 \text{ m}}{18 \text{ s}}$ and $\frac{9 \text{ m}}{12 \text{ s}}$ 6. $\frac{14 \text{ km}}{15 \text{ h}}$ and $\frac{42 \text{ km}}{45 \text{ h}}$ 7. $\frac{\$3}{51 \text{ cm}}$ and $\frac{\$1}{17 \text{ cm}}$

8. $\frac{13 \text{ bolts}}{16 \text{ nails}}$ and $\frac{52 \text{ bolts}}{68 \text{ nails}}$ 9. $\frac{27 \text{ books}}{2 \text{ shelves}}$ and $\frac{135 \text{ books}}{5 \text{ shelves}}$ 10. $\frac{14 \text{ chairs}}{3 \text{ tables}}$ and $\frac{110 \text{ chairs}}{24 \text{ tables}}$

If one of the four terms of a proportion is unknown, solve for the unknown by using *cross products* or *equivalent fractions*.

If 5 computers were shared by 8 people, how many people could share 15 computers, at the same rate?

Method 1: Cross products

$$\frac{\text{computers:}}{\text{people:}} \quad \frac{5}{8} = \frac{15}{x} \qquad \frac{5}{8} \times \frac{15}{x} \qquad \begin{matrix} 5x = 8 \times 15 \\ 5x = 120 \\ x = 24 \end{matrix}$$

← 24 people could share 15 computers

Method 2: Equivalent fractions

$$\frac{\text{computers:}}{\text{people:}} \quad \frac{5}{8} = \frac{15}{x} \qquad \frac{5 \boxed{\times 3}}{8 \boxed{\times 3}} = \frac{15}{24} \qquad x = 24$$

Consider each of the following as a proportion. Solve for the unknown term using the method of your choice.

11. $\frac{1}{3} = \frac{4}{n}$ 12. $\frac{2}{5} = \frac{x}{10}$ 13. $\frac{a}{5} = \frac{16}{20}$ 14. $\frac{5}{6} = \frac{25}{y}$

15. $\frac{3}{t} = \frac{6}{8}$ 16. $\frac{p}{2} = \frac{12}{24}$ 17. $\frac{3}{8} = \frac{9}{w}$ 18. $\frac{r}{7} = \frac{14}{49}$

19. $\frac{28 \text{ cm}}{4 \text{ h}} = \frac{x \text{ (cm)}}{1 \text{ h}}$ 20. $\frac{g \text{ (m)}}{8 \text{ s}} = \frac{9 \text{ m}}{12 \text{ s}}$ 21. $\frac{\$12}{20 \text{ m}} = \frac{y \text{ (\$)}}{5 \text{ m}}$

22. $\frac{25 \text{ km}}{3 \text{ h}} = \frac{a \text{ (km)}}{6 \text{ h}}$ 23. $\frac{20 \text{ kg}}{\$4} = \frac{15 \text{ kg}}{k \text{ (\$)}}$ 24. $\frac{\$96}{8 \text{ h}} = \frac{z \text{ (\$)}}{3 \text{ h}}$

SKILL 20 Writing Decimals as Percents

Percent (%) is a rate per 100. The diagram at the right shows that 45 of 100 squares are shaded, that is, 45% of the large square is shaded.

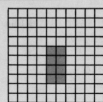

Express the shaded region as a percent.

1.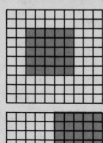

2.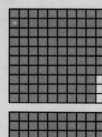

3.

4.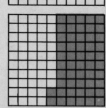

5.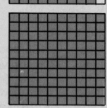

To express a decimal number as a percent, move the decimal point two places to the right, and annex the percent symbol.

    0.45 = 45%   0.7 = 70%   0.003 = 0.3%   6 = 600%

Write each as a percent.

6. 0.26	**7.** 0.19	**8.** 0.39	**9.** 0.52
10. 5.10	**11.** 4.80	**12.** 0.30	**13.** 7.7
14. 2.25	**15.** 0.13	**16.** 0.47	**17.** 0.9
18. 0.2	**19.** 6.5	**20.** 28.0	**21.** 8
22. 10.4	**23.** 11.5	**24.** 8.3	**25.** 0.6
26. 0.20	**27.** 0.64	**28.** 1.23	**29.** 0.347
30. 8.103	**31.** 0.006	**32.** 0.05	**33.** 0.16
34. 72	**35.** 3.212	**36.** 0.491	**37.** 0.325
38. 0.109	**39.** 0.001	**40.** 0.25	**41.** 0.83
42. 6.753	**43.** 0.194	**44.** 10.28	**45.** 0.007
46. 0.504	**47.** 3	**48.** 6.1	**49.** 12.0

SKILL 21 Writing Fractions and Mixed Numbers as Percents

A fraction is changed to a percent by first writing it as a decimal.

$$\frac{3}{8} = \begin{array}{r} 0.375 \\ 8\overline{)3.000} \end{array} = 37.5\%$$ ← Since the numerator is less than the denominator, the percent is less than 100%.

$\frac{3}{8}$ means 3 ÷ 8.

Write each fraction as a percent.

1. $\frac{7}{10}$

2. $\frac{7}{20}$

3. $\frac{7}{40}$

4. $\frac{2}{5}$

5. $\frac{1}{8}$

6. $\frac{1}{4}$

7. $\frac{7}{50}$

8. $\frac{5}{8}$

9. $\frac{12}{25}$

10. $\frac{11}{16}$

11. $\frac{4}{32}$

12. $\frac{8}{8}$

Write each fraction as a percent to the nearest tenth.

13. $\frac{1}{3}$

14. $\frac{5}{6}$

15. $\frac{7}{12}$

16. $\frac{7}{9}$

A mixed number is changed to a percent by first changing it to a decimal.

$3\frac{1}{2}$ = 3.5 = 350%   ← Note that the percent
$2\frac{5}{8}$ = 2.625 = 262.5%   equivalents of mixed numbers are over 100%.

5 ÷ 8 = 0.625

Write each as a percent.

17. $2\frac{1}{4}$

18. $4\frac{7}{40}$

19. $3\frac{5}{8}$

20. $12\frac{3}{16}$

21. $2\frac{11}{50}$

22. $11\frac{3}{10}$

23. $6\frac{7}{40}$

24. $3\frac{3}{8}$

Write each as a percent. Round to the nearest tenth.

25. $5\frac{1}{6}$

26. $4\frac{5}{9}$

27. $8\frac{2}{3}$

28. $6\frac{4}{15}$

29. $\frac{5}{9}$

30. $\frac{17}{30}$

31. $21\frac{5}{6}$

32. $5\frac{11}{15}$

33. $\frac{5}{12}$

34. $21\frac{5}{14}$

35. $\frac{8}{9}$

36. $\frac{5}{7}$

SKILL 22 Writing Percents as Decimals

If 25 of the 100 squares are shaded at the right, we say that 25%, or 0.25, is shaded.

To express a percent as a decimal number, move the decimal point two places to the left and remove the percent symbol.

25% = 0.25   2.5% = 0.025

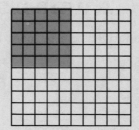

Write each percent as a decimal.

1. 85%	**2.** 60%	**3.** 40%	**4.** 14.5%
5. 120%	**6.** 146%	**7.** 3.6%	**8.** 8.1%
9. 7.9%	**10.** 16.2%	**11.** 98.5%	**12.** 100%
13. 75%	**14.** 219%	**15.** 8.6%	**16.** 4.25%
17. 5.4%	**18.** 165%	**19.** 124.3%	**20.** 0.56%

If the percent is written as a mixed number, write the mixed number in decimal form first. Then change the percent to a decimal.

$37\frac{1}{2}\%$ = 37.5% = 0.375 $\frac{3}{4}\%$ = 0.75% = 0.0075

Write each percent as a decimal.

21. $10\frac{1}{2}\%$	**22.** $21\frac{1}{4}\%$	**23.** $5\frac{3}{8}\%$	**24.** $5\frac{3}{5}\%$
25. $6\frac{1}{10}\%$	**26.** $\frac{7}{10}\%$	**27.** $20\frac{2}{5}\%$	**28.** $\frac{5}{8}\%$
29. $72\frac{13}{20}\%$	**30.** $14\frac{21}{25}\%$	**31.** $11\frac{3}{4}\%$	**32.** $9\frac{4}{5}\%$

Write each percent as a decimal rounded to the nearest thousandth.

33. $11\frac{1}{3}\%$	**34.** $10\frac{3}{7}\%$	**35.** $22\frac{1}{9}\%$	**36.** $\frac{1}{40}\%$
37. $2\frac{3}{11}\%$	**38.** $28\frac{2}{3}\%$	**39.** $37\frac{6}{7}\%$	**40.** $\frac{8}{9}\%$
41. 4.5673%	**42.** 78.2569%	**43.** 0.5267%	**44.** 1.259%
45. 6.25%	**46.** $6\frac{1}{4}\%$	**47.** $6\frac{7}{28}\%$	**48.** $14\frac{9}{21}\%$
49. $127\frac{1}{2}\%$	**50.** 18.96%	**51.** $12\frac{24}{25}\%$	**52.** $\frac{1}{100}\%$

SKILL 23 Writing Percents as Fractions

The diagram at the right shows that 28%, or $\frac{28}{100}$ of the square region, is shaded.

To change a percent to a fraction, write the percent as a fraction with a denominator of 100 and simplify.

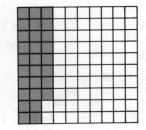

$$28\% = \frac{28}{100} = \frac{7}{25} \quad \leftarrow \text{simplest form}$$

$$28.5\% = \underset{\uparrow}{\frac{28.5}{100}} = \underset{\uparrow}{\frac{285}{1000}} = \frac{57}{200} \quad \leftarrow \text{simplest form}$$

$$\underset{\text{equivalent fractions}}{}$$

$$16\tfrac{2}{3}\% = \frac{16\frac{2}{3}}{100} = \frac{\frac{50}{3}}{100} = \frac{50}{3} \times \frac{1}{100} = \frac{50}{300} = \frac{1}{6} \quad \leftarrow \text{simplest form}$$

Write each as a fraction in simplest form.

1. 42.5%   **2.** 70%   **3.** 125%   **4.** 300%

5. 12.7%   **6.** 0.1%   **7.** 0.12%   **8.** 54.3%

9. 11.6%   **10.** 65.44%   **11.** 27.63%   **12.** 61.24%

13. 50%   **14.** 95%   **15.** 9.1%   **16.** 0.13%

17. 40.9%   **18.** 43.4%   **19.** 5.25%   **20.** 87.5%

21. $43\frac{3}{4}\%$ **22.** $2\frac{1}{2}\%$ **23.** $9\frac{1}{4}\%$ **24.** $11\frac{1}{3}\%$

25. $16\frac{1}{2}\%$ **26.** $10\frac{1}{2}\%$ **27.** $25\frac{1}{2}\%$ **28.** $25\frac{3}{4}\%$

29. $16\frac{1}{3}\%$ **30.** $33\frac{2}{3}\%$ **31.** 20%   **32.** $12\frac{1}{3}\%$

33. 35%   **34.** $5\frac{3}{4}\%$ **35.** 46.2%   **36.** $11\frac{1}{9}\%$

37. 100%   **38.** 48.55%   **39.** $12\frac{3}{4}\%$ **40.** 132%

41. 200%   **42.** 5.8%   **43.** $68\frac{1}{2}\%$ **44.** 75%

45. 23.2%   **46.** 95%   **47.** 55.5%   **48.** 97.3%

49. $\frac{3}{10}\%$ **50.** $10\frac{3}{8}\%$ **51.** $2\frac{8}{10}\%$ **52.** $5\frac{1}{2}\%$

SKILL 24 Finding the Percent

In diagram *A*, 12 out of 60 squares are shaded while in diagram *B*, 20 out of 100 squares are shaded. This means that 20 *per hundred* are shaded in diagram *B*, or 20%.

The percent of shaded squares in diagram *A* can be found by two methods.

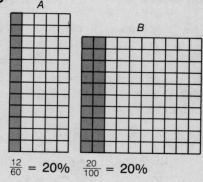

Method 1: Division

$$\frac{12}{60} \text{ means } 60\overline{)12.00}^{\,0.20} = 20\%$$

To help you remember this method for finding the *percent*, place your finger over the word *percent* in the diagram. The uncovered portion reminds you that the *part* is divided by the *whole*.

$$\frac{12}{60} = 20\% \qquad \frac{20}{100} = 20\%$$

Method 2: Proportion

Twelve out of 60 is the same as ___?___ per hundred.

$$\frac{12}{60} = \frac{n}{100} \rightarrow \begin{array}{l} 60n = 12(100) \\ 60n = 1200 \\ n = 20 \end{array} \rightarrow \frac{12}{60} = 20\%$$

You can see that in both diagram *A* and diagram *B* above, 20% of the squares are shaded.

Solve. Round the result to the nearest tenth of a percent where needed.

1. 30 out of 60 = ___?___ %

2. 10 out of 40 = ___?___ %

3. 15 out of 90 = ___?___ %

4. 9 out of 60 = ___?___ %

5. 25 out of 80 = ___?___ %

6. 50 out of 90 = ___?___ %

7. 48 out of 55 = ___?___ %

8. 65 out of 74 = ___?___ %

9. 69 out of 103 = ___?___ %

10. 110.5 out of 230 = ___?___ %

11. 92 out of 104 = ___?___ %

12. 239.76 out of 296 = ___?___ %

13. 15.5 out of 100 = ___?___ %

14. 63 out of 21 = ___?___ %

15. 112 out of 222 = ___?___ %

Solve by using a proportion. Round the result to the nearest tenth of a percent where needed.

16. 30 is ___?___ % of 20

17. 7.5 is ___?___ % of 48

18. 130 is ___?___ % of 110

19. 0.10 is ___?___ % of 0.60

20. 13.4 is ___?___ % of 54

21. 2.5 is ___?___ % of 90

22. ___?___ % of 18 is 6

23. ___?___ % of 60.5 is 52

24. ___?___ % of 15 is 2.5

25. ___?___ % of 2.5 is 0.2

26. ___?___ % of 12.3 is 5

27. ___?___ % of 100 is 125

SKILL 25 Finding the Part

To help you remember the method for finding
the *part* in a problem involving *percent*, place
your finger over *part* in the diagram. The uncovered portion shows you that the *percent* is
multiplied by the *whole* to find the *part*.

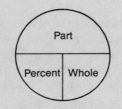

What is 10.5% of 70?

10.5% of 70 means 10.5% × 70. $0.105 \times 70 = 7.35$
                                             $= 7.4$ (rounded to the nearest tenth)

You can also use the proportion method to find the *part*.
What is 70% of 140?

$$\frac{70}{100} = \frac{x}{140} \rightarrow \begin{array}{l} 100x = 70(140) \\ 100x = 9800 \\ \quad\; x = 98 \end{array} \rightarrow 98 \text{ is } 70\% \text{ of } 140.$$

Find the *part*.

1. 12% of 50	**2.** 40% of 80	**3.** 3% of 86
4. 5% of 130	**5.** 2.5% of 110	**6.** 7.3% of 150
7. 12.8% of 70	**8.** 21.3% of 130	**9.** 4.72% of 70
10. 1.05% of 45	**11.** 37.7% of 87	**12.** 8.35% of 55
13. $6\frac{1}{4}$% of 120	**14.** $\frac{3}{5}$% of 50	**14.** $7\frac{1}{2}$% of 75

Find the *part* by using a proportion. Round the result to the nearest tenth
where needed.

16. 2.25% of 80	**17.** 35% of 72	**18.** 125% of 54
19. 8.1% of 70	**20.** 6.2% of 224	**21.** 90% of 100
22. 1% of 1000	**23.** 8.55% of 200	**24.** 99.9% of 80
25. 9% of 85.6	**26.** 12% of 125.05	**27.** 7.2% of 42.58
28. 34.6% of 60	**29.** $\frac{3}{4}$% of 800	**30.** $\frac{1}{2}$% of 400
31. 110% of 100	**32.** 4% of 4	**33.** 53% of 50
34. 119% of 109	**35.** 225% of 12	**36.** 0.25% of 61
37. $6\frac{1}{4}$% of 500	**38.** $10\frac{1}{2}$% of 750	**39.** 1.9% of 7050
40. $10\frac{3}{8}$% of 2500	**41.** 3.6% of 259	**42.** 100% of 53

SKILL 26 Finding the Whole

To help you remember the method for finding the *whole* in a problem involving percent, place your finger over *whole* in the diagram. The uncovered portion shows you that the *part* is divided by the *percent* to find the *whole*.

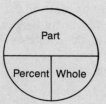

12.18 is 20.3% of what number?

$$\frac{12.18}{20.3\%} \rightarrow \frac{12.18}{0.203} = 60 \rightarrow 12.18 \text{ is } 20.3\% \text{ of } 60.$$

A proportion can also be used to find the *whole*.

42 is 37.5% of what number?

$$\frac{42}{x} = \frac{37.5}{100} \rightarrow \begin{array}{l} 37.5x = 42(100) \\ 37.5x = 4200 \\ x = 112 \end{array} \rightarrow 42 \text{ is } 37.5\% \text{ of } 112.$$

Find the *whole*. Round the result to the nearest tenth where needed.

1. 12% of n = 400

2. 37.5% of n = 25

3. 3.25% of n = 122

4. 3% of n = 4.5

5. 6.75% of n = 82

6. 8% of n = 7.2

7. 11% of n = 27

8. 1% of n = 3.7

9. 9% of n = 2

10. 52 is 12% of what number?

11. 28.2 is 15% of what number?

12. 942 is 95.2% of what number?

13. 31 is 11.5% of what number?

14. 40 is 27.3% of what number?

15. 24 is 71% of what number?

16. 70% of what number is 60?

17. 20% of what number is 3.9?

18. 35% of what number is 35?

19. 1.2% of what number is 3.6?

20. 43% of what number is 80?

21. 55% of what number is 19?

Find the number by using a proportion. Round the result to the nearest tenth.

22. 70% of n = 60

23. 110% of n = 3.5

24. 6% of n = 5.2

25. 1.234 is 8.08% of what number?

26. 62 is 90% of what number?

27. 82 is 8% of what number?

28. 12 is 20% of what number?

29. 110% of what number is 3.5?

30. 7.5% of what number is 29?

31. 0.5% of what number is 0.7?

32. 1.2% of what number is 3.6?

SKILL 27 Solving Equations by Adding or Subtracting

In order to solve for an unknown in an equation, the variable representing the unknown needs to be isolated. Often we need to isolate a variable in an equation by adding the same number to or subtracting the same number from *each side* of an equation.

$$x + 3.1 = 7.4$$
$$x + 3.1 - 3.1 = 7.4 - 3.1$$
$$x = 4.3$$

Do the *opposite* of "add 3.1".
← Subtract 3.1 from each side to isolate x.

Check: $4.3 + 3.1 = 7.4$
$$7.4 = 7.4$$

← The solution, $x = 4.3$, is correct since the left side and the right side are equal.

$$n - 46 = 52$$
$$n - 46 + 46 = 52 + 46$$
$$n = 98$$

Do the *opposite* of "subtract 46".
← Add 46 to each side to isolate n.

Check: $98 - 46 = 52$
$$52 = 52$$

← The solution, $n = 98$, is correct since the left side and the right side are equal.

Solve each equation by doing the *opposite* operation. Then check your solution.

1. $y + 3 = 5$

2. $y + 7 = 12$

3. $x + 3 = 4$

4. $x + 1.1 = 3.6$

5. $w + 0.3 = 5.6$

6. $w + 1.15 = 7.7$

7. $0.02 + t = 3.77$

8. $10.5 + t = 90$

9. $2.5 + r = 3.75$

10. $4\frac{1}{3} + r = 10$

11. $p + \frac{4}{3} = 5\frac{1}{6}$

12. $q + 1\frac{1}{2} = \frac{17}{8}$

13. $x - 7 = 5$

14. $x - 8 = 22$

15. $r - 6 = 12$

16. $s - 4.2 = 5.5$

17. $y - 3.6 = 24.7$

18. $y - 1.1 = 0.5$

19. $7.2 - t = 4.01$

20. $5.3 - t = 1.02$

21. $4.5 - w = 0.5$

22. $q - 2\frac{2}{3} = 8\frac{5}{6}$

23. $z - \frac{9}{4} = 10\frac{3}{8}$

24. $x - \frac{1}{2} = 7\frac{3}{5}$

25. $x + 3.5 = 7.2$

26. $y - 0.1 = 3.5$

27. $11.5 + r = 36$

28. $m + 8.67 = 123.5$

29. $p - 16.3 = 31.44$

30. $n - 2.5 = 14.7$

31. $1.5 + y = 10$

32. $12.25 + r = 88.75$

33. $s - 2.3 = 8.3$

34. $d + 35\frac{1}{2} = 57\frac{3}{4}$

35. $b - 5750 = 275$

36. $1\frac{7}{28} + z = 1.25$

SKILL 28 Solving Equations by Multiplying or Dividing

Sometimes when we solve for the unknown in an equation, the variable is isolated by multiplying or dividing *each side* by the same number.

$$15x = 75$$

Do the *opposite* of "multiply by 15".

$$\frac{15x}{15} = \frac{75}{15}$$

← Divide each side by 15 to isolate x.

$$x = 5$$

Check: $15(5) = 75$

The solution, $x = 5$, is correct since the

$$75 = 75$$

← left side and the right side are equal.

$$\frac{w}{1.2} = 6.0$$

$$\frac{\overset{1}{\cancel{w}}}{\underset{1}{\cancel{1.2}}} \times 1.2 = 6.0 \times 1.2 \quad \leftarrow \quad \text{Do the } \textit{opposite} \text{ of "divide by 1.2".}$$

$$w = 7.2 \qquad \text{Multiply each side by 1.2 to isolate } w.$$

Check: $\frac{7.2}{1.2} = 6.0$ ← The solution, $w = 7.2$, is correct since

$$6.0 = 6.0 \qquad \text{the left side and the right side are equal.}$$

Solve each equation by doing the *opposite* operation. Then check your solution. Round to the nearest tenth where needed.

1. $\dfrac{x}{5} = 4$

2. $\dfrac{y}{12} = 20$

3. $\dfrac{t}{15} = 15$

4. $\dfrac{r}{30} = 1.7$

5. $\dfrac{n}{100} = 21$

6. $\dfrac{p}{18} = 170$

7. $\dfrac{t}{5.2} = 12.7$

8. $\dfrac{w}{1.5} = 12$

9. $\dfrac{s}{0.05} = 21.6$

10. $\dfrac{z}{0.02} = 12.7$

11. $\dfrac{y}{35} = 7$

12. $\dfrac{x}{0.7} = 1.05$

13. $3r = 93$

14. $8x = 800$

15. $3y = 240$

16. $25t = 325$

17. $16w = 240$

18. $12y = 360$

19. $9w = 450$

20. $7r = 24$

21. $5s = 230$

22. $14z = 65$

23. $24t = 648$

24. $1.2x = 60$

25. $9x = 17.1$

26. $2.5x = 7.50$

27. $1.2y = 74.15$

28. $\dfrac{t}{3} = 21$

29. $\dfrac{w}{8} = 4$

30. $\dfrac{d}{18} = 66$

SKILL 29 Solving Equations by More Than One Step

There are times when the solution of an equation requires more than one operation to isolate the variable.

$$\frac{x}{2} - 15 = 70$$

$$\frac{x}{2} - 15 + 15 = 70 + 15 \quad \leftarrow \text{ Do the } opposite. \text{ Add 15 to each side.}$$

$$\frac{x}{2} = 85$$

$$\frac{x}{2} \times 2 = 85 \times 2 \quad \leftarrow \text{ Do the } opposite. \text{ Multiply each side by 2.}$$

$$x = 170$$

Check: $\quad \frac{170}{2} - 15 = 70 \quad \leftarrow \text{ The solution, } x = 170, \text{ is correct since}$

$$70 = 70 \qquad\qquad \text{the right side and the left side are equal.}$$

Sometimes **like terms** are combined before solving equations.

$$5t - 2t + 7 = 70$$

$$3t + 7 = 70 \qquad \leftarrow \text{ Combine like terms: } 5t - 2t = 3t.$$

$$3t + 7 - 7 = 70 - 7 \quad \leftarrow \text{ Subtract 7 from each side.}$$

$$3t = 63$$

$$\frac{3t}{3} = \frac{63}{3} \qquad \leftarrow \text{ Divide each side by 3.}$$

$$t = 21$$

Check: $\quad 5(21) - 2(21) + 7 = 70$

$$70 = 70 \qquad \leftarrow \text{ The solution, } t = 21, \text{ is correct.}$$

Solve each equation.

1. $3x + 4 = 19$

2. $6y + 12 = 30$

3. $8x - 10 = 30$

4. $7y - 15 = 20$

5. $19t + 12 = 88$

6. $21r - 7 = 35$

7. $4w - 22 = 78$

8. $15z + 13 = 62$

9. $16x + 18 = 146$

10. $4(b + 2) = 20$

11. $2(n - 3) = 16$

12. $2x + 5 = 15\frac{1}{2}$

13. $5(x + 4) = 35$

14. $3x + 2\frac{1}{4} = 22$

15. $3.1x + 5.2 = 11.4$

Solve each equation by first combining like terms. Round to the nearest tenth where needed.

16. $5x - 3x + 8 = 24$

17. $7y - 2y - 7 = 33$

18. $12t - 5t - 5 = 51$

19. $24r - 18r + 4 = 58$

20. $4.2x - 1.1x + 5.2 = 12.8$

21. $0.3m + 1.4m + 1.05 = 15$

22. $14.6y - 3.8y - 10.5 = 7.6$

23. $1\frac{1}{4} + \frac{1}{2}d + \frac{1}{2}d = 3$

24. $\frac{1}{4}t + \frac{1}{4}t - 3 = 5$

25. $\frac{1}{4}k + \frac{1}{2}k - 4 = 2$

SKILL 30   Using Exponents

An exponent tells how many times a number is used as a factor.

$7^2 = 7 \times 7$
$\quad\ = 49$

base   exponent

← Since the exponent is 2, there are 2 factors of 7. We say, "seven to the second power" or "seven squared".

$(1.2)^3 = 1.2 \times 1.2 \times 1.2$
$\qquad\ = 1.728$

← The exponent is 3 and there are 3 factors. We say, "1.2 to the third power" or "1.2 cubed".

Notice the patterns in *powers of ten*.

$10^5 = 10 \times 10 \times 10 \times 10 \times 10$
$\quad\ = 100,000$

← 10 to the 5th power has *five* factors and there are *five* zeroes in the product.

$0.1^5 = 0.1 \times 0.1 \times 0.1 \times 0.1 \times 0.1$
$\quad\ = 0.00001$

← There are *five* decimal factors and the product has *five* decimal places.

Simplify.

1. 5^2	**2.** 8^2	**3.** 9^2	**4.** 1^2	**5.** 4^2
6. 3^3	**7.** 2^5	**8.** 10^3	**9.** 10^6	**10.** 2^7
11. 3^4	**12.** 7^3	**13.** 2^{10}	**14.** 9^7	**15.** 8^3
16. 7^4	**17.** 5^3	**18.** 4^4	**19.** 11^3	**20.** 14^2
21. 25^2	**22.** 10^2	**23.** 12^3	**24.** 10^4	**25.** 30^2
26. 100^2	**27.** $\left(\frac{1}{2}\right)^2$	**28.** $\left(\frac{2}{3}\right)^3$	**29.** 6^5	**30.** $\left(\frac{1}{4}\right)^4$
31. $\left(\frac{2}{5}\right)^3$	**32.** 10^7	**33.** $(1.01)^2$	**34.** $(0.05)^3$	**35.** $\left(\frac{1}{3}\right)^5$

Simplify. Work with the exponent first. Then round the result to the nearest hundredth.

36. $2.6 \times (3.9)^2$	**37.** $3.14 \times (1.3)^2$	**38.** $12.2 \times (1.4)^2$
39. $(2.1)^2 \times 1.4$	**40.** $(3.1)^2 \times 2.72$	**41.** $(2.1)^2 \times (1.1)^2$
42. $(0.6)^3 \times (1.3)^2$	**43.** $(7.5)^2 \times (1.2)^2$	**44.** $(1.1)^3 \times (3.7)^2$
45. 2.72×7^2	**46.** $(3.14)^2 \times 8$	**47.** $3.14 \times (1.7)^2$
48. $18 \times (5.2)^2$	**49.** $3.14 \times 2 \times (4.1)^2$	**50.** $5 \times (1.2)^2 \times 3.1$
51. $6^2 \times 3.14 \times (2.5)^2$	**52.** $(1.1)^3 \times 2.72$	**53.** $3.14 \times 2^3 \times (3.1)^2$
54. $0.75 \times 3.14 \times (4)^3$	**55.** $4 \times 3.14 \times (1.2)^2$	**56.** $3.14 \times (1.1)^2 \times 2$

SKILL 31 Multiplying by Powers of 10 and Metric Conversions

Notice the short cuts you can take when multiplying an amount by a power of 10.

$9 \times 10^3 = 9000$ ← Since the exponent is 3, there are 3 *zeros* in the product. The decimal point moves three places to the right.

$4.3 \times 10^2 = 430$ ← The decimal point moves 2 places to the right.

$26 \times 10^{-3} = 0.026$ ← Since the exponent is -3, there are 3 decimal places in the product. The decimal point moves three places to the left.

$1.7 \times 10^{-2} = 0.017$ ← The decimal point moves 2 places to the left.

Multiply each by a power of ten.

1. 4×10^2	**2.** 12×10^3	**3.** 21×10^2	**4.** 12×10^1
5. 22×10^4	**6.** 123×10^1	**7.** 14×10^3	**8.** 25×10^2
9. 8.1×10^5	**10.** 1.5×10^{-3}	**11.** 6.2×10^{-4}	**12.** 0.5×10^2
13. 0.3×10^{-2}	**14.** 12.1×10^{-3}	**15.** 32.2×10^2	**16.** 123.5×10^{-5}
17. 75.6×10^3	**18.** 0.32×10^6	**19.** 0.175×10^{-1}	**20.** 4.15×10^{-3}
21. 2.5×10^2	**22.** 6.7×10^{-1}	**23.** 7.6×10^4	**24.** 5.3×10^{-6}

We multiply by powers of ten as we change metric units.

$4.5 \text{ km} = 4500 \text{ m}$ ← 1 km = 1000 m, so multiply 4.5 by 1000 (or 10^3). The decimal moves 3 places to the right,

$65 \text{ mL} = 0.065 \text{ L}$ ← 1 mL = 0.001 L, so multiply 65 by 0.001 (or 10^{-3}). The decimal moves 3 places to the left.

$16 \text{ cm} = 0.16 \text{ m}$ ← 1 cm = 0.01 m, so multiply 16 by 0.01 (or 10^{-2}). The decimal moves 2 places to the left.

Change each of the following units by using powers of ten.

25. 15 m = ___ mm	**26.** 129 mL = ___ cL	**27.** 5.3 kg = ___ g
28. 236 dg = ___ g	**29.** 10.2 dm = ___ mm	**30.** 8.6 L = ___ mL
31. 13 mL = ___ L	**32.** 2.2 km = ___ m	**33.** 3.4 mg = ___ g
34. 2.5 dm = ___ cm	**35.** 3.4 mg = ___ cg	**36.** 3.6 kL = ___ mL

SKILL 32   Finding Square Roots

A square best illustrates the meaning of
square root. The area of the square at the right
is 25 m². We find it by saying 5 × 5 = 25, or
5² = 25.

If the length of a side were unknown, we could
take the *square root* of 25 to find it. The square
root of 25 is written as $\sqrt{25}$

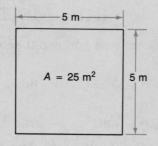

$\sqrt{25}$ = 5, because 5² = 25

radical   radicand

To isolate *d* in the following equation, we must do the *opposite*
operation.

$$d^2 = 1.21$$
$$\sqrt{d^2} = \sqrt{1.21}$$
$$d = 1.1$$

← Do the opposite. Find the square root of each
side of the equation.

Check: 1.1² = 1.21
         1.21 = 1.21

← The check shows the solution, *d* = 1.1, is correct.

Find the square root.

1. $\sqrt{4}$ 2. $\sqrt{25}$ 3. $\sqrt{100}$ 4. $\sqrt{9}$ 5. $\sqrt{1}$

6. $\sqrt{36}$ 7. $\sqrt{196}$ 8. $\sqrt{144}$ 9. $\sqrt{121}$ 10. $\sqrt{49}$

11. $\sqrt{225}$ 12. $\sqrt{81}$ 13. $\sqrt{256}$ 14. $\sqrt{64}$ 15. $\sqrt{169}$

16. $\sqrt{16}$ 17. $\sqrt{289}$ 18. $\sqrt{361}$ 19. $\sqrt{900}$ 20. $\sqrt{400}$

21. $\sqrt{324}$ 22. $\sqrt{484}$ 23. $\sqrt{441}$ 24. $\sqrt{576}$ 25. $\sqrt{529}$

26. $\sqrt{0.01}$ 27. $\sqrt{0.0001}$ 28. $\sqrt{1.69}$ 29. $\sqrt{5.76}$ 30. $\sqrt{1.44}$

31. $\sqrt{1.96}$ 32. $\sqrt{2.25}$ 33. $\sqrt{2.89}$ 34. $\sqrt{3.24}$ 35. $\sqrt{0.81}$

Solve for *x* in each equation.

36. $x^2 = 1024$ 37. $x^2 = 484$ 38. $x^2 = 2500$ 39. $x^2 = 961$

40. $x^2 = 1.00$ 41. $x^2 = 0.09$ 42. $x^2 = 3600$ 43. $x^2 = 0.64$

44. $x^2 = 0.81$ 45. $x^2 = 4900$ 46. $x^2 = 6400$ 47. $x^2 = 56.25$

48. $x^2 = 4.84$ 49. $x^2 = 9.61$ 50. $x^2 = 16.81$ 51. $x^2 = 34.81$

52. $x^2 = 53.29$ 53. $x^2 = 571.21$ 54. $x^2 = 462.25$ 55. $x^2 = 156.25$

Mathematical Formulas

Area of a circle	$A = \pi r^2$
Area of a parallelogram	$A = bh$
Area of a rectangle	$A = lw$
Area of a square	$A = s^2$
Area of a trapezoid	$A = 0.5\, h(b_1 + b_2)$
Area of a triangle	$A = 0.5\, bh$
Circumference of a circle	$C = \pi d$, or $2\pi r$
Diameter of a circle	$d = 2r$
Perimeter of a polygon	$P =$ sum of sides
Perimeter of a rectangle	$P = 2l + 2w$
Perimeter of a square	$P = 4s$
Surface area of prism, pyramid, cylinder, or cone	$A =$ Lateral area + Area of base(s)
Surface area of a sphere	$A = 4\pi r^2$
Volume of a cube	$V = s^3$
Volume of a prism or cylinder	$V =$ (Area of base) $\times$ height
Volume of a pyramid or cone	$V = \dfrac{\text{(Area of base)} \times \text{height}}{3}$
Volume of a sphere	$V = \dfrac{4\pi r^3}{3}$

$$\text{Mean} = \frac{\text{sum of values}}{\text{number of values}} \qquad \text{Percent} = \frac{\text{part}}{\text{whole}} \times 100$$

Trigonometric ratios

$$\sin A = \frac{\text{opposite}}{\text{hypotenuse}} \qquad \cos A = \frac{\text{adjacent}}{\text{hypotenuse}} \qquad \tan A = \frac{\text{opposite}}{\text{adjacent}}$$

Symbols

$=$	is equal to	$:$	the ratio of
$>$	is greater than	$\angle$	angle
$\geq$	is greater than or equal to	$\circ$	degrees
$<$	is less than	$\triangle$	triangle
$\leq$	is less than or equal to	$\sqrt{\ }$	the square root of
$\cong$	is congruent to	ϕ	diameter measurement
$\sim$	is similar to	Ω	ohms
$\perp$	is perpendicular to	π	pi, (approximately 3.1415926536 to
$\approx$	is approximately equal to		ten decimal places)
$\pm$	plus or minus	$\overline{AB}$	line segment AB
		$\overrightarrow{AB}$	ray A to B

Units of Measurement (U.S. System)

Measurement	Unit	Abbreviation
Length	inch	in.
	foot	ft
	yard	yd
	mile	mi
Weight	ounce	oz
	pound	lb
	ton	t
Area	square foot	ft^2
	acre	acre
Volume	cubic foot	ft^3
Capacity	fluid ounce	oz
	gallon	gal
Time	second	s
Temperature	degrees Fahrenheit	°F
Force	pound	lb
Pressure	pounds per square inch	psi
Power	watt	W
Electrical current	ampere	A
Electrical energy	kilowatt hour	kW·h
Electrical voltage	volt	V
Electrical resistance	ohm	Ω

Units of Measurement (Metric System)

Measurement	Unit	Abbreviation
Length	meter	m
Mass	gram	g
	tonne	t
Area	square meter	m^2
	hectare	ha
Volume	cubic meter	m^3
Capacity	liter	L
Time	second	s
Temperature	degrees Celsius	°C
Force	newton	N
Pressure	Pascal	Pa
Power	watt	W
Electrical current	ampere	A
Electrical energy	kilowatt hour	kW·h
Electrical voltage	volt	V
Electrical resistance	ohm	Ω

SI Metric Prefixes

Prefix	Symbol	Meaning	
mega	M	one million;	1,000,000
kilo	k	one thousand;	1,000
hecto	h	one hundred;	100
deca	da	ten;	10
deci	d	one tenth;	0.1
centi	c	one hundredth;	0.01
milli	m	one thousandth;	0.001
micro	μ	one millionth;	0.000001
nano	η	one billionth;	0.000000001

Measurement Equivalents, and Conversion Tables

U.S. System Equivalents

Length
1 ft = 12 in.
1 yd = 3 ft
1 mi = 5280 ft

Area
$1 ft^2 = 144 in.^2$
$1 yd^2 = 9 ft^2$
$1 acre = 4840 yd^2$

Volume
$1 ft^3 = 1728 in.^3$
$1 yd^3 = 27 ft^3$

Capacity
16 fl oz = 1 pt.
2 pt = 1 qt
4 qt = 1 gal

Weight
16 oz = 1 lb
2000 lb = 1 t

Metric System Equivalents

Length
1 cm = 100 mm
1 m = 100 cm
1 km = 1000 m

Area
$1 cm^2 = 100 mm^2$
$1 m^2 = 10,000 cm^2$
$1 ha = 10,000 m^2$

Volume
$1 cm^3 = 1000 mm^3$
$1 m^3 = 1,000,000 cm^3$

Capacity
1 L = 1000 mL
1 kL = 1000 L

Mass
1 g = 1000 mg
1 kg = 1000 g
1 t = 1000 kg

Tables

Approximate U.S. and Metric Equivalents

Length
1 in. = 2.540 cm
1 ft = 0.305 m
1 yd = 0.914 m
1 mi = 1.609 km

Area
1 in.^{2} = 6.452 cm^2
1 ft^2 = 0.093 m^2
1 yd^2 = 0.836 m^2
1 mi^2 = 2.590 km^2

Volume
1 in.^{3} = 16.387 cm^3
1 ft^3 = 0.028 m^3
1 yd^3 = 0.765 m^3

Capacity
1 fl oz = 29.573 mL
1 pt = 0.473 L
1 gal = 3.785 L
1 qt = 0.946 L

Weight
1 oz = 28.350 g
1 lb = 0.454 kg
1 t = 0.907 t (tonne)

Set Squares

45°–45°–90° Set Square

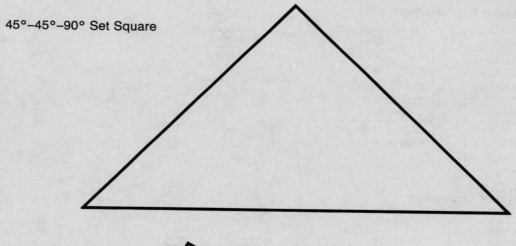

30°–60°–90° Set Square

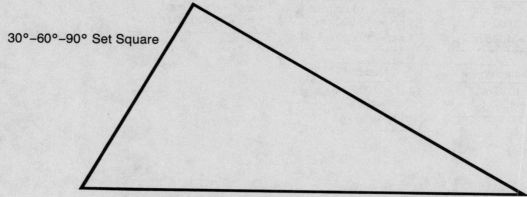

Pica Rule

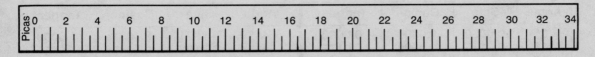

Metric Triangular Rules

Scale: 1 : 100 (1 mm : 100 mm)

Scale: 1 : 50 (1 mm : 50 mm)

Scale: 1 : 20 (1 mm : 20 mm)

Architect's Triangular Rules (U.S. system)

Scale: 1 in. : 1 ft

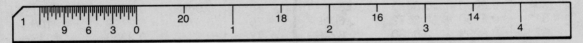

Scale: $\frac{1}{2}$ in. : 1 ft

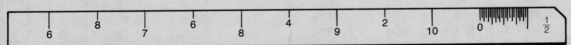

Scale: $\frac{1}{4}$ in. : 1 ft

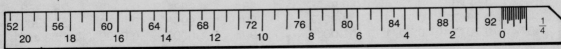

Tables

389

Table of Trigonometric Ratios

Angle	Sine	Cosine	Tangent	Angle	Sine	Cosine	Tangent
0°	0.0000	1.0000	0.0000	45°	0.7071	0.7071	1.0000
1°	0.0175	0.9998	0.0175	46°	0.7193	0.6947	1.0355
2°	0.0349	0.9994	0.0349	47°	0.7314	0.6820	1.0724
3°	0.0523	0.9986	0.0524	48°	0.7431	0.6691	1.1106
4°	0.0698	0.9976	0.0699	49°	0.7547	0.6561	1.1504
5°	0.0872	0.9962	0.0875	50°	0.7660	0.6428	1.1918
6°	0.1045	0.9945	0.1051	51°	0.7771	0.6293	1.2349
7°	0.1219	0.9925	0.1228	52°	0.7880	0.6157	1.2799
8°	0.1392	0.9903	0.1405	53°	0.7986	0.6018	1.3270
9°	0.1564	0.9877	0.1584	54°	0.8090	0.5878	1.3764
10°	0.1736	0.9848	0.1763	55°	0.8193	0.5736	1.4281
11°	0.1908	0.9816	0.1944	56°	0.8290	0.5592	1.4826
12°	0.2079	0.9781	0.2126	57°	0.8387	0.5446	1.5399
13°	0.2250	0.9744	0.2309	58°	0.8480	0.5299	1.6003
14°	0.2419	0.9703	0.2493	59°	0.8572	0.5150	1.6643
15°	0.2588	0.9659	0.2679	60°	0.8660	0.5000	1.7321
16°	0.2756	0.9613	0.2867	61°	0.8746	0.4848	1.8040
17°	0.2924	0.9563	0.3057	62°	0.8829	0.4695	1.8807
18°	0.3090	0.9511	0.3249	63°	0.8910	0.4540	1.9626
19°	0.3256	0.9455	0.3443	64°	0.8988	0.4384	2.0503
20°	0.3420	0.9397	0.3640	65°	0.9063	0.4226	2.1445
21°	0.3584	0.9336	0.3839	66°	0.9135	0.4067	2.2460
22°	0.3746	0.9272	0.4040	67°	0.9205	0.3907	2.3559
23°	0.3907	0.9205	0.4245	68°	0.9272	0.3746	2.4751
24°	0.4067	0.9135	0.4452	69°	0.9336	0.3584	2.6051
25°	0.4226	0.9063	0.4663	70°	0.9397	0.3420	2.7475
26°	0.4384	0.8988	0.4877	71°	0.9455	0.3256	2.9042
27°	0.4540	0.8910	0.5095	72°	0.9511	0.3090	3.0777
28°	0.4695	0.8829	0.5317	73°	0.9563	0.2924	3.2709
29°	0.4848	0.8746	0.5543	74°	0.9613	0.2756	3.4874
30°	0.5000	0.8660	0.5774	75°	0.9659	0.2588	3.7321
31°	0.5150	0.8572	0.6009	76°	0.9703	0.2419	4.0108
32°	0.5299	0.8480	0.6249	77°	0.9744	0.2250	4.3315
33°	0.5446	0.8387	0.6494	78°	0.9781	0.2079	4.7046
34°	0.5592	0.8290	0.6745	79°	0.9816	0.1908	5.1446
35°	0.5736	0.8192	0.7002	80°	0.9848	0.1736	5.6713
36°	0.5878	0.8090	0.7265	81°	0.9877	0.1564	6.3138
37°	0.6018	0.7986	0.7536	82°	0.9903	0.1392	7.1154
38°	0.6157	0.7880	0.7813	83°	0.9925	0.1219	8.1443
39°	0.6293	0.7771	0.8098	84°	0.9945	0.1045	9.5144
40°	0.6428	0.7660	0.8391	85°	0.9962	0.0872	11.4301
41°	0.6561	0.7547	0.8693	86°	0.9976	0.0698	14.3007
42°	0.6691	0.7431	0.9004	87°	0.9986	0.0523	19.0811
43°	0.6820	0.7314	0.9325	88°	0.9994	0.0349	28.6363
44°	0.6947	0.7193	0.9657	89°	0.9998	0.0175	57.2900
45°	0.7071	0.7071	1.0000	90°	1.0000	0.0000	undefined

Glossary

Acute angle An angle measuring less than 90°. (6-1)

Adjacent angles Two angles in a plane having a common vertex and a common side but no common interior points. (6-4)

Aligned dimensioning Dimensions in a technical drawing that are readable from either the bottom or the right side of the drawing. (1-8)

Allowance The difference between the maximum value and the minimum value of a dimension or the range of allowable error. (1-10)

Ammeter A device used to measure electrical current flow in a circuit. (3-6)

Ampere (A) The basic unit of measure of electrical current. (3-6)

Angle of depression The angle determined by the line representing the horizontal and the line of sight to some point below the observer. (10-5)

Angle of elevation The angle determined by the line representing the horizontal and the line of sight to some point above the observer. (10-5)

Arc A part of a circle's circumference. (1-9)

Average A single value that is used to typify and/or summarize a set of data. (11-2)

Axis (horizontal) A horizontal number line used to locate points on a coordinate plane. (11-5)

Axis (vertical) A vertical number line used to locate points on a coordinate plane. (11-5)

Bar graph A display of data in which each quantity is represented by the proportional length of a bar. (11-5)

Bevel square A simple tool used in cutting wood to a specific angle size. (6-3)

Bisector (of an angle) A ray that divides an angle into two congruent adjacent angles. (6-1)

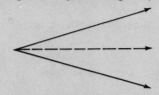

Board foot A unit of measure for sawn lumber that is 1 in. thick, 1 ft wide, and 1 ft long. (9-2)

Bore (cylinder) The inside diameter of a cylinder. (7-6)

Box plot A display of data which includes a horizontal number line that highlights the median, extremes, and hinges. (11-4)

Boyle's Law The absolute pressure of a confined body of gas varies inversely as its volume, provided its temperature remains constant. (7-9)

Broken-line graph A visual display of data consisting of a broken line connecting points. This graph is used to show continuous data that change over time. (11-6)

Capacity The measure of the volume of material, usually liquid, a vessel can hold. (7-5)

Carpenter's square A simple tool used to confirm that two pieces of wood fit together at a 90° angle. It is also used to calculate the rise and pitch of a roof as well as other angled cuts in carpentry work. (1-5)

Celsius scale A temperature scale with the freezing point of water set at 0°C to the boiling point of water set at 100°C.

Centering square A tool used to help locate the center of a circle quickly. (6-3)

Central angle An angle with its vertex at the center of a circle and two radii making its sides. (6-6)

Chord (of a circle) A line segment that joins two points on a circle's circumference. (6-5)

Circle A closed curve whose points are all the same distance from one fixed point (the center). (4-8)

Circle graph A graph that displays the ratios of parts of a set of data to the whole set as proportional pieces of a whole pie. (11-7)

Circumference The distance around the rim of a circle. (4-2)

Combined variation A relation between one variable and a combination of other variables. (11-11)

Compass An instrument used for drawing arcs and circles. (6-1)

Complementary angles Two angles whose measures have the sum of 90°. (6-4)

Compression ratio The ratio of the maximum space to the minimum space in the cylinder of an engine. (7-7)

Cone A solid bounded by a circle and all the line segments from a point outside the plane of the circle to all the points of the circle. (7-2)

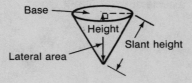

Congruent Having the same size and shape. (6-3)

Constant of variation A quantity having a fixed value in a variation. (11-9)

Coordinates Two numbers in an ordered pair that locate a point in a plane. (11-9)

Cosine ratio For a given angle of a right-angled triangle, the cosine is the ratio of the length of the adjacent side to the length of the hypotenuse. (10-2)

Cube A regular solid with six congruent square faces. (7-1)

Current A flow of electric charge, measured in amperes. (3-6)

Cylinder A three-dimensional figure with two congruent and parallel circular bases joined by a curved surface. (7-2)

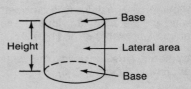

Data Facts. (11-1)

Density The mass of a substance divided by its volume. For example, the density of water is about 62 lb/ft^3. (3-7)

Diameter A chord that passes through the center of a circle. (4-2)

Dimension lines Lines in a technical drawing with arrowheads at both ends showing the specific lengths of a part of the object. (1-8)

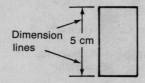

Direct variation A relation between two variables that has a constant ratio. (11-9)

Discount rate The difference between the sale price and the regular price of an item expressed as a percent of the regular selling price. (2-11)

Displacement (car engine) The amount of space through which a cylinder piston travels in one stroke. (7-6)

Efficiency ratio (of an engine) The ratio of useful power output by an engine to the power input. (5-2)

Em (in typesetting) A unit of measure of area in the printing trade equal to the square space occupied by the capital letter M in any given typeface. (8-1)

Equation A mathematical sentence stating that two expressions are equal. (3)

Equilateral triangle A triangle having three congruent sides. (10-4)

Estimate A reasonable guess. (2-4)

Exponent A small number written above and to the right of another number telling how many times the base number is to be used as a factor. (3)

Expression A combination of mathematical symbols, variables, and numerals. (3)

Extension lines Lines extending beyond the edges of an object between which the dimension lines are drawn in a technical drawing. (1-8)

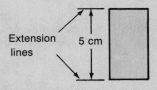

Face A plane figure that forms one of the sides of a three-dimensional figure. (7-1)

Fahrenheit scale A temperature scale with the freezing point of water set at 32° to the boiling point of water set at 212°.

Frequency The number of times a particular event occurs in a set of events. (11)

Frequency table A table showing the frequencies of the occurrences of certain events. (11-1)

Gear ratio The ratio of the speed of the driver gear to the speed of the driven gear.

Gravity The force which accelerates falling bodies towards Earth's center. On Earth, the acceleration due to gravity is 32.2 ft/s². (7-5)

Gravure printing A method of printing in which the impression is obtained from etched plates. (8-4)

Hidden feature lines Dashed lines in a technical drawing which indicate a part of the figure that cannot be seen in a particular view. (1-7)

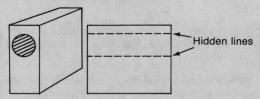

Hinge (of a box plot) The median of the upper half of the data or the lower half of the data. (11-4)

Histogram A type of bar graph used to display the frequencies of data. (11-5)

Hydraulic power Power transmitted through a confined fluid. (7-8)

Hypotenuse The side of a right-angled triangle opposite the right angle. (10-1)

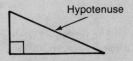

Imposition The arrangement of several pages of a book on a large sheet of paper so that the pages will be in the proper order when the sheet is folded after printing. (8-6)

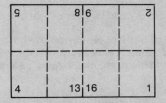

Inclined plane A simple machine, consisting of a plane inclined to the horizontal. (5-7)

Interpolate To approximate a value between two known values. (1-5)

Intersecting lines Two lines that meet at the same point. (6-2)

Inverse proportion A proportion in which the ratios are based on reciprocal relationships. (5-8)

Inverse variation A relation between two variables such that the ratio of one to the reciprocal of the other is constant. (11-10)

Isosceles triangle A triangle having at least two sides congruent. (10-3)

Joint variation Joint variation occurs when a quantity varies directly as the product of two or more other quantities. (11-11)

Kilowatt hour (kW·h) The basic unit of measure of the consumption of electricity. (1-1)

Lateral area The sum of the areas of the lateral faces of a solid. (7-1)

Leader A line, in a technical drawing, used to direct information and symbols to a place in the drawing. (1-8)

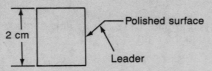

Lever A simple machine consisting of a rigid body pivoted on a fixed fulcrum. (5-7)

Lineal foot The running length of a board. (9-3)

Mass The amount of material of which an object is composed. Mass is usually expressed in grams. (7-5)

Mean A measure of central tendency that is the sum of the values divided by the number of values. (11-2)

Measure of central tendency A statistic that represents a central value in a set of data. The mean, median, and mode are measures of central tendency. (11-2)

Mechanical advantage The capacity for a machine to multiply a small input force into a large output force. (5-6)

Mechanical advantage (actual) Mechanical advantage calculated considering the force due to friction. (5-6)

Mechanical advantage (ideal) Mechanical advantage calculated ignoring the effects of friction. (5-6)

Median A measure of central tendency that is the middle value of a set of data when the data are listed in numerical order. (11-2)

Micrometer An instrument used in making precise measurements to the nearest thousandth of an inch or hundredth of a millimeter. (2-6)

Mode A measure of central tendency that is the most frequently occurring number in a set of data. (11-2)

Obtuse angle An angle measuring between 90° and 180°. (6-1)

Offset lithography A method of printing in which the page is photographically reproduced on a thin flexible metal plate. (8-4)

Ohm (Ω) The basic unit of measure of electrical resistance. (2-12)

Ohmmeter An instrument used to measure the amount of electrical resistance in a circuit. (3-6)

Origin The fixed point (0,0) on a coordinate grid where the horizontal and vertical axes intersect. (11-9)

Orthographic projection A method of drawing the top, front, and side views of a three-dimensional figure. (1-6)

Parallel lines Lines on a flat surface that do not meet no matter how far they are extended. (6-3)

Parallelogram A quadrilateral with both pairs of opposite sides parallel. (4-5)

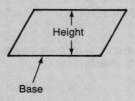

Pascal's Law The pressure on a confined fluid is transmitted undiminished in all directions and acts with equal force on equal areas at right angles to them. (7-8)

Percent The ratio of a number to one hundred.

Perimeter The distance around the rim of a figure. (3-1)

Perpendicular lines Lines that intersect at right angles. (6-2)

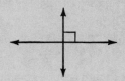

Pi (π) The ratio of the circumference of a circle to its diameter ($\pi \approx 3.14159265$). (4-2)

Pica (in typesetting) A unit used to measure distances in printed matter equal to about $\frac{1}{6}$ inch. (8-1)

Pictograph A visual display that uses pictures or symbols to represent large values. (11-6)

Pitch (of a roof) The ratio of the rise of a roof to its span. (5-1)

Pneumatic power Power transmitted through a compressed column of gas. (7-9)

Point (in typesetting) A unit used to measure the size of printing type equal to 0.01384 in. or $\frac{1}{12}$ pica. (8-1)

Polygon A closed figure whose sides are three or more line segments. (6-6)

Polygon (regular) A polygon with all sides of equal length. (6-6)

Power rating The rate at which a device uses electrical energy. (3-6)

Pressure The force per unit area exerted by a fluid or gas. It is usually expressed in pounds per square inch (psi).

Prism A three-dimensional figure which has two congruent and parallel bases that are polygons and three or more rectangular faces. (7-1)

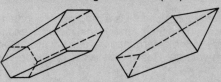

Proportion A statement of equality between two ratios or rates. (5-4)

Pyramid A three-dimensional figure that has one polygonal base and three or more lateral triangular faces. (7-1)

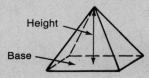

Quadrilateral A four-sided polygon.

Radius A line segment from any point on the circumference of a circle to its center. (4-2)

Range The difference between the smallest and largest values of a set of data. (11-2)

Ray Part of a line extending without end in one direction only. (6-1)

Ream (of paper) A standard quantity of paper equal to 500 sheets of a basic size. (8-5)

Reciprocals Two numbers whose product is one. For example, $\frac{5}{4}$ and $\frac{4}{5}$ are reciprocals, since $\frac{5}{4} \times \frac{4}{5} = 1$.

Rectangle A parallelogram with four right angles. (4-1)

Relief printing A printing method in which the image to be printed is raised on the printing plate and makes a depression on the paper, much like a rubber stamp. (8-4)

Representative sample A number of randomly selected items that are to represent the characteristics of an entire population. (2-9)

Resistance Resistance is a property of conductors, depending upon their dimensions, material, and temperature which determines the current produced by a given voltage. Resistance is measured in ohms. (3-6)

Right angle An angle whose measure is 90°. (6-1)

Rule of Pythagoras The rule of Pythagoras states that in every right-angled triangle, the square of the hypotenuse, c, equals the sum of the squares of the other two sides, a and b. (10-1)

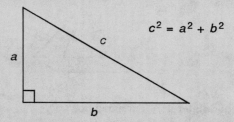

$$c^2 = a^2 + b^2$$

Sale price A purchase price that is lower than the regular selling price. (2-11)

Scale The ratio of the length of a line in a technical drawing to the corresponding length of the actual object. (5-3)

Set square A triangular frame used in making technical drawings. (6-2)

Sextant An instrument that measures angles in navigation. (10-6)

Sheet-fed press A press that prints on large sheets of paper. (8-4)

Signature (of a book) A large sheet of paper that is folded to form one section of a book. (8-5)

Significant digits The digits in a measurement that are used to determine its accuracy. (2-1)

Silk-screen printing An adaptable form of printing done with a stencil. (8-4)

Similar polygons Polygons having the same shape but not necessarily the same size. (6-7)

Sine ratio For a given angle of a right-angled triangle, the sine is the ratio of the length of the opposite side to the length of the hypotenuse. (10-2)

Solid lines Unbroken lines in a technical drawing that are used to show the visible edges of a particular view of an object. (1-7)

Speed (of the cutting edge) The rate at which a cutting edge of a circular tool moves across a surface, usually expressed in feet per second (ft/s). (4-4)

Sphere The set of all points in space that are a given distance from a fixed point. (7-2)

Square A rectangle having four right angles and four congruent sides. (4-1)

Square (of shingles) The amount of shingles that will cover an area of 100 ft². (9-4)

Square root The square root of a number is that number which, when multiplied by itself, results in the given number. (10-1)

Statistics The branch of mathematics involving the collection and analysis of data. (11-1)

Stem-and-leaf plot A visual display of data that highlights the range, extremes, and the median. (11-4)

Straight angle An angle whose measure is 180°. (6-1)

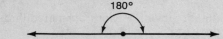

Stroke The distance through which a piston moves within a cylinder. (7-6)

Supplementary angles Two angles whose measures have the sum of 180°. (6-4)

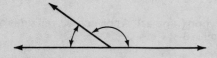

Surface area The total outer area of the faces of a three-dimensional object. (7-1)

Systolic (blood) pressure The upper number in a blood pressure reading measuring the force of the blood pushing against the artery walls. (3-7)

T-square A T-shaped drafting instrument used for drawing parallel lines. (6-2)

Tangent (to a circle) A line in the plane of a circle that meets the circle in exactly one point, called the point of tangency. (6-5)

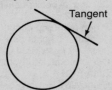

Tangent ratio For a given angle of a right-angled triangle, the tangent is the ratio of the length of the opposite side to the length of the adjacent side. (10-2)

Tolerance The upper and lower limits of a dimension. (1-10)

Transmission ratio The ratio of the engine speed to the drive shaft speed. (5-9)

Transversal A line that intersects two or more other lines. (6-3)

Trapezoid A quadrilateral having exactly one pair of parallel sides, called the bases. (4-6)

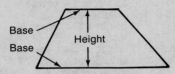

Triangle A polygon having three sides. (4-6)

Triangle (right) A triangle containing a right angle. (10-1)

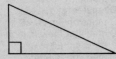

Triangle (right-isosceles) A triangle with one right angle and two sides of equal length. (10-3)

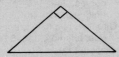

Trigonometric ratios In a right-angled triangle, a trigonometric ratio is the ratio formed by the lengths of any two sides. (10-2)

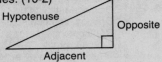

Trim size (of a book) The width and depth of a finished page of a book. (8-1)

Type page (of a book) The width and depth of a page including text, illustrations, headings, and footnotes. (8-1)

Typeface (in typesetting) The style of the letter or character on the page. (8-1)

Unidirectional dimensioning Dimensions in a technical drawing that can be read from the bottom of the drawing. (1-8)

Unit rate A comparison of quantities in which the second term is one. (5-4)

Variable An unknown value that is represented in a mathematical expression or equation by a letter. (3-5)

Vertex The common endpoint of the two sides of an angle. (6-1)

Vertical angles Two angles whose sides form two pairs of opposite rays. (6-4)

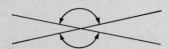

Volt (V) The basic unit of measure of electrical potential. Voltage is analogous to pressure in that it causes electrical current to flow. (3-6)

Voltage A measure of the energy given in a circuit. Voltage is the product of current and resistance. (3-6)

Voltmeter A device used to measure the voltage of a source of energy. (3-6)

Volume The measure of space occupied by an object. (7-3)

Web-fed press A press that unwinds a continuous roll of paper, prints on both sides, and then folds the paper as it outputs. (8-4)

Weight The amount of force that gravity exerts on an object. It is usually expressed in pounds (lb). (7-5)

INDEX

Photograph Credits

Every effort has been made to ascertain proper ownership of copyrighted materials and obtain permission for their use. Any omission is unintentional and will be corrected in future printing upon proper notification.

Cover

Design: The Dragon's Eye Press
Photograph: Michael Stuckey/Miller Comstock

Chapter One

0	Tom Grill/Miller Comstock
1	E.I. DuPont DeNemour and Company
4	Stewart Warner Corporation
7	New Britian Hand Tools
23	General Electric
29	Canapress Photo Service
30 (T)	Donald Dietz - Dietz/Hamlin
30 (B)	Peter B. Typer
31	H. Armstrong Roberts

Chapter Two

34	Peter Coney/The Stock Market Inc.
37	Chapman Photography
38	The Stanley Works
39	National Machinery Company
40	The Seven-Up Company
42 (T)	Sikorsky Aircraft
42 (B)	The Lamson and Sessions Company
44	Penn Central Railroad
46	L. B. Foster Company
47	Owens Corning Fiberglass Corp.
48 (T)	Kemsies/Norm Scudellari Photography
48 (B)	Kemsies/Norm Scudellari Photography
49	Kemsies/Norm Scudellari Photography
50	Jon Riley/The Stock Shop
52	Ontario Robotic Centre
53	Kemsies/Norm Scudellari Photography
54	Toronto Transit Commission
55	Canapress Photo Service
56 (T)	Miller Comstock
56 (B)	Gabriel V. Guillén
57	The Seven-Up Company
60	Ryerson Polytechnical Institute
65	Portland Cement

66 (T)	General Electric Company
66 (B)	Miller Comstock
67	George Brown College
68	Kemsies/Norm Scudellari Photography
69	The Lamson and Sessions Company
71	L. S. Starrett Company

Chapter Three

72	Gerry Wright/Canapress Photo Service
75 (T)	L. S. Starrett Company
75 (B)	NASA
76	The Bank of Nova Scotia
78	Jerry Davey/Ryerson Polytechnical Inst.
80	IBM Corporation
83	Armstrong Brothers Tool Company
84	Canapress Photo Service
85	The Lamson and Sessions Company
86	Gabriel V. Guillén
88	General Electric Company
90	George Brown College
93	Canapress Photo Service
95	Earl Roberge/Photo Researchers, Inc.

Chapter Four

98	Sir Sandford Fleming College
105	Chevrolet Motor Division, General Motors Corp.
106	Peter B. Tepper
107 (T)	Simonds Saw and Steel
107 (B.L.)	The Black and Decker Manufacturing Company
107 (B.R.)	Busy Bee Machine Tools Ltd.
108 (T)	Peter Seidenberg
108 (B)	George Brown College
120	Billy E. Barnes/Click/Chicago
121	Ontario Ministry of Natural Resources

Chapter Five

Chapter Six

Chapter Seven

Chapter Eight

Chapter Nine

Chapter Ten

Chapter Eleven

Photo Research   Debbie Brewer

Answers to Selected Exercises

Chapter 1 - Reading Industrial Measurements

1–1, Exercises, pages 2–3
1. 9824 kW•h   **3.** 6645 kW•h   **5.** 17,006 kW•h   **7.**   **a.** 6149 kW•h   **b.** 2618 kW•h
c. 3531 kW•h   **9.** 5234 kW•h   **11.** 24,828 kW•h
13. a. $45.19   **b.** 1863, $104.33   **c.** 3572, $200.03

1–2, Exercises, pages 4–5
1. 3500 r/min   **7.**
3. 5300 r/min
5. 4300 r/min

Self-Analysis Test, page 5
1. 595 kW•h   **3.** 1200 r/min

1-3, Exercises, page 7
1. $2\frac{1}{2}$ in.   **3. a.** 0.3 cm   **b.** 1.0 cm   **c.** 2.9 cm
11. a. $AB = 5$ mm $BC = 22$ mm $CD = 52$ mm
    b. $AB = 76$ mm $CD = 6$ mm $EF = 10$ mm $GH = 9$ mm $JK = 40$ mm

1–4, Exercises, page 9
1. $2\frac{7}{8}$ in.   **3.** $1\frac{3}{8}$ in.   **5.** 86 mm, 8.6 cm, 0.86 dm   **7.** 108 mm, 10.8 cm, 1.08 dm
9. Yes   **11.** Yes   **13.** 1.2 cm

1–5, Exercises, pages 11–12
1. $C = 3\frac{27}{32}$ in.   **3.** $E = 2\frac{7}{32}$ in.   **5.** $M = 9.85$ cm   **7.** $O = 3.45$ cm
9. $A = 1$ in. $B = 1$ in. $C = 1$ in. $D = \frac{3}{4}$ in. $E = 2\frac{3}{10}$ in. $F = \frac{11}{20}$ in. $G = \frac{1}{2}$ in. $H = 1\frac{1}{20}$ in.
    $J = 2\frac{1}{10}$ in. $K = \frac{2}{5}$ in. $L = 2\frac{3}{10}$ in. $M = \frac{2}{5}$ in. $N = 3\frac{1}{20}$ in.
11. $A = 6.25$ cm $B = 0.25$ cm $C = 5.35$ cm $D = 5.5$ cm $E = 1.1$ cm $F = 8.25$ cm $G = 7.6$ cm

Self-Analysis Test, page 13
3. Yes   **5. a.** $A = 1\frac{3}{32}$ in. $B = 1\frac{3}{4}$ in. $C = 1\frac{19}{32}$ in. $D = \frac{15}{16}$ in. $E = 1\frac{17}{32}$ in.
    $F = \frac{1}{2}$ in. $G = \frac{13}{16}$ in. $H = 3\frac{17}{32}$ in.
b. $A = 2.8$ cm $B = 4.45$ cm $C = 4$ cm $D = 2.4$ cm $E = 3.9$ cm $F = 1.25$ cm $G = 2.1$ cm $H = 9$ cm

1–6, Exercises, page 15
1. B   **3.** C
5.   **7.**   **9.**

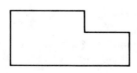

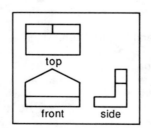

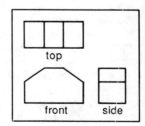

1–7, Exercises, pages 18–19
1. B, front   **3.** B, top

5. 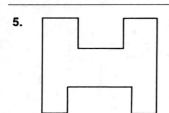　　**7.** 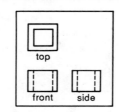　　**9.**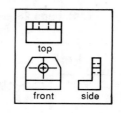

Check Your Skills, page 19
1. a. 104.7　**b.** 3.2　　　**c.** 46.0　　　**3.** 59,889　**5.** 47,403　**7.** 37,741
9. 58.34　　**11.** 11.153　**13.** 4　　　**15.** 8,496,000 mm, 84,960 dm, 8.496 km
17. 21,300 mm, 213 dm, 0.0213 km　　　　**19.** 456,123 mm, 4561.23 dm, 0.456123 km

1–8, Exercises, pages 21–23
1. 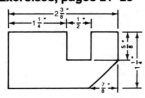　　**3.**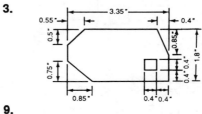

5.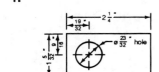

Drawing is not correct size.

9.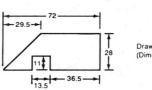

Drawing is not correct size.
(Dimensions are in millimeters.)

1–9, Exercises, page 25
1. 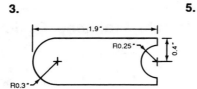　　**3.**　　**5.**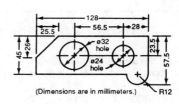

(Dimensions are in millimeters.)

1–10, Exercises, page 27

	Maximum Size	Minimum Size		Maximum Size	Minimum Size
1.	$2\frac{5}{8}$ in.	$2\frac{3}{8}$ in.	**3.**	2.447 in.	2.443 in.
5.	1.875 in.	1.775 in.	**7.**	45.5 mm	45.0 mm
9.	83.5 mm	83.0 mm			

		Basic Size	Tolerance	Maximum Size	Minimum Size
11.	**a.**	A = 3.50 in.	± 0.03 in.	3.53 in.	3.47 in.
		B = 0.28 in.	± 0.05 in.	0.33 in.	0.23 in.
		C = 1.25 in.	± 0.03 in.	1.28 in.	1.22 in.
		D = 4.00 in.	± 0.04 in.	4.04 in.	3.96 in.
		E = 0.26 in.	± 0.02 in.	0.28 in.	0.24 in.
	b.	A = 90.0 mm	± 0.7 mm	90.7 mm	89.3 mm
		B = 7.0 mm	± 0.1 mm	7.1 mm	6.9 mm
		C = 31.0 mm	± 0.7 mm	31.7 mm	30.3 mm
		D = 100.0 mm	± 0.8 mm	100.8 mm	99.2 mm
		E = 3.0 mm	± 0.5 mm	3.5 mm	2.5 mm

13. Maximum Size 858 mm　　　Minimum Size 854 mm　　　Allowance 4 mm

Self-Analysis Test, page 28

1. a.

top

b.

top

c.

top

3.

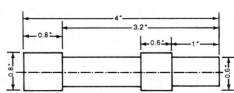

5.

	Maximum Size	Minimum Size	Allowance
length	4.850 in.	4.800 in.	0.050 in.
diameter	0.650 in.	0.600 in.	0.050 in.

Focus on Industry, page 31

1. Answers will vary.

Chapter 1 Test, pages 32–33

1. 1004 kW•h   **3.** 7600 r/min   **5.** $F = 11$ mm $G = 51$ mm $H = 13$ mm

7. a. $2\frac{29}{32}$ in.   **b.** $1\frac{13}{32}$ in.   **c.** 30.5 mm   **d.** 73.5 mm

9.

top

11.

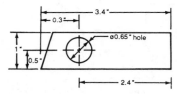

Chapter 2 - Measurements, Fractions, Decimals, and Percent

2–1, Exercises, pages 36–37

1. exact   **3.** approximate   **5.** exact   **7.** 4   **9.** 2

11. 2   **13.** 4   **15. a.** 3.94 in.   **b.** 3.9 in.   **17. a.** 75,400 mi

b. 75,000 mi   **19. a.** 0.825 in.   **b.** 0.83 in.   **21.** 13.5 in.   **23.** 78.1 lb

25. 5 m   **27.** 20 ft   **29.** 0.0003 in.   **31.** 0.02 km

2–2, Exercises, page 39

1. $\frac{13}{16}$ in.   **3.** 41.8 lb   **5.** $2\frac{11}{16}$ in.   **7.** $1\frac{15}{32}$ in.   **9.** $6\frac{1}{4}$ ft

2–3, Exercises, page 41

1. $34\frac{1}{8}$ pages   **3.** $62\frac{1}{3}$ mi   **5.** 2.63 kg   **7.** 66¢   **9. a.** $55\frac{1}{2}$ in.

b. 4 ft $7\frac{1}{2}$ in.   **11.** 3.81 cm   **13.** 13 ft 5 in.   **15.** 1300 L

2–4, Exercises, pages 43–44

1. $1\frac{3}{10}$ gal   **3.** 2.80 ft   **5.** 8 recipes   **7.** 32 boards   **9.** $\frac{1}{8}$ in.

11. 42¢   **13.** 9 strips   **15.** 15 bars   **17.** 19 strips   **19.** $1.29

21. 21¢/ft   **23. a.** 5.72 m   **b.** 0.28 m

Self-Analysis Test, page 45

1. a. 3   **b.** 2   **c.** 3   **d.** 2   **3.** $3\frac{9}{16}$ in.

5. a. 18 bricks   **b.** 294 layers   **7.** 6 pieces

2–5, Exercises, page 47
1. $23\frac{3}{4}$ in.　　**3.** 7 ft $1\frac{1}{8}$ in.　　**5.** 7 ft $8\frac{1}{8}$ in.　　**7.** 19 ft 9 in.　　**9.** 6 ft $1\frac{1}{2}$ in.
11. 10 ft $7\frac{7}{8}$ in.　　**13.** 37 ft 9 in.　　**15. a.** 176 ft 2 in.　　**b.** 173 ft 10 in.　　**17.** 30 in.
19. 37 in.

2–6, Exercises, page 49
1. 0.147 in.　　**3.** 0.477 in.　　**5.** 0.612 in.　　**7.** 0.076 in.　　**9.** 0.991 in.

2–7, Exercises, pages 51–52
1. $\frac{2}{3}$ h　　**3.** 44 baskets　　**5.** $2.63　　**7.** 43 ft 2 in.　　**9.** No
11. a. 13.02 cm　　**b.** 3.33 cm

Self-Analysis Test, page 53
1. 102 ft 4 in.　　**3.** $94.79　　**5.** 0.343 in.　　**7.** 0.21 oz　　**9.** No

2–8, Exercises, page 55
1. Reduced, 7 in.　　**3.** Enlarged, 63 L　　**5.** Reduced, 1 km　　**7.** Enlarged, 14 ft
9. Reduced, 6.0 m　　**11.** 19.5 ft by 41.6 ft　　**13.** 39 faucets　　**15.** 18.0 in. by 28.8 in.
17. 2.9 cm

2–9, Exercises, page 57
1. 2%　　**3.** 72 jars　　**5.** 5000 movies　　**7.** 0.06%　　**9.** 15 parts
11. 0.5%　　**13.** Yes

2–10, Exercises, page 59
1. 12%　　**3.** $125.00　　**5.** $8.44　　**7.** 34,400　　**9.** 28,750　　**11.** 12%

2–11, Exercises, pages 60–61
1. $672.00　　**3.** $70.00　　**5.** 73.0 lb　　**7.** 30%　　**9.** $19,270　　**11.** b.

2–12, Exercises, pages 63–64
1. 12.6 Ω and 8.4 Ω　　**3.** 12.54 Ω and 10.26 Ω　　**5.** 12,000 Ω and 8000 Ω
7. 68,250 Ω and 61,750 Ω　　**9.** 0.627 Ω and 0.513 Ω　　**11.** 143 Ω and 117 Ω
13. 4,100,000 Ω ± 10%　　**15.** 600,000,000 Ω ± 5%　　**17.** 4,920,000 Ω and 3,280,000 Ω
19. a. 7,600 Ω ± 5%　　**b.** 7,980 Ω and 7,220 Ω
21. a. 35,000,000 Ω ± 20%　　**b.** 42,000,000 Ω and 28,000,000 Ω
23. a. 98 Ω ± 10%　　**b.** 107.8 Ω and 88.2 Ω
25. a. 81,000 Ω ± 5%　　**b.** 85,050 Ω and 76,950 Ω
27. a. 81,000 Ω ± 10%　　**b.** 89,100 Ω and 72,900 Ω
29. a. 800,000 Ω ± 20%　　**b.** 960,000 Ω and 640,000 Ω

Check Your Skills, page 64
1. a. 100　　**b.** 900　　**c.** 700　　**3.** 9.5　　**5.** 418.77　　**7.** 0.1884
9. 8　　**11.** $5\frac{3}{4}$　　**13.** $\frac{7}{20}$　　**15.** $5\frac{5}{9}$　　**17.** $4\frac{4}{15}$　　**19.** $1\frac{3}{5}$
21. 521%　　**23.** 0.09　　**25.** 0.0062　　**27.** 1.7724　　**29.** 520　　**31.** 120 cm; 1.2 m; 1200 mm

Self-Analysis Test, page 65
1. 1.40 in. by 3.20 in.　　**3.** 5 light bulbs　　**5.** 12%　　**7.** $41.65
9. a. 6.48 Ω; 4.32 Ω　　**b.** 3,885,000 Ω; 3,515,000 Ω
11. 9.2 Ω ± 20%　　11.04 Ω and 7.36 Ω　　**13.** 16 Ω ± 10%; 17.6 Ω and 14.4 Ω

Focus on Industry, page 67
1. a. $181.50　　**b.** $435.60　　**c.** $1210.00　　**d.** $1.53

Chapter 2 Test, pages 68–69
1. a. 0.62 in.　　**b.** 730,000 mi　　**c.** 31 m　　**d.** 1.8 km
3. $G = 375.0$ mm; $H = 625.0$ mm　　**5.** 4 mm　　**7.** $13.50　　**9.** 0.498 in.
11. a. 93　　**b.** 45　　**c.** 272　　**13.** 8 fasteners　　**15.** $7025.25
17. 73 Ω ± 5%　　76.65 Ω and 69.35 Ω

1. 473 kW·h   3. $P = 1.9$ in. $Q = 0.7$ in. $R = 0.3$ in. $S = 1.1$ in. $T = 0.8$ in.
5. *A:* Upper Limit 2.467 in.; Lower Limit 2.407 in.; Allowance 0.060 in.
   B: Upper Limit 0.737 in.; Lower Limit 0.637 in.; Allowance 0.100 in.
7. 4.8 in. by 6.0 in.   9. $6.23   11. 6 ft 8.7 in.   13. 0.2%
15. $70,000 \, \Omega \pm 5\%$ $73,500 \, \Omega$ and $66,500 \, \Omega$

Chapter 3 - Expressions and Equations

3–1, Exercises, pages 73–75
1. $5a$ 3. $2x$ 5. t 7. $7x + 1$ 9. $7t + 8$ 11. a
13. $2s + 6$ 15. $6d$ 17. a. $8.9, 2y + 8.9, 2x + 2y + 8.9$ b. $3z$ 19. $40x$

3–2, Exercises, page 77
1. $x + 37 = 198$ 3. $x - 6.33 = 27.96$ 5. $89.95 + x = 295.44$ 7. $x - 1784 = 14{,}693$
9. $x - 1\frac{3}{4} = 22\frac{1}{2}$; $x = 24\frac{1}{4}$ in.   11. $y + 2.55 = 2.75$; $y = 0.20$ in.
13. $x - 0.1250 = 0.4375$; $x = 0.5625$ in.

3–3, Exercises, page 79
1. $6x = 24$ 3. $0.07x = 5600$ 5. $\frac{x}{5} = 0.83$ 7. $3x = 1.2$; $x = 0.40$ in.   9. a. $5.28
b. $20.87   c. $239.63   11. a. 0.4 kg   b. 3.6 kg   c. 1.6 kg   d. 2.8 kg

3–4, Exercises, pages 81–82
1. $25n + 78 = 378$ 3. $5.50n - 13.50 = 166.50$ 5. $2h + 1.96 = 2.32$; $h = 0.18$ in.
7. $2x + 2\frac{3}{8} = 4\frac{1}{2}$; $x = 1\frac{1}{16}$ in. 9. $x = 60$ cm; $1.1x = 66$ cm; $y = 29$ cm 11. 96 plants 13. 160 customers

Check Your Skills, page 82
1. 49   3. 5   5. 0.889   7. 8.124   9. 19.55   11. 46.27   13. $11\frac{1}{4}$
15. $7\frac{7}{8}$ 17. $7\frac{1}{32}$ 19. 18.07   21. 1.5   23. 0.6615   25. 1   27. 0.5975

Self-Analysis Test, page 83
1. $11n$ 3. $3x$ 5. $4.5x$ 7. $101x$ 9. 25 mm   11. 56.3 ft   13. 38 h

3–5, Exercises, pages 84–85
1. $i = 12f$ 3. $k = 0.454p$ 5. $d = \frac{c}{100}$ 7. $d = st$ 9. $x = \frac{y}{3}$ 11. $w = tr$ 13. $t = \frac{1}{p}$
15. Let x represent efficiency in percent, let y represent input, and let z represent output; $x = 100 \, \frac{y}{z}$
17. Let f be the flow rate, v be the total infusion volume, d be the drop factor, and t be the infusion time
   in minutes; $f = \frac{vd}{t}$

3–6, Exercises, pages 88–89
1. 120 V   3. 130 V   5. 1200 W   7. 900 W   9. $22 \, \Omega$ 11. 10 A
13. $22 \, \Omega$ 15. 6.3 A   17. a. $9.0 \, \Omega$ b. 110 V

3–7, Exercises, pages 90–91
1. a. 11 cm   b. 9.2 cm   3. a. 23.6 mi/gal   b. 9990 mi   c. 27 gal
5. 45 in.   7. 510 gal   9. a. 0.28 in.   b. 7.75 cm   11. 0.05 cm

Self-Analysis Test, page 92
1. $D = \frac{m}{v}$ 3. $I = prt$ 5. a. $x = R + r$ b. 3.68 cm   c. 9.150 cm
d. 15.2 cm   e. 5.9 cm   7. 700 W   9. a. 109   b. 112.5
c. 120   d. 132.5   10. 56 lb/ ft³

Focus on Industry, page 95
1. a. 40.1%   b. No   3. 3,300,000 kW

Chapter 3 Test, pages 96–97
1. $11m$
3. $4.50 + 1.25n$
5. $2.50 + 0.75n$
7. $5.4 + y = 18.15$; $y = 12.8$ cm
9. 5.33 cm
11. \$14.79/h
13. 2 h 6 min
15. 5.0 h
17. $A = \frac{h}{b}$
19. 1400 W
21. **a.** \$16,000
b. 300 items

Chapter 4 - Working With Measurement Formulas

4–1, Exercises, page 100
1. $22\frac{1}{2}$ in.
3. 56 in.
5. $5\frac{5}{12}$ in.
7. 16 ft 6 in.
9. 39.65 in.
11. 6.0 ft
13. 62 plants
15. 8.65 in.

Tricks of the Trade, page 101
1.

3. 316 m

4–2, Exercises, pages 102–103
1. 15.7 in.
3. 75.4 ft
5. 81.6 in.
7. 13.0 ft
9. 3.00 in.
11. 54.9 in.

4–3, Exercises, page 105
1. 240 ft lb/s
3. 190 ft lb/s
5. 510 ft lb/s
7. 700 ft lb/s
9. 200 ft lb/s

4–4, Exercises, pages 107–108
1. 3.14 ft
3. 1.18 ft
5. 45.5 ft/s
7. 1.14 ft/s
9. 2.94 ft/s
11. 41.4 ft/s
13. 3670 r/min
15. 153 r/min

Self-Analysis Test, page 109
1. 162 ft
3. 13 in.
5. 0.96 m
7. 4800 ft lb/s
9. 9.1 lb
11. 51.0 ft/s

4–5, Exercises, pages 110–111
1. 15,490 ft²
3. 4.65 cm
5. \$997.50
7. \$4,950

Tricks of the Trade, page 111
1. 200 yd²; 190 m²

4–6, Exercises, page 113
1. 70 in.²
3. 27.06 cm
5. 27 in.
7. 2 qt
9. 500 in.²

4–7, Exercises, pages 114–115
1. 200 ft²
3. 196 ft²
5. 220 ft²
7. 12.0 ft²
9. \$33.66

4–8, Exercises, pages 117–118
1. 50 square units
3. 113 square units
5. 21.5 in.
7. 2100 mm²
9. **a.** 6400 mm²
b. 2600 mm²
11. \$49,144
13. **a.** 62.6 lb
b. \$69
15. **a.** 12 in. size
b. \$0.01/in.²

Check Your Skills, page 118
1. 15.24
3. 7.5
5. 0.5
7. 1
9. 0.2
11. 0.5
13. 1.23 m
15. 16,453.2 m
17. 630 mm
19. 13
21. 25
23. 12.08
25. 11.66
27. 18.57

Self-Analysis Test, page 119
1. 12 m
3. \$2607
5. 13.0 in.
7. \$1499
9. 2 *whole* bags

Focus on Industry, page 121
1. **a.** \$285.00 **b.** \$935.00
c. \$1440.00
d. \$8400.00
3. 6200 trees

1. 200 ft **3.** 300 in. **5.** The circumference is doubled. **7.** 470 hp **9.** 147 ft/min
11. 135 lb **13.** 76 in.² **15.** 600 mm² **17.** 10.6 m

Chapter 5 - Ratio, Scale, and Proportion

5–1, Exercises, pages 126–127
1. $\frac{3}{4}$ **3.** $\frac{1}{10}$ **5.** $\frac{1}{4}$ **7.** $\frac{4}{7}$ **9.** $\frac{2}{9}$ **11.** $\frac{4}{3}$ **13.** $\frac{1}{3}$ **15.** $\frac{1}{2}$ **17.** $\frac{1}{2}$
19. 3 : 2 **21.** 1 : 4 **23.** 6 : 11 **25.** 5 : 4 **27.** 20 times **29. a.** $\frac{245}{13}$ **b.** $\frac{425}{156}$

5–2, Exercises, pages 128–129
1. Yes **3.** $\frac{4}{5}$; 80% **5.** $\frac{3}{4}$; 75% **7.** 80% **9.** 63% **11.** 3740 W
13. 450,000 hp **15.** 12 copies/min; 33%

5–3, Exercises, pages 132–134
1. a. a line measuring $1\frac{1}{2}$ in. **b.** a line measuring 5 in.
   c. a line measuring $3\frac{1}{2}$ in. **d.** a line measuring $2\frac{3}{4}$ in.
3. a. a line measuring 100 mm **b.** a line measuring 75 mm
   c. a line measuring 40 mm **d.** a line measuring 62.5 mm
5. $8\frac{3}{4}$ ft **7.** 5000 mm **9. a.** a line measuring $1\frac{1}{2}$ in. **b.** a line measuring 5 in.
   c. a line measuring $4\frac{1}{4}$ in. **17.**
11. length = 82 mm; width = 40 mm
13. 3 in.; 3 in.; 1 ft $4\frac{1}{4}$ in.; $2\frac{1}{4}$ in.; 1 ft $9\frac{1}{2}$ in.
15. Answers will vary.

Self-Analysis Test, page 135
1. 2 : 21 **3.** $\frac{3}{1}$ **5.** 37 % **7.** a line measuring $4\frac{1}{4}$ in. **9.** a line measuring 47 mm
11. a line measuring 69 mm **13.** a line measuring 30 mm
15. a. 12 m **b.** 6 m **c.** 2.4 m

5–4, Exercises, pages 136–137
1. 2000 km **3.** 0.384 Ω **5.** $67.57 **7.** 9740 widgets **9. a.** 740 ft **b.** 5.9 in.

5–5, Exercises, page 139
1. a. 4800 pick-up trucks **b.** 249,600 pick-up trucks **3.** 3.75 gal
5. 400,000 s **7. a.** 240 loaves **b.** 720 loaves **9.** 0.60 lb

5–6, Exercises, page 141
1. 25 **3.** 5.0 **5.** 210 lb **7.** 210 lb

5–7, Exercises, page 143
1. 160 lb **3. a.** 28 lb **b.** 39 lb **5.** 62 lb **7.** 100 lb

5–8, Exercises, pages 145–147
1. 1, 24, 100, 450 **3.** 40, 20, 120 **5.** 14 teeth **7.** 40 r/min **9.** 1560 r/min
11. 3130 r/min

5–9, Exercises, page 149
1. 3.0 : 1.0 **3.** 1.1 : 1.0 **5.** 3.0 : 1.0 **7.** 1.6 : 1.0

Check Your Skills, page 149
1. 77 **3.** 36 **5.** $\frac{4}{3}$ **7.** 4 **9.** 21
11. 0.1875 **13.** 450% **15.** 0.45

Self-Analysis Test, page 150
1. about 18.5 h   3. 3 h 20 min   5. 115 lb   7. 600 r/min   9. 1.8 : 1.0

Computer Connection, page 151
1. 23 mi/gal   3. 33 mi/gal

Focus on Industry, page 153
1. 2 : 3   3. 34 : 35

Chapter 5 Test, pages 154–155
1. 3 : 1   3. $2893.71; $3858.29   5. 3.8 hp   7. **a.** 12 ft   **b.** 6 ft   **c.** 3 ft
9. **a.** 18 ft 6 in. **b.** 9 ft 3 in.   **c.** 4 ft 7.5 in.   11. **a.** 5.5 m   **b.** 2.75 m   **c.** 1.10 m
13. **a.** a line measuring 5.5 in.   **b.** a line measuring 6 in.   **c.** a line measuring 3.25 in.
15. 22,000 lb   17. 360,000 revolutions   19. 64,800 units   21. 20,000 lb   23. 130 lb
25. 1350 r/min

Cumulative Review, pages 156–157
1. $5x + 7$ 3. Let x be the total. $0.04x = 18$; $x = 450$ 5. $V_{air} = V_{ground} - V_{headwind}$
7. 12 V   9. 19.0 m   11. 710 hp   13. 10 cm   15. about 14 bags
17. 177 : 151   19. **a.** a line measuring $\frac{1}{2}$ in.   **b.** a line measuring 3 in.
c. a line measuring $2\frac{1}{8}$ in.   21. 6.7 gal   23. 89.29 r/min

---

Chapter 6 - Plane Geometry

6–1, Exercises, page 161
1. right; 90°   3. acute; 30°

6–2, Exercises, pages 163–165
1. No   3. No

6–4, Exercises, page 171
1. **a.** 55°   **b.** 6°   **c.** 23.7°   3. **a.** ∠1 = 152°; ∠2 = 28°; ∠3 = 152°
b. ∠1 = 70°; ∠2 = 110°; ∠3 = 70° **c.** ∠1 = 117.5°; ∠2 = 62.5°; ∠3 = 117.5°   5. 98°

6–5, Exercises, page 173
1. x 3. Answers will vary.

6–6, Exercises, pages 175–176
1. 120°   7. **a.** 76°, 64°, 40°   **b.** 180°   3. 72°   5. 45°
9. **a.** Each angle measures 108°.   **b.** 540°
11.   13. 60°   15.   **a.** 90°   17.
b. 90°
c. Any angle on a circle that is subtended by a diameter is 90°.

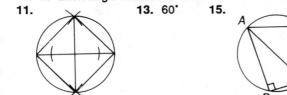

Self-Analysis Test, page 177
7. ∠1 = 120°; ∠2 = 150°

6–8, Exercises, page 181
1. 21.6 in.   3. 8.5 in.   5. 6.2 in.   7. 12 in.   9. 3.75 ft; 3.00 ft
11. 1.50 cm; 2.25 cm   13. 35 cm; 25 cm

6–9, Exercises, pages 183–184
1. 2 : 3   3. 15 : 17   5. **a.** 4 : 9   **b.** 4 : 9   **c.** Yes   7. **a.** 25 : 121
b. 25 : 121   **c.** Yes   9. **a.** 1 : 2   **b.** 2 : 3   **c.** 1 : 16   **d.** 49 : 16
11. **a.** 5 : 7   **b.** 6 : 5   **c.** 1 : 9   **d.** 1 : 9   13. **a.** 16 ft²   **b.** 36 ft²
c. 784 ft²   **d.** 374 ft²

Check Your Skills, page 188

1. 800   **3.** 18,750   **5.** 215,072   **7.** $19\frac{2}{7}$ **9.** $62\frac{3}{7}$ **11.** $175\frac{2}{3}$

13. $\frac{9}{23}$ **15.** $\frac{1}{2}$ **17.** 1 : 9   **19.** 15   **21.** 5   **23.** 6

25. 81   **27.** 324   **29.** 2.25   **31.** $\frac{64}{125}$ **33.** 0.729

Self-Analysis Test, page 185

5. 48 ft   **7.** area P : area $Q = 49 : 81$

Focus on Industry, page 187

1. 100; 570; 850; 1010 (Answers will vary.)   **3.** 1945 to 1950

Chapter 6 Test, pages 188–189

7. a. 45°   **b.** 45°   **c.** 45°   **d.** 92°   **11.**

9. Answers will vary.

13. Answers will vary.

15. 640 cm²

Chapter 7 - Solid Geometry

7–1, Exercises, pages 192–193

1. 72 in.²   **3.** 300 in.²   **5.** 32 m²   **7.** 47 ft²   **9.** 9000 ft²   **11.** 16 cans of paint

7–2, Exercises, page 196

1. 470 in.²   **3.** 44 cm²   **5.** 200 in.²   **7.** 1100 cm²   **9.** 44 in.²   **11.** 58,300 cm²

13. 35,600 ft²   **15.** 30,600 mm²   **17.** 6.30 gal

Self-Analysis Test, page 197

1. 710 in.²   **3.** 2000 in.²   **5.** 9 squares   **7.** 1200 in.²   **9.** 1.5 in.²   **11.** 170 ft²

7–3, Exercises, page 199

1. 240 in.³   **3.** 0.13 m³   **5.** 1555 in.³   **7.** 58,000 in.³   **9.** 1400 in.³   **11.** 200 yd³

13. 3,100,000 yd³

7–4, Exercises, page 201

1. 2500 in.³   **3.** 98.1 cm³   **5.** 630 in.³   **7.** 12 cm³   **9.** 7230 in.³   **11.** 0.1 m³

13. 2280 ft³   **15.** 265 ft³   **17.** 2160 in.³   **19.** 200 cm³

7–5, Exercises, pages 203–204

1. 5500 lb   **3.** 5.9 lb   **5.** 0.025 L   **7.** 2.6 qt   **9.** 3.05 L   **11.** $0.32/L

13. 0.525 L   **15.** 1000 kg

Self-Analysis Test, page 205

1. 3 yd³   **3.** 65 cm³   **5.** 7150 yd³   **7.** 180 in.³   **9.** $4.96/kg   **11.** 449 kL; 119,000 gal

7–6, Exercises, page 207

1. 1.5 L   **3.** 5.7 L   **5.** 2000 cm³   **7.** 8.62 cm; 8.36 cm   **9.** 6

11. 3.67 L   **13.** 3.5 L   **15.** 4.7 L   **17.** 2.3 L   **19.** 10 cm

7–7, Exercises, page 209

1. 6.13 : 1   **3.** 6.3 : 1   **5.** 560 cm³   **7.** 520 cm³

7–8, Exercises, page 211

1. 38 psi   **3.** 3.0 in.²   **5.** 24 in.²   **7.** 2.00 in.² ; 5.00 in.²   **9.** 3.500 in.²; 32.09 in.²

7–9, Exercises, pages 213–214

1. a. 185.3 psi   **b.** 345.3 psi   **c.** 110.3 psi   **3.** 453.3 psi   **5.** 1012 psi   **7.** 174 psi

9. 3.75 ft³   **11.** 58.0 psi

Check Your Skills, page 215
1. 9.3   **3.** 1.95471   **5.** 50.5   **7.** 4.575   **9.** 16.3   **11.** 17.5 paces/min
13. $x = 16$ **15.** $q = 34$ **17.** $m = 7$ **19.** $x = 7.091$ **21.** $w = \frac{1}{3}$ **23.** $n = 0.8755$
25. 216   **27.** 0.5625   **29.** 564.898

Self-Analysis Test, page 216
1. 5600 cm³   **3.** 2.0 L   **5.** 700 cm³   **7.** 8.0 psi   **9.** 14 psi   **11.** 392.1 psi

Focus on Industry, page 219
1.

Rank	State	Production Percentage	Rank	State	Production Percentage
1	California	9.013%	6	Michigan	5.802%
2	Texas	7.460%	7	Georgia	4.072%
3	Florida	6.711%	8	Pennsylvania	3.461%
4	Arizona	6.672%	9	Missouri	3.164%
5	Minnesota	6.663%	10	New York	2.828%

3. $79.32

Chapter 7 Test, pages 220–221
1. 250 in.²   **3.** 640 in.²   **5.** 84 ft²   **7.** 20 in.³   **9.** 1700 in.³   **11.** $0.35/L
13. 4500 cm³   **15.** 6.6 L   **17.** 27 in.³   **19.** 640 lb   **21.** 338 lb   **23.** 683 psi

Cumulative Review, pages 222–223
1. $\angle BAC = 115°$; AD bisects $\angle BAC$ **3.** $DF \parallel CB$ **5.** 27.7 °
7. Each central angle is 72°.   **9. a.** $DE = 57$ m; $EF = 43$ m   **b.** area of $\triangle DEF = 1000$ m²
11. 1620 in.²   **13. a.** 404 m²   **b.** 763 t (tonnes)   **15.** 2100 kg; 4600 lb
17. 1900 cm³   **19.** 6.0 ft³

Chapter 8 - Publishing a Book

8–1, Exercises, pages 228–229
1. a. bold   **b.** 18   **3. a.** roman   **b.** 26.5   **5. a.** bold   **b.** 29
7. 20.5 picas; 3.4 in.   **9.** 14 picas; 2.3 in.   **11. a.** 4 picas   **b.** 8 picas
   c. 9 picas   **d.** 6 picas   **13. a.** 30 picas   **b.** 12 picas   **c.** 48 picas   **d.** 18 picas
15. 19.5 picas; 23.4 ems   **17.** 31 ems   **19.** 210 points
21. 12 picas wide by 3 picas deep; 52 ems   **23.** 10,738 lines   **25.** 9 in.

8–2, Exercises, page 231
1. 80%   **3.** 171%   **5.** 25%   **7.** 60.5 picas   **9.** 21.5 picas   **11.** 12.0 picas
13. 51.5 picas   **15.** 200%

8–3, Exercises, pages 233–234
1. $4864.00   **3.** $2.27   **5. a.** $5.56   **b.** $3.51
7. $5.11   **9. a.** $36,360.00   **b.** $28,360.00

Self-Analysis Test, page 235
1. 11.5 picas   **3. a.** 36.0 picas   **b.** 65 picas   **5.** $14,116.00; $15,916.00

8–4, Exercises, pages 238–239
1. letterpress   **3.** offset lithography   **5.** web-fed   **7.** 36,000 sheets
9. $12,840.00   **11.** $17,920.00   **13.** 320,000 sheets   **15.** 144,000 fliers and 720,000 brochures

8–5, Exercises, page 241
1. 100 reams   **3.** $2625.00   **5.** No   **7. a. i.** 300,000 sheets
ii. 150,000 sheets   **iii.** 100,000 sheets   **iv.** 50,000 sheets   **b.** 24 pages   **c.** $112,500.00
9. 8 pages

8–6, Exercises, pages 243–244

1.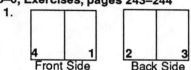
Front Side   Back Side

3.
Front Side   Back Side

5. **a.** 64 pages
    b. 19 signatures
7. 56,000 signatures
9. $120.00

11. **a.** 1152 pages   **b.** $9600.00   **c.** Yes; $2400.00

Check Your Skills, page 244
1. 1.04   3. 12.92   5. 966   7. 1,037,680   9. 38.56   11. $4.63
13. $4.12   15. 15.701   17. $336.66   19. 81.150472   21. 3.7

Self-Analysis Test, page 245
1. 4 h   3. 3000 sheets   5. $6970.00; $12,750.00   7. 75,000; 150
9. 384,000 pages   11. **a.** 768 pages   **b.** $124,800.00   **c.** Yes   **d.** $62,400.00

Focus on Industry, page 247
1. **a.** $18,480,000,000   **b.** $23,760,000,000   **c.** $6,160,000,000   **d.** $2,640,000,000
3. $4,200,000,000

Chapter 8 Test, pages 248–249
1. 22.5 picas   3. 36 picas wide by 39 picas deep   5. **a.** 50%   **b.** 30 picas
7. **a.** $8580.00   **b.** $0.86   **c. i.** $0.62   **ii.** $1.23
9. 2,400,000 bookmarks and 288,000 fliers   11. **a.** 130 reams   **b.** $5850.00
13. 4166 copies   15. $444.00

Chapter 9 - Building a House

9–1, Exercises, pages 252–253
1. **a.** 120 ft²   **b.** 48 ft²
3. **a.**

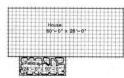

5. **a.**

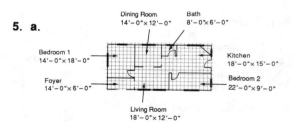

    b. $86,000   **b.** 1300 ft²   **c.** $75,000

9–2, Exercises, pages 256–257
1. **a.** 898 ft³   **b.** 33 yd³   **c.** $2000   3. $5250   5. **a.** 9 yd³   **b.** $500
7. **a.** $68.04   **b.** $1624   **c.** $2913.44   **d.** $275.90   **e.** $19.44   **f.** 172.50
   g. $5073.32   9. $300

9–3, Exercises, pages 259–260
1. 2000 lin ft   3. Yes   5. $730   7. 20   9. $179

Self-Analysis Test, page 261
1. $43,200   3. $13,500   5. $1340

9–4, Exercises, page 263
1. **a.** 18 squares   **b.** 4 rolls   **c.** 36 lb   3. **a.** 8 squares   **b.** 2 rolls
   c. 16 lb   5. $520   7. **a.** $670   **b.** $120

9–5, Exercises, pages 265–267
1. $520 **3.** 672 bricks **9. a.** 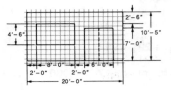 **b.** 130 ft²
5. a. 13 rolls **b.** 12 rolls **c.** $241
7. $14,400 **d.** 5 rolls
11. $280 **13.** Answers will vary.

9–6, Exercises, pages 268–269
1. 275 ft²; 30.6 yd²
3. a. 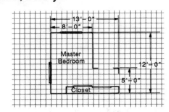 **b.** 13 yd² **5. a.** 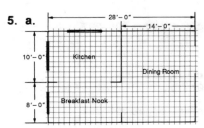 **b.** 40 yd²
c. $207 **c.** $638

7. $310

9–7, Exercises, pages 271–273
1. $33.25 **3.** $160 **5.** $1350 **7.** $676.67 **9.** $200 **11.** $60
13. $1310 **15.** Answers will vary.

Check Your Skills, page 273
1. 95.46 **3.** 243.10 **5.** 36 **7.** 439 **9.** 12.4753 **11.** 10.007
13. 14.1352 **15.** 7367 **17.** 33.423 **19.** $79,821.50 **21.** 117.35 **23.** 0.03663

Self-Analysis Test, page 274
1. 3 squares **3.** 7 lb **5.** 160 ft² **7.** 1400 bricks **9.** $860 **11.** 56 yd³

Focus on Industry, page 277
1. Number of Stores: 148,800; Store Sales: $52,078,000,000; Annual Payroll: $7,008,000,000;
    Paid Employees: 597,000.
3. $12,187

Chapter 9 Test, pages 278–279
1. $115,000 **3.** $281.50 **5.** $570 **7.** 18 squares **9.** 130 ft² **11.** four gallon-cans
13. $610 **15.** $380

Cumulative Review, pages 280–281
1. a. 17.5 picas **b.** 15.5 picas **3.** $22,732 **5.** $216.00 **7.** 3 yd³
9. 47 **11.** $3675 **13.** $1500

Chapter 10 - Right Angle Trigonometry

10–1, Exercises, pages 284–285
1. 6 cm **3.** 6 in. **5.** 9.3 in. **7.** 5.92 m **9.** 588 cm **11.** 9 ft
13. $1\frac{1}{2}$ in. **15.** 14 ft **17.** 268.3 m **19.** AC = 100 m; BC = 300 m; AB = 400 m
21. 12 ft **23.** 4.1 ft **25.** 280 ft

10–2, Exercises, page 287
1. 0.8513 **3.** 0.8513 **5.** 0.6167 **7.** 0.4564 **9.** 0.8899 **11.** 0.5128
13. a. 2.5000 **b.** 8.6 ft **15.** 0.1640

10–3, Exercises, page 289
1. 0.7071 **3.** 1.0000 **5.** 0.5736 **7.** 5.7 cm **9.** 34.1 cm **11.** 6.5 cm
13. 34.12 in. **15.** 13 in. **17.** 48.8 cm **19. a.** 109.5 ft **b.** 146.7 ft

10–4, Exercises, pages 291–292
1. 0.8660 **3.** 1.7321 **5.** 84 cm **7.** 2.8 cm **9.** 9.9 cm **11.** 43 ft
13. 16.5 m **15.** 15.6 m **17.** 10,000 ft **19.** 83 in. **21.** 14.07 in. **23.** 2030 in.²

Self-Analysis Test, page 293
1. 20.6 ft **3.** 87.2 in. **5.** 0.7071 **7.** 1.0000 **9.** 0.7071 **11.** 0.0508
13. 27.2 cm **15.** 290.1 m

10–5, Exercises, pages 295–297
1. 291 m **3.** 63 m **5.** 50 m **7.** 11° **9.** 38 ft **11.** 56 m
13. 1.72° **15.** 176 yd **17.** 2700 ft

10–6, Exercises, page 299
1. a. 1300 m **b.** 1900 m **3.** 16 km **5.** 1.4 mi **7.** 3.0° **9. a.** 6.2 mi **b.** 39 mi

10–7, Exercises, pages 300–301
1. 60 m **3.** 188 ft **5.** 110 m **7.** 9.45° **9.** 2.2 m **11.** 12.7 ft

Check Your Skills, page 302
1. 9.9 **3.** 8.89 **5.** 1091.44 **7.** 428.3264 **9.** 415.3408 **11.** 194.64 **13.** 413 : 110
15. 12 : 413 **17.** 17.35 **19.** 30.056 **21.** 54.792 **23.** 1331 **25.** 29 **27.** 37

Self–Analysis Test, page 303
1. 1.62° **3.** 64.9° **5.** 2030 m **7.** 32.5 mi **9.** 16° **11.** 30°

Focus on Industry, page 305
1. a. 24,200,000,000 barrels **b.** 41,700,000,000,000 ft³ **c.** 7,900,000,000 barrels

Chapter 10 Test, pages 306–307
1. Yes **3.** 3.63 ft **5.** $\sin A \approx 0.2415$; $\cos A \approx 0.9707$; $\tan A \approx 0.2488$
7. 32 in. **9.** 16.3 ft **11.** 1.3 mi **13.** 18.8° **15.** 19.2 km **17.** 258 ft **19.** 66°

Chapter 11 - Statistics

11–1, Exercises, pages 310–311
1. Adult males 29; Adult females 43; Fawns 17; Total 89
3. 7; 8% **5.** Friday
9. 56%

7.

Direction	Tally	Frequency
L	ᚋ ᚋ ᚋ ᚋ	20
R	ᚋ ᚋ ᚋ ᚋ ᚋ II	27
S	ᚋ ᚋ ᚋ ᚋ ᚋ ᚋ ᚋ II	37
total		84

13. Location 1
15. 70%

11.

Hour (P.M.)	Location 1	Location 2	Location 3
4:00 - 5:00	24.6%	8.3%	11.0%
5:00 - 6:00	18.4%	17.3%	18.6%
6:00 - 7:00	9.5%	33.8%	18.1%
7:00 - 8:00	17.6%	10.1%	18.6%
8:00 - 9:00	20.7%	15.1%	16.5%
9:00 - 10:00	9.2%	15.4%	17.2%

11–2, Exercises, page 313
1. False **3.** False **5.** 44 **7.** Yes; 38
9. 0.253 in.; 0.254 in.; No mode **11.** 0.247 in.

11–3, Exercises, pages 315–316

1. The mean, 2051, supports the claim that the five boats averaged more than 2000 salmon caught. The median, 1496, is less than 1500 and therefore supports the claim that more boats caught under 1500 salmon.

3. The median and the mode, $30,000, best describe the salaries of the six officers.

5. a. The mean, $0.20, would be the best measure of central tendency to promote the sale of this stock.
b. The mode, $0.14, would be the best measure of central tendency to discredit the stock.

Self-Analysis Test, page 317

1. a. 23 **b.** 35 **c.** 52 **d.** 31 **e.** 17 **f.** 9
g. 3 **3.** 72,000 **5.** 39.2¢/lb **7.** The mean, 913,000, is just under 1,000,000.

11–4, Exercises, pages 320–321

1. 55 cm; 63 cm **3.** $58; $32
5. 32 cm; 51 cm; 19 cm **7.** 148; 406
11.

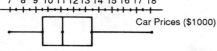

Car Prices ($1000)

15. 4th quarter (between the upper hinge and the highest amount)

9.

Books Sold (1,000,000s)

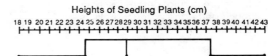

13. Heights of Seedling Plants (cm)

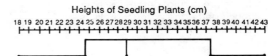

11–5, Exercises, pages 323–325

1. The horizontal axis shows an evenly-divided scale ranging the year 1982 to the year 1988.
3. 1988
5. about 50,000,000
7. about 320,000,000
9. about 20%
11. about 35%
13. Under 18 years **19.** 14%

15.

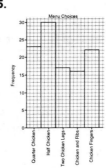

17.

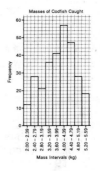

11–6, Exercises, pages 327–329

1. 1,800,000; 2,700,000 **3.** 2,550,000 **5.** 101.0° F; the morning of June 22
7. 99.3° F **9.** Brand E **11.** 65 beats/min **13.** 81 beats/min
15. about $1.26 **17.** 35% **19.** 70 : 79 **21.** 15.3 in.
23. 13%

11–7, Exercises, page 331

1. 80%
3. 2105 mi **5.** 1610 mi
9.

Expense	Central Angle
Apartment	144°
Transportation	36°
Living	54°
Loans/Savings	108°
Other	18°

11. $600

7. 162°

13. Favorite Courier Service

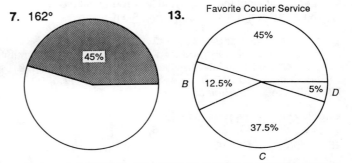

11–8, Exercises, pages 332–334

1. The bar graph shows only some of the data, but the line graphs show the data changing throughout the years.
3. By giving the bars depth, the information seems more substantial.
5.

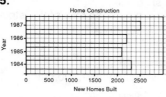

9. This graph is of no real value because a vertical scale is not given.
   The consumer is not able to read the graph accurately and therefore cannot determine how much better the product really is.
11. 2 times

7.

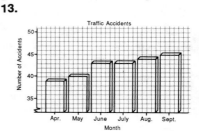

13.

Self-Analysis Test, page 335

1. median = 27,900,000;
   lower hinge = 21,300,000;
   upper hinge = 31,900,000.
3. bar graph
5. a. Canada catches half as many pounds of fish as the U.S.
      Peru catches half as many pounds of fish as Japan.
   b. Peru and Norway

7.

11–9, Exercises, page 337

1. Yes; $k = 4$ 3. a. $y = 291x$ b. $y = 8x$
5. The quotients (ratios) of the two variables in each graph are constant. Also, each graph is a straight line passing through the origin.
7. Let y represent the cost in dollars and let x represent the length in yards, $y = 12.95x$. $64.75; $155.40
9. $k = 1.80$; $9.00; $14.40

11–10, Exercises, page 339

1. inverse variation   3. inverse variation   5. inverse variation
7. As the number of people doing the job increases, the amount of time needed to do the job decreases, such that $xy = 60$.
9. 3750, 3000
11. Let y represent the length and let x represent the width, such that $xy = 48$. 1 cm, 48 cm; 2 cm, 24 cm; 3 cm, 16 cm; 4 cm, 12 cm; 6 cm, 8 cm

11–11, Exercises, pages 342–343

1. The number of nails produced doubles.   3. The number of bricks laid remains the same.
5. The amount of painted wall surface is increased 6 times.   7. $80   9. $90
11. $360   13. 1000   15. 3000   17. 900   19. 7 min   21. 250 cm²
23. 300 cm²   25. 900 cm²

Check Your Skills, page 343

1. 89   3. 308   5. $298\frac{2}{3}$ 7. 1 : 2   9. 9 : 103   11. 2782.1%
13. 0.00023   15. 19.4%   17. 1.8   19. 0.2

Self-Analysis Test, page 344

1. $y = 24x$ **3.** 2.25; 4.50; 9.00; 11.25; 13.5; 15.75; 18.00

5. As the number of hours traveled increases, the speed decreases, such that $xy = 216$.

7. Direct variation; If mass varies directly with age, then when Tom is 32 years old, he will have a mass of 100 kg. Similarly at age 64, he will have a mass of 200 kg. This relationship is unrealistic.

9. Inverse variation; This variation does not make sense since the average life of a light bulb does not depend greatly on the power of the bulb.

Computer Connection, page 345

1.
3.

From	To	Departure Time	Arrival Time	Carrier	Flight	Class of Service
New York	Toronto	3:50p	5:17p	AC	712	FJYBV
New York	Toronto	6:59a	8:24a	AA	171	FYBQM
		7:10a	8:44a	AC	701	YBV
Toronto	New York	5:00p	6:35p	OC	609	YB
		6:20p	7:41p	AC	714	YBV
		7:10p	8:45p	OU	611	YB

Focus on Industry, page 347

1.

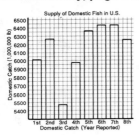

3. These data fall into distinct categories, and may be visually represented better by either a bar graph or a pictograph.

Chapter 11 Test, page 348

1.

Flower	Tally	Frequency
Asters(A)	︴︴ ︴︴ ‖	12
Daisies (D)	︴︴ ∣	6
Marigolds (M)	︴︴ ‖‖	9
Petunias (P)	︴︴ ︴︴	10
Roses (R)	︴︴ ︴︴ ︴︴	15
Total		52

3. All measures of central tendency support the claim that the pulse rates of the workers fall within a normal range of 72 beats/min to 80 beats/min.

5.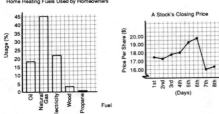

7.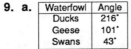

9. a.

Waterfowl	Angle
Ducks	216°
Geese	101°
Swans	43°

b.

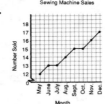

11. $72; $216; $288; $360

13. 15 people

Cumulative Review, pages 350–351

1. 4.0 m

3. 15' – 7"

5. 95.95 in.

7. 326 m

9. a. 8,000,000

b. 16,000,000

c. In 1988, there was a shortage of 12,000,000 t of cement.

11. a.

b.

Skills File

Skill 1 – Rounding
1. 30   **3.** 430   **5.** 1880   **7.** 700   **9.** 1000   **11.** 10,600
13. 6000   **15.** 26,000   **17.** 1000   **19.** 12.2   **21.** 0.3   **23.** 5.5
25. 0.123   **27.** 0.500   **29.** 0.901   **31.** 10   **33.** 51   **35.** 68
37. $1.69   **39.** $75.75   **41.** $0.26   **43.** $6   **45.** $23   **47.** $123

Skill 2 – Adding and Subtracting Whole Numbers
1. 5139   **3.** 2680   **5.** 58,788   **7.** 8021   **9.** 95,581   **11.** $21,993
13. 359   **15.** 1138   **17.** 2664   **19.** 604   **21.** 2365   **23.** 42,228
25. 21,469   **27.** 87,440   **29.** $993   **31.** $12,074   **33.** $369   **35.** 173
37. 3996   **39.** 414

Skill 3 – Multiplying Whole Numbers
1. 3876   **3.** 64   **5.** 4118   **7.** 1458   **9.** 3538   **11.** 305,547
13. 155,061   **15.** 891,990   **17.** 171,000   **19.** 4,629,966   **21.** 21,270,384   **23.** 475,272
25. $138   **27.** $13,488   **29.** $3276   **31.** 480   **33.** 672   **35.** 6144

Skill 4 – Dividing Whole Numbers
1. $1\frac{41}{43}$ **3.** $2\frac{5}{28}$ **5.** $12\frac{46}{59}$ **7.** $53\frac{32}{33}$ **9.** $54\frac{38}{53}$ **11.** $31\frac{36}{43}$
13. $650\frac{5}{8}$ **15.** $12\frac{13}{15}$ **17.** 13.5   **19.** 35.5   **21.** 35.2   **23.** 10.5   **25.** 5.6
27. 13.3   **29.** 2.0   **31.** 4.5   **33.** 9.5   **35.** 56   **37.** 53

Skill 5 – Using Order of Operations
1. 78   **3.** 59   **5.** 35   **7.** 19   **9.** 48   **11.** 11   **13.** 11
15. 14   **17.** 8   **19.** 680   **21.** 78   **23.** 1431   **25.** 87   **27.** 10
29. 26   **31.** 10   **33.** 1

Skill 6 – Adding and Subtracting Decimals
1. 18.3   **3.** 114.595   **5.** 298.77   **7.** 2.296   **9.** 81.85   **11.** 83.353   **13.** 30.6
15. 36.558   **17.** 4811.865   **19.** 7.957   **21.** 97.67   **23.** 2.75513   **25.** 1445.023   **27.** 482.987
29. 7.15   **31.** 41.69   **33.** 13.28   **35.** 9.66   **37.** 15.73   **39.** 7.4   **41.** 9.7

Skill 7 – Multiplying Decimals
1. 6.3   **3.** 77.724   **5.** $15,198.81   **7.** 0.2952   **9.** 472.0645   **11.** 0.2184
13. $4261   **15.** 85,999.806   **17.** 0.11111   **19.** 11.943   **21.** 22.835623   **23.** $90.88
25. 110.4192   **27.** 412.56   **29.** 11,884.581   **31.** 44.61   **33.** 62.62   **35.** 10.41
37. 54.352   **39.** 37.18   **41.** 3217.5   **43.** 218.4   **45.** 160.212

Skill 8 – Dividing Decimals
1. 3.29   **3.** 4.35   **5.** 13.83   **7.** 11.86   **9.** 34.33   **11.** 0.21   **13.** 352.20
15. 0.29   **17.** 240.62   **19.** 142.83   **21.** 3.50   **23.** 1.83   **25.** 115.5   **27.** 0.01
29. 0.10   **31.** 178.8   **33.** 280   **35.** 80   **37.** 41.05   **39.** 3.75   **41.** 24.63
43. $0.51   **45.** $11.21   **47.** $84.77

Skill 9 – Changing Mixed Numbers and Decimals
1. $\frac{21}{4}$ **3.** $\frac{75}{8}$ **5.** $\frac{23}{3}$ **7.** $\frac{44}{5}$ **9.** $\frac{33}{10}$ **11.** $\frac{177}{32}$ **13.** $\frac{33}{8}$
15. $\frac{19}{10}$ **17.** $\frac{37}{16}$ **19.** $\frac{133}{32}$ **21.** $\frac{201}{32}$ **23.** $\frac{25}{2}$ **25.** $\frac{133}{64}$ **27.** 4
29. $1\frac{1}{16}$ **31.** $4\frac{4}{7}$ **33.** $3\frac{3}{4}$ **35.** $3\frac{2}{15}$ **37.** $8\frac{1}{3}$ **39.** $6\frac{5}{7}$ **41.** $5\frac{7}{8}$
43. $30\frac{1}{2}$ **45.** $17\frac{1}{4}$ **47.** $10\frac{1}{3}$ **49.** $14\frac{2}{11}$

Skill 10 – Equivalent and Simplest Form Fractions
1. $\frac{10}{15}$; $\frac{12}{18}$ **3.** $\frac{20}{45}$; $\frac{24}{54}$ **5.** $\frac{15}{50}$; $\frac{18}{60}$ **7.** 50   **9.** 72   **11.** 14   **13.** 21
15. 51   **17.** $\frac{1}{5}$ **19.** $\frac{7}{10}$ **21.** $\frac{8}{9}$ **23.** $\frac{3}{8}$ **25.** $\frac{1}{4}$ **27.** $\frac{7}{9}$ **29.** $\frac{7}{16}$
31. $\frac{13}{16}$ **33.** $\frac{23}{64}$ **35.** $\frac{21}{32}$ **37.** $\frac{11}{12}$ **39.** $\frac{1}{2}$ **41.** $\frac{15}{32}$ **43.** $\frac{3}{16}$ **45.** $\frac{3}{16}$

Skill 11 – Changing Fractions and Decimals
1. 0.6   **3.** 1.38   **5.** 3.13   **7.** 6.5   **9.** 0.7   **11.** 3.38
13. 5.08   **15.** 0.03   **17.** 7.13   **19.** 2.22   **21.** 21.67   **23.** 1.94
25. 175.8   **27.** $\frac{3}{100}$ **29.** $\frac{13}{50}$ **31.** $1\frac{1}{8}$ **33.** $5\frac{3}{8}$ **35.** $5\frac{5}{8}$
37. $\frac{21}{50}$ **39.** $6\frac{1}{2}$ **41.** $3\frac{1}{8}$ **43.** $1\frac{411}{1000}$ **45.** $\frac{21}{200}$ **47.** $1\frac{29}{1000}$
49. $5\frac{11}{20}$ **51.** $1\frac{3}{5}$ **53.** $6\frac{1}{8}$ **55.** $1\frac{5}{16}$ **57.** $15\frac{1}{4}$ **59.** $1\frac{473}{500}$

Skill 12 – Comparing Fractions
1. <   **3.** =   **5.** >   **7.** <   **9.** >   **11.** <   **13.** >
15. <   **17.** <   **19.** <   **21.** <   **23.** <   **25.** =   **27.** <

Skill 13 – Adding Fractions
1. $\frac{4}{5}$ **3.** $\frac{1}{2}$ **5.** $\frac{4}{5}$ **7.** 1   **9.** $\frac{3}{8}$ **11.** $1\frac{7}{16}$ **13.** $\frac{11}{16}$
15. $\frac{45}{64}$ **17.** $10\frac{7}{8}$ **19.** $15\frac{37}{64}$ **21.** $23\frac{17}{33}$ **23.** $3\frac{1}{32}$ **25.** $25\frac{37}{40}$

Skill 14 – Subtracting Fractions with Like Denominators
1. $\frac{1}{4}$ **3.** $\frac{1}{4}$ **5.** $\frac{1}{2}$ **7.** $\frac{2}{5}$ **9.** $\frac{1}{2}$ **11.** $\frac{4}{5}$
13. $3\frac{1}{2}$ **15.** $4\frac{1}{4}$ **17.** $3\frac{3}{4}$ **19.** $4\frac{3}{5}$ **21.** $11\frac{1}{6}$ **23.** $18\frac{25}{32}$
25. $1\frac{3}{4}$ **27.** $4\frac{2}{5}$ **29.** $22\frac{1}{8}$ **31.** $3\frac{1}{4}$ **33.** 93

Skill 15 – Subtracting Fractions with Unlike Denominators
1. $\frac{1}{4}$ **3.** $\frac{1}{2}$ **5.** $\frac{3}{8}$ **7.** $\frac{57}{110}$ **9.** $\frac{1}{5}$ **11.** $\frac{8}{21}$ **13.** $2\frac{1}{18}$
15. $7\frac{1}{8}$ **17.** $9\frac{53}{60}$ **19.** $1\frac{37}{70}$ **21.** $2\frac{7}{8}$ **23.** $8\frac{7}{8}$ **25.** $2\frac{5}{24}$ **27.** $1\frac{3}{8}$

Skill 16 – Multiplying Fractions
1. $\frac{3}{8}$ **3.** $\frac{5}{64}$ **5.** $\frac{5}{12}$ **7.** $\frac{7}{8}$ **9.** $\frac{1}{15}$ **11.** $\frac{15}{128}$
13. $8\frac{1}{3}$ **15.** $\frac{19}{40}$ **17.** $43\frac{3}{4}$ **19.** $4\frac{125}{128}$ **21.** $6\frac{7}{8}$ **23.** $33\frac{1}{3}$
25. $1\frac{11}{64}$ **27.** $7\frac{7}{15}$ **29.** $2\frac{4}{9}$ **31.** $18\frac{3}{4}$ **33.** 1

Skill 17 – Dividing Fractions
1. $\frac{2}{3}$ **3.** $\frac{9}{10}$ **5.** 11   **7.** 6   **9.** 12   **11.** $\frac{3}{16}$
13. $1\frac{1}{2}$ **15.** $5\frac{1}{3}$ **17.** $3\frac{1}{9}$ **19.** $\frac{1}{2}$ **21.** $\frac{3}{10}$ **23.** $12\frac{1}{2}$
25. $4\frac{33}{47}$ **27.** 6   **29.** $9\frac{258}{263}$ **31.** $1\frac{1}{728}$ **33.** $\frac{3}{16}$

Skill 18 – Writing Ratios and Rates
1. 11 : 3   **3.** 4 : 3   **5.** 6 : 11 : 3 : 4   **7.** $1\frac{2}{3}$ **9.** $\frac{7}{16}$
11. 200 students/ 8 instructors; 25 students/instructor   **13.** 96 m/16 s; 6 m/s
15. $1280/40 h; $32/h   **17.** 500,000 people/ 200 km²; 2500 people/km²
19. 32 L/400 km; 0.08 L/km   **21.** 4950 words/9 min; 550 words/min

Skill 19 – Writing Proportions
1. Yes   **3.** Yes   **5.** No   **7.** Yes   **9.** No   **11.** $n = 12$ **13.** $a = 4$
15. $t = 4$ **17.** $w = 24$ **19.** $x = 7$ cm   **21.** $y = 3.00 **23.** $K = 3

Skill 20 – Writing Decimals as Percents
1. 25%   **3.** 8%   **5.** 115%   **7.** 19%   **9.** 52%   **11.** 480%   **13.** 770%
15. 13%   **17.** 90%   **19.** 650%   **21.** 800%   **23.** 1150%   **25.** 60%   **27.** 64%
29. 34.7%   **31.** 0.6%   **33.** 16%   **35.** 321.2%   **37.** 32.5%   **39.** 0.1%   **41.** 83%
43. 19.4%   **45.** 0.7%   **47.** 300%   **49.** 1200%

Skill 21 – Writing Fractions and Mixed Numbers as Percents
1. 70%   **3.** 17.5%   **5.** 12.5%   **7.** 14%   **9.** 48%   **11.** 12.5%   **13.** 33.3%
15. 58.3%   **17.** 225%   **19.** 362.5%   **21.** 222%   **23.** 617.5%   **25.** 516.7%   **27.** 866.7%
29. 55.6%   **31.** 2183.3%   **33.** 41.7%   **35.** 88.9%

Skill 22 – Writing Percents as Decimals
1. 0.85   **3.** 0.40   **5.** 1.20   **7.** 0.036   **9.** 0.079   **11.** 0.985   **13.** 0.75
15. 0.086   **17.** 0.054   **19.** 1.243   **21.** 0.105   **23.** 0.05375   **25.** 0.061   **27.** 0.204
29. 0.7265   **31.** 0.1175   **33.** 0.113   **35.** 0.221   **37.** 0.023   **39.** 0.379   **41.** 0.046
43. 0.005   **45.** 0.063   **47.** 0.063   **49.** 1.275   **51.** 0.130

Skill 23 – Writing Percents as Fractions
1. $\frac{17}{40}$ **3.** $1\frac{1}{4}$ **5.** $\frac{127}{1000}$ **7.** $\frac{3}{2500}$ **9.** $\frac{29}{250}$ **11.** $\frac{2763}{10\,000}$ **13.** $\frac{1}{2}$
15. $\frac{91}{1000}$ **17.** $\frac{409}{1000}$ **19.** $\frac{21}{400}$ **21.** $\frac{7}{16}$ **23.** $\frac{37}{400}$ **25.** $\frac{33}{200}$ **27.** $\frac{51}{200}$
29. $\frac{49}{300}$ **31.** $\frac{1}{5}$ **33.** $\frac{7}{20}$ **35.** $\frac{231}{500}$ **37.** 1   **39.** $\frac{51}{400}$ **41.** 2
43. $\frac{137}{200}$ **45.** $\frac{29}{125}$ **47.** $\frac{111}{200}$ **49.** $\frac{3}{1000}$ **51.** $\frac{7}{250}$

Skill 24 – Finding the Percent
1. 50%   **3.** 16.7%   **5.** 31.3%   **7.** 87.3%   **9.** 67.0%   **11.** 88.5%   **13.** 15.5%
15. 50.5%   **17.** 15.6%   **19.** 16.7%   **21.** 2.8%   **23.** 86.0%   **25.** 8%   **27.** 125%

Skill 25 – Finding the Part
1. 6   **3.** 2.58   **5.** 2.75   **7.** 8.96   **9.** 3.304   **11.** 32.799   **13.** 7.5
15. 5.625   **17.** 25.2   **19.** 5.7   **21.** 90   **23.** 17.1   **25.** 7.7   **27.** 3.1
29. 6   **31.** 110   **33.** 26.5   **35.** 27   **37.** 31.3   **39.** 134.0   **41.** 9.3

Skill 26 – Finding the Whole
1. 3333.3   **3.** 3753.8   **5.** 1214.8   **7.** 245.5   **9.** 22.2   **11.** 188   **13.** 269.6
15. 33.8   **17.** 19.5   **19.** 300   **21.** 34.5   **23.** 3.2   **25.** 15.3   **27.** 1025
29. 3.2   **31.** 140

Skill 27 – Solving Equations by Adding or Subtracting
1. $y = 2$ **3.** $x = 1$ **5.** $w = 5.3$ **7.** $t = 3.75$ **9.** $r = 1.25$ **11.** $p = 3\frac{5}{6}$
13. $x = 12$ **15.** $r = 18$ **17.** $y = 28.3$ **19.** $t = 3.19$ **21.** $w = 4$ **23.** $z = 12\frac{5}{8}$
25. $x = 3.7$ **27.** $r = 24.5$ **29.** $p = 47.74$ **31.** $y = 8.5$ **33.** $s = 10.6$ **35.** $b = 6025$

Skill 28 – Solving Equations by Multiplying or Dividing
1. $x = 20$ **3.** $t = 225$ **5.** $n = 2100$ **7.** $t = 8.84$ **9.** $s = 1.08$ **11.** $y = 245$
13. $r = 31$ **15.** $y = 80$ **17.** $w = 15$ **19.** $w = 50$ **21.** $s = 46$ **23.** $t = 27$
25. $x = 1.9$ **27.** $y = 61.8$ **29.** $w = 32$

Skill 29 – Solving Equations by More Than One Step
1. $x = 5$ **3.** $x = 5$ **5.** $t = 4$ **7.** $w = 25$ **9.** $x = 8$ **11.** $n = 11$ **13.** $x = 3$
15. $x = 2$ **17.** $y = 8$ **19.** $r = 9$ **21.** $m = 8.2$ **23.** $d = 1\frac{3}{4}$ **25.** $k = 8$

Skill 30 – Using Exponents
1. 25   **3.** 81   **5.** 16   **7.** 32   **9.** 1,000,000   **11.** 81
13. 1024   **15.** 512   **17.** 125   **19.** 1331   **21.** 625   **23.** 1728   **25.** 900
27. $\frac{1}{4}$ **29.** 7776   **31.** $\frac{8}{125}$ **33.** 1.0201   **35.** $\frac{1}{243}$ **37.** 5.31   **39.** 6.17
41. 5.34   **43.** 81   **45.** 133.28   **47.** 9.07   **49.** 105.57   **51.** 706.5   **53.** 241.40
55. 18.09

Skill 31 – Multiplying by Powers of 10 and Metric Conversion
1. 400   **3.** 2100   **5.** 220,000   **7.** 14,000   **9.** 810,000   **11.** 0.00062   **13.** 0.003
15. 3220   **17.** 75,600   **19.** 0.0175   **21.** 250   **23.** 76,000   **25.** 15,000   **27.** 5300
29. 1020   **31.** 0.013   **33.** 0.0034   **35.** 0.34

Skill 32 – Finding Square Roots
1. 2   **3.** 10   **5.** 1   **7.** 14   **9.** 11   **11.** 15   **13.** 16
15. 13   **17.** 17   **19.** 30   **21.** 18   **23.** 21   **25.** 23   **27.** 0.01
29. 2.4   **31.** 1.4   **33.** 1.7   **35.** 0.9   **37.** 22   **39.** 31   **41.** 0.3
43. 0.8   **45.** 70   **47.** 7.5   **49.** 3.1   **51.** 5.9   **53.** 23.9   **55.** 12.5

Printed In Canada

Donald G. McCombs
772-2730